Vector Analysis

SCHAUM'S
outlines

Vector Analysis

and an introduction to TENSOR ANALYSIS

Second Edition

Seymour Lipschutz, Ph.D.

Mathematics Department, Temple University

Dennis Spellman, Ph.D.

Mathematics Department, Temple University

Murray R. Spiegel, Ph.D.

Former Professor and Chairman, Mathematics Department
Rensselaer Polytechnic Institute, Hartford Graduate Center

Schaum's Outline Series

New York   Chicago   San Francisco
Lisbon   London   Madrid   Mexico City
Milan   New Delhi   San Juan
Seoul   Singapore   Sydney   Toronto

SEYMOUR LIPSCHUTZ is on the faculty of Temple University and formerly taught at the Polytechnic Institute of Brooklyn. He received his Ph.D. from New York University and is one of Schaum's most prolific authors. In particular he has written, among others, *Linear Algebra, Probability, Discrete Mathematics, Set Theory, Finite Mathematics*, and *General Topology*.

DENNIS SPELLMAN is on the faculty of Temple University and he formerly taught at the University of the East in Venezuela. He received his Ph.D. from New York University where he wrote his thesis with Wilhelm Magnus. He is the author of over twenty-five journal articles in pure and applied mathematics.

The late **MURRAY R. SPIEGEL** received the MS degree in Physics and the Ph.D. degree in Mathematics from Cornell University. He had positions at Harvard University, Columbia University, Oak Ridge, and Rensselaer Polytechnic Institute, and served as a mathematical consultant at several large companies. His last position was Professor and Chairman of Mathematics at the Rensselaer Polytechnic Institute, Hartford Graduate Center. He was interested in most branches of mathematics, especially those that involve applications to physics and engineering problems. He was the author of numerous journal articles and 14 books on various topics in mathematics.

Schaum's Outline of VECTOR ANALYSIS

8 9 0 CUS CUS 0 1 4

ISBN 978-0-07-161545-7
MHID 0-07-161545-8

Sponsoring Editor: Charles Wall
Production Supervisor: Tama Harris McPhatter
Editing Supervisor: Maureen Dennehy
Interior Designer: Jane Tenenbaum
Project Manager: Hannah Turner

Library of Congress Cataloging-in-Publication Data

Schaum's outline of complex variables : with an introduction to conformal mapping and its applications / Murray R. Spiegel . . . [et al.]. -- 2nd ed.
        p. cm. -- (Schaum's outlines series)
    Includes bibliographical references and index.
    ISBN-13: 978-0-07-161569-3 (alk. paper)
    ISBN-10: 0-07-161569-5 (alk. paper)
    1. Functions of complex variables. 2. Functions of several complex variables. I. Spiegel, Murray R. II. Title: Complex variables.
    QA331.7.S33 2009
    515′. 9--dc22

                                                                    2009007907

Preface

The main purpose of this second edition is essentially the same as the first edition with changes noted below. Accordingly, first we quote from the preface by Murray R. Spiegel in the first edition of this text.

"This book is designed to be used either as a textbook for a formal course in vector analysis or as a useful supplement to all current standard texts."

"Each chapter begins with a clear statement of pertinent definitions, principles and theorems together with illustrated and other descriptive material. This is followed by graded sets of solved and supplementary problems. . . . Numerous proofs of theorems and derivations of formulas are included among the solved problems. The large number of supplementary problems with answers serve as complete review of the material of each chapter."

"Topics covered include the algebra and the differential and integral calculus of vectors, Stokes' theorem, the divergence theorem, and other integral theorems together with many applications drawn from various fields. Added features are the chapters on curvilinear coordinates and tensor analysis . . . ."

"Considerable more material has been included here than can be covered in most first courses. This has been done to make the book more flexible, to provide a more useful book of reference, and to stimulate further interest in the topics."

Some of the changes we have made to the first edition are as follows: (a) We expanded many of the sections to make it more accessible for out readers. (b) We reformatted the text, such as, the chapter number is included in the label of all problems and figures. (c) Many results are restated formally as Propositions and Theorems. (d) New material was added, such as, a discussion of linear dependence and linear independence, and a discussion of $\mathbf{R}^n$ as a vector space.

Finally, we wish to express our gratitude to the staff of McGraw-Hill, particularly to Charles Wall, for their excellent cooperation at every stage in preparing this second edition.

Seymour Lipschutz

Dennis Spellman

Temple University

Contents

CHAPTER 8 TENSOR ANALYSIS   189

INDEX   235

Vector Analysis

Vectors and Scalars

1.1 Introduction

The underlying elements in vector analysis are vectors and scalars. We use the notation **R** to denote the real line which is identified with the set of real numbers, $\mathbf{R}^2$ to denote the Cartesian plane, and $\mathbf{R}^3$ to denote ordinary 3-space.

Vectors

There are quantities in physics and science characterized by both magnitude and direction, such as displacement, velocity, force, and acceleration. To describe such quantities, we introduce the concept of a *vector* as a directed line segment $\overrightarrow{PQ}$ from one point P to another point Q. Here P is called the *initial point* or *origin* of $\overrightarrow{PQ}$, and Q is called the *terminal point*, *end*, or *terminus* of the vector.

We will denote vectors by bold-faced letters or letters with an arrow over them. Thus the vector $\overrightarrow{PQ}$ may be denoted by **A** or $\vec{A}$ as in Fig. 1-1(a). The magnitude or length of the vector is then denoted by $|\overrightarrow{PQ}|$, $|\mathbf{A}|$, $|\vec{A}|$, or A.

The following comments apply.

(a) Two vectors **A** and **B** are *equal* if they have the same magnitude and direction regardless of their initial point. Thus **A** = **B** in Fig. 1-1(a).

(b) A vector having direction opposite to that of a given vector **A** but having the same magnitude is denoted by −**A** [see Fig. 1-1(b)] and is called the *negative* of **A**.

(a)   (b)

Fig. 1-1

Scalars

Other quantities in physics and science are characterized by magnitude only, such as mass, length, and temperature. Such quantities are often called *scalars* to distinguish them from vectors. However, it must be emphasized that apart from units, such as feet, degrees, etc., scalars are nothing more than real numbers. Thus we can denote them, as usual, by ordinary letters. Also, the real numbers 0 and 1 are part of our set of scalars.

1.2   Vector Algebra

There are two basic operations with vectors: (a) Vector Addition; (b) Scalar Multiplication.

(a)   Vector Addition

Consider vectors **A** and **B**, pictured in Fig. 1-2(a). The *sum* or *resultant* of **A** and **B**, is a vector **C** formed by placing the initial point of **B** on the terminal point of **A** and then joining the initial point of **A** to the terminal point of **B**, pictured in Fig. 1-2(b). The sum **C** is written **C = A + B**. This definition here is equivalent to the Parallelogram Law for vector addition, pictured in Fig. 1-2(c).

     (a)   (b)   (c)

Fig. 1-2

Extensions to sums of more than two vectors are immediate. Consider, for example, vectors **A, B, C, D** in Fig. 1-3(a). Then Fig. 1-3(b) shows how to obtain the sum or resultant **E** of the vectors **A, B, C, D**, that is, by connecting the end of each vector to the beginning of the next vector.

           (a)   (b)

Fig. 1-3

The *difference* of vectors **A** and **B**, denoted by **A − B**, is that vector **C**, which added to **B**, gives **A**. Equivalently, **A − B** may be defined as **A + (−B)**.

If **A = B**, then **A − B** is defined as the *null* or *zero* vector; it is represented by the symbol **0** or 0. It has zero magnitude and its direction is undefined. A vector that is not null is a *proper* vector. All vectors will be assumed to be proper unless otherwise stated.

(b)   Scalar Multiplication

Multiplication of a vector **A** by a scalar m produces a vector m**A** with magnitude $|m|$ times the magnitude of **A** and the direction of m**A** is in the same or opposite of **A** according as m is positive or negative. If $m = 0$, then m**A = 0**, the null vector.

Laws of Vector Algebra

The following theorem applies.

THEOREM 1.1: Suppose **A**, **B**, **C** are vectors and m and n are scalars. Then the following laws hold:

[A_1]	$(A + B) + C = (A + B) + C$	Associative Law for Addition
[A_2]	There exists a zero vector **0** such that, for every vector **A**,	
	$A + 0 = 0 + A = A$	Existence of Zero Element
[A_3]	For every vector **A**, there exists a vector $-A$ such that	
	$A + (-A) = (-A) + A = 0$	Existence of Negatives
[A_4]	$A + B = B + A$	Commutative Law for Addition
[M_1]	$m(A + B) = mA + mB$	Distributive Law
[M_2]	$(m + n)A = mA + nA$	Distributive Law
[M_3]	$m(nA) = (mn)A$	Associative Law
[M_4]	$1(A) = A$	Unit Multiplication

The above eight laws are the axioms that define an abstract structure called a *vector space*.

The above laws split into two sets, as indicated by their labels. The first four laws refer to vector addition. One can then prove the following properties of vector addition.

(a) Any sum $A_1 + A_2 + \cdots + A_n$ of vectors requires no parentheses and does not depend on the order of the summands.

(b) The zero vector **0** is unique and the negative $-A$ of a vector **A** is unique.

(c) (Cancellation Law) If $A + C = B + C$, then $A = B$.

The remaining four laws refer to scalar multiplication. Using these additional laws, we can prove the following properties.

PROPOSITION 1.2: (a) For any scalar m and zero vector **0**, we have $m0 = 0$.

(b) For any vector **A** and scalar 0, we have $0A = 0$.

(c) If $mA = 0$, then $m = 0$ or $A = 0$.

(d) For any vector **A** and scalar m, we have $(-m)A = m(-A) = -(mA)$.

1.3 Unit Vectors

Unit vectors are vectors having unit length. Suppose **A** is any vector with length $|A| > 0$. Then $A/|A|$ is a unit vector, denoted by **a**, which has the same direction as **A**. Also, any vector **A** may be represented by a unit vector **a** in the direction of **A** multiplied by the magnitude of **A**. That is, $A = |A|a$.

EXAMPLE 1.1 Suppose $|A| = 3$. Then $a = |A|/3$ is a unit vector in the direction of **A**. Also, $A = 3a$.

1.4 Rectangular Unit Vectors i, j, k

An important set of unit vectors, denoted by **i**, **j**, and **k**, are those having the directions, respectively, of the positive x, y, and z axes of a three-dimensional rectangular coordinate system. [See Fig. 1-4(a).]

The coordinate system shown in Fig. 1-4(a), which we use unless otherwise stated, is called a *right-handed coordinate system*. The system is characterized by the following property. If we curl the fingers of the right hand in the direction of a 90° rotation from the positive x-axis to the positive y-axis, then the thumb will point in the direction of the positive z-axis.

Generally speaking, suppose nonzero vectors **A**, **B**, **C** have the same initial point and are not coplanar. Then **A**, **B**, **C** are said to form a *right-handed system* or *dextral system* if a right-threaded screw rotated through an angle less than 180° from **A** to **B** will advance in the direction **C** as shown in Fig. 1-4(b).

(a)　　　　　　　　　(b)　　　　　　　　　(c)

Fig. 1-4

Components of a Vector

Any vector **A** in three dimensions can be represented with an initial point at the origin $O = (0, 0, 0)$ and its end point at some point, say, (A_1, A_2, A_3). Then the vectors $A_1\mathbf{i}$, $A_2\mathbf{j}$, $A_3\mathbf{k}$ are called the *component vectors* of **A** in the x, y, z directions, and the scalars A_1, A_2, A_3 are called the *components* of **A** in the x, y, z directions, respectively. (See Fig. 1-4(c).)

The sum of $A_1\mathbf{i}$, $A_2\mathbf{j}$, and $A_3\mathbf{k}$ is the vector **A**, so we may write

$$\mathbf{A} = A_1\mathbf{i} + A_2\mathbf{j} + A_3\mathbf{k}$$

The magnitude of **A** follows:

$$|\mathbf{A}| = \sqrt{A_1^2 + A_2^2 + A_3^2}$$

Consider a point $P(x, y, z)$ in space. The vector **r** from the origin O to the point P is called the *position* vector (or *radius* vector). Thus **r** may be written

$$\mathbf{r} = x\mathbf{i} + y\mathbf{j} + z\mathbf{k}$$

It has magnitude $|\mathbf{r}| = \sqrt{x^2 + y^2 + z^2}$.

The following proposition applies.

PROPOSITION 1.3:　Suppose $\mathbf{A} = A_1\mathbf{i} + A_2\mathbf{j} + A_3\mathbf{k}$ and $\mathbf{B} = B_1\mathbf{i} + B_2\mathbf{j} + B_3\mathbf{k}$. Then

(i)　$\mathbf{A} + \mathbf{B} = (A_1 + B_1)\mathbf{i} + (A_2 + B_2)\mathbf{j} + (A_3 + B_3)\mathbf{k}$

(ii)　$m\mathbf{A} = m(A_1\mathbf{i} + A_2\mathbf{j} + A_3\mathbf{k}) = (mA_1)\mathbf{i} + (mA_2)\mathbf{j} + (mA_3)\mathbf{k}$

EXAMPLE 1.2　Suppose $\mathbf{A} = 3\mathbf{i} + 5\mathbf{j} - 2\mathbf{k}$ and $\mathbf{B} = 4\mathbf{i} - 8\mathbf{j} + 7\mathbf{k}$.

(a)　To find $\mathbf{A} + \mathbf{B}$, add corresponding components, obtaining $\mathbf{A} + \mathbf{B} = 7\mathbf{i} - 3\mathbf{j} + 5\mathbf{k}$

(b)　To find $3\mathbf{A} - 2\mathbf{B}$, first multiply by the scalars and then add:

$$3\mathbf{A} - 2\mathbf{B} = (9\mathbf{i} + 15\mathbf{j} - 6\mathbf{k}) + (-8\mathbf{i} + 16\mathbf{j} - 14\mathbf{k}) = \mathbf{i} + 31\mathbf{j} - 20\mathbf{k}$$

(c)　To find $|\mathbf{A}|$ and $|\mathbf{B}|$, take the square root of the sum of the squares of the components:

$$|\mathbf{A}| = \sqrt{9 + 25 + 4} = \sqrt{38} \quad \text{and} \quad |\mathbf{B}| = \sqrt{16 + 64 + 49} = \sqrt{129}$$

1.5 Linear Dependence and Linear Independence

Suppose we are given vectors $\mathbf{A}_1, \mathbf{A}_2, \ldots, \mathbf{A}_n$ and scalars $a_1, a_2, \ldots, a_n$. We can multiply the vectors by the corresponding scalars and then add the corresponding scalar products to form the vector

$$\mathbf{B} = a_1\mathbf{A}_1 + a_2\mathbf{A}_2 + \cdots + a_n\mathbf{A}_n$$

Such a vector $\mathbf{B}$ is called a *linear combination* of the vectors $\mathbf{A}_1, \mathbf{A}_2, \ldots, \mathbf{A}_n$.

The following definition applies.

DEFINITION   Vectors $\mathbf{A}_1, \mathbf{A}_2, \ldots, \mathbf{A}_n$ are *linearly dependent* if there exist scalars $a_1, a_2, \ldots, a_n$, not all zero, such that

$$a_1\mathbf{A}_1 + a_2\mathbf{A}_2 + \cdots + a_n\mathbf{A}_n = \mathbf{0}$$

Otherwise, the vectors are *linearly independent*.

The above definition may be restated as follows. Consider the vector equation

$$x_1\mathbf{A}_1 + x_2\mathbf{A}_2 + \cdots + x_n\mathbf{A}_n = \mathbf{0}$$

where $x_1, x_2, \ldots, x_n$ are unknown scalars. This equation always has the zero solution $x_1 = 0$, $x_2 = 0, \ldots, x_n = 0$. If this is the only solution, the vectors are linearly independent. If there is a solution with some $x_j \neq 0$, then the vectors are linearly dependent.

Suppose $\mathbf{A}$ is not the null vector. Then $\mathbf{A}$, by itself, is linearly independent, since

$$m\mathbf{A} = 0 \quad \text{and} \quad \mathbf{A} \neq 0, \text{ implies } m = 0$$

The following proposition applies.

PROPOSITION 1.4:   Two or more vectors are linearly dependent if and only if one of them is a linear combination of the others.

COROLLARY 1.5:   Vectors $\mathbf{A}$ and $\mathbf{B}$ are linearly dependent if and only if one is a multiple of the other.

EXAMPLE 1.3

  (a) The unit vectors $\mathbf{i}, \mathbf{j}, \mathbf{k}$ are linearly independent since neither of them is a linear combination of the other two.

  (b) Suppose $a\mathbf{A} + b\mathbf{B} + c\mathbf{C} = a'\mathbf{A} + b'\mathbf{B} + c'\mathbf{C}$ where $\mathbf{A}, \mathbf{B}, \mathbf{C}$ are linearly independent. Then $a = a'$, $b = b'$, $c = c'$.

1.6 Scalar Field

Suppose that to each point (x, y, z) of a region D in space, there corresponds a number (scalar) $\phi(x, y, z)$. Then ϕ is called a *scalar function of position*, and we say that a *scalar field* ϕ has been defined on D.

EXAMPLE 1.4

  (a) The temperature at any point within or on the Earth's surface at a certain time defines a scalar field.

  (b) The function $\phi(x, y, z) = x^3y - z^2$ defines a scalar field. Consider the point $P(2, 3, 1)$. Then
     $\phi(P) = 8(3) - 1 = 23$.

A scalar field ϕ, which is independent of time, is called a *stationary* or *steady-state scalar field*.

1.7 Vector Field

Suppose to each point (x, y, z) of a region D in space there corresponds a vector $\mathbf{V}(x, y, z)$. Then $\mathbf{V}$ is called a *vector function of position*, and we say that a *vector field* $\mathbf{V}$ has been defined on D.

EXAMPLE 1.5

(a) Suppose the velocity at any point within a moving fluid is known at a certain time. Then a vector field is defined.
(b) The function $\mathbf{V}(x, y, z) = xy^2\mathbf{i} - 2yz^3\mathbf{j} + x^2z\mathbf{k}$ defines a vector field. Consider the point P(2, 3, 1). Then $\mathbf{V}(P) = 18\mathbf{i} - 6\mathbf{j} + 4\mathbf{k}$.

A vector field $\mathbf{V}$ which is independent of time is called a *stationary* or *steady-state vector field*.

1.8 Vector Space $\mathbf{R}^n$

Let $\mathbf{V} = \mathbf{R}^n$ where $\mathbf{R}^n$ consists of all *n*-element sequences $\mathbf{u} = (a_1, a_2, \ldots, a_n)$ of real numbers called the components of $\mathbf{u}$. The term *vector* is used for the elements of $\mathbf{V}$ and we denote them using the letters $\mathbf{u}$, $\mathbf{v}$, and $\mathbf{w}$, with or without a subscript. The real numbers we call *scalars* and we denote them using letters other than $\mathbf{u}$, $\mathbf{v}$, or $\mathbf{w}$.

We define two operations on $\mathbf{V} = \mathbf{R}^n$:

(a) Vector Addition

Given vectors $\mathbf{u} = (a_1, a_2, \ldots, a_n)$ and $\mathbf{v} = (b_1, b_2, \ldots, b_n)$ in $\mathbf{V}$, we define the vector sum $\mathbf{u} + \mathbf{v}$ by

$$\mathbf{u} + \mathbf{v} = (a_1 + b_1, a_2 + b_2, \ldots, a_n + b_n)$$

That is, we add corresponding components of the vectors.

(b) Scalar Multiplication

Given a vector $\mathbf{u} = (a_1, a_2, \ldots, a_n)$ and a scalar k in $\mathbf{R}$, we define the scalar product k$\mathbf{u}$ by

$$k\mathbf{u} = (ka_1, ka_2, \ldots, ka_n)$$

That is, we multiply each component of $\mathbf{u}$ by the scalar **k**.

PROPOSITION 1.6: $\mathbf{V} = \mathbf{R}^n$ satisfies the eight axioms of a vector space listed in Theorem 1.1.

SOLVED PROBLEMS

1.1. State which of the following are scalars and which are vectors:

(a) specific heat, (b) momentum, (c) distance, (d) speed, (e) magnetic field intensity

Solution

(a) scalar, (b) vector, (c) scalar, (d) scalar, (e) vector

1.2. Represent graphically: (a) a force of 10 lb in a direction 30° north of east,
(b) a force of 15 lb in a direction 30° east of north.

Solution

Choosing the unit of magnitude shown, the required vectors are as indicated in Fig. 1-5.

Fig. 1-5

1.3. An automobile travels 3 miles due north, then 5 miles northeast. Represent these displacements graphically and determine the resultant displacement: (a) graphically, (b) analytically.

Solution

Figure 1.6 shows the required displacements.
Vector **OP** or **A** represents displacement of 3 miles due north.
Vector **PQ** or **B** represents displacement of 5 miles north east.
Vector **OQ** or **C** represents the resultant displacement or sum of vectors **A** and **B**, i.e. **C** = **A** + **B**. This is the *triangle law* of vector addition.

The resultant vector **OQ** can also be obtained by constructing the diagonal of the parallelogram *OPQR* having vectors **OP** = **A** and **OR** (equal to vector **PQ** or **B**) as sides. This is the *parallelogram law* of vector addition.

(a) *Graphical Determination of Resultant.* Lay off the 1 mile unit on vector **OQ** to find the magnitude 7.4 miles (approximately). Angle *EOQ* = 61.5°, using a protractor. Then vector **OQ** has magnitude 7.4 miles and direction 61.5° north of east.

(b) *Analytical Determination of Resultant.* From triangle *OPQ*, denoting the magnitudes of **A**, **B**, **C** by A, B, C, we have by the law of cosines

$$C^2 = A^2 + B^2 - 2AB \cos \angle OPQ = 3^2 + 5^2 - 2(3)(5) \cos 135° = 34 + 15\sqrt{2} = 55.21$$

and C = 7.43 (approximately).

By the law of sines, $\dfrac{A}{\sin \angle OQP} = \dfrac{C}{\sin \angle OPQ}$. Then

$$\sin \angle OQP = \frac{A \sin \angle OPQ}{C} = \frac{3(0.707)}{7.43} = 0.2855 \quad \text{and} \quad \angle OQP = 16°35'.$$

Thus vector **OQ** has magnitude 7.43 miles and direction (45° + 16°35′) = 61°35′ north of east.

Fig. 1-6   Fig. 1-7

1.4. Find the sum (resultant) of the following displacements:
A: 10 ft northwest,   **B**: 20 ft 30° north of east,   **C**: 35 ft due south.

Solution

Figure 1-7 shows the resultant obtained as follows (where one unit of length equals 5 feet).

Let **A** begin at the origin. At the terminal point of **A**, place the initial point of **B**. At the terminal point of **B**, place the initial point of **C**. The resultant **D** is formed by joining the initial point of **A** to the terminal point of **C**, that is, **D** = **A** + **B** + **C**. Graphically, the resultant **D** is measured to have magnitude 4.1 units = 20.5 ft and direction 60° south of east.

1.5. Show that addition of vectors is commutative, that is, $\mathbf{A} + \mathbf{B} = \mathbf{B} + \mathbf{A}$. (Theorem 1.1 [$A_4$].)

Solution

As indicated by Fig. 1-8,

$$OP + PQ = OQ \text{ or } \mathbf{A} + \mathbf{B} = \mathbf{C} \quad \text{and} \quad OR + RQ = OQ \text{ or } \mathbf{B} + \mathbf{A} = \mathbf{C}$$

Thus $\mathbf{A} + \mathbf{B} = \mathbf{B} + \mathbf{A}$.

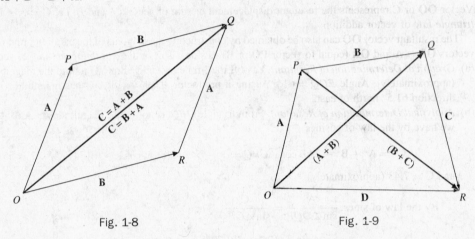

Fig. 1-8   Fig. 1-9

1.6. Show that addition of vectors is associative, that is, $\mathbf{A} + (\mathbf{B} + \mathbf{C}) = (\mathbf{A} + \mathbf{B}) + \mathbf{C}$. (Theorem 1.1 [$A_1$].)

Solution

As indicated by Fig. 1-9,

$$OP + PQ = OQ = (\mathbf{A} + \mathbf{B}) \quad \text{and} \quad PQ + QR = PR = (\mathbf{B} + \mathbf{C})$$
$$OP + PR = OR = \mathbf{D} \text{ or } \mathbf{A} + (\mathbf{B} + \mathbf{C}) = \mathbf{D} \quad \text{and} \quad OQ + QR = OR = \mathbf{D} \text{ or } (\mathbf{A} + \mathbf{B}) + \mathbf{C} = \mathbf{D}$$

Then $\mathbf{A} + (\mathbf{B} + \mathbf{C}) = (\mathbf{A} + \mathbf{B}) + \mathbf{C}$.

1.7. Forces $\mathbf{F}_1, \mathbf{F}_2, \ldots, \mathbf{F}_6$ act on an object P as shown in Fig. 1-10(a). Find the force that is needed to prevent P from moving.

Solution

Since the order of addition of vectors is immaterial, we may start with any vector, say $\mathbf{F}_1$. To $\mathbf{F}_1$ add $\mathbf{F}_2$, then $\mathbf{F}_3$, and so on as pictured in Fig. 1-10(b). The vector drawn from the initial point of $\mathbf{F}_1$ to the terminal point of $\mathbf{F}_6$ is the resultant $\mathbf{R}$, that is, $\mathbf{R} = \mathbf{F}_1 + \mathbf{F}_2 + \cdots + \mathbf{F}_6$.

The force needed to prevent P from moving is $-\mathbf{R}$, sometimes called the *equilibrant*.

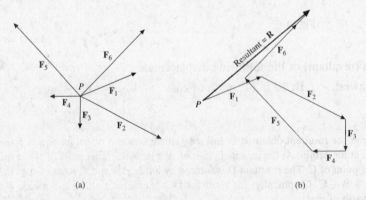

(a)   (b)

Fig. 1-10

1.8. Given vectors **A**, **B**, and **C** in Fig. 1-11(a), construct $\mathbf{A} - \mathbf{B} + 2\mathbf{C}$.

Solution

Beginning with **A**, we add $-\mathbf{B}$ and then add $2\mathbf{C}$ as in Fig. 1-11(b). The resultant is $\mathbf{A} - \mathbf{B} + 2\mathbf{C}$.

(a) (b)

Fig. 1-11

1.9. Given two non-collinear vectors **a** and **b**, as in Fig. 1-12. Find an expression for any vector **r** lying in the plane determined by **a** and **b**.

Solution

Non-collinear vectors are vectors that are not parallel to the same line. Hence, when their initial points coincide, they determine a plane. Let **r** be any vector lying in the plane of **a** and **b** and having its initial point coincident with the initial points of **a** and **b** at O. From the terminal point R of **r**, construct lines parallel to the vectors **a** and **b** and complete the parallelogram $ODRC$ by extension of the lines of action of **a** and **b** if necessary. From Fig. 1-12,

$$\mathbf{OD} = x(\mathbf{OA}) = x\mathbf{a}, \text{ where } x \text{ is a scalar}$$
$$\mathbf{OC} = y(\mathbf{OB}) = y\mathbf{b}, \text{ where } y \text{ is a scalar.}$$

But by the parallelogram law of vector addition

$$\mathbf{OR} = \mathbf{OD} + \mathbf{OC} \quad \text{or} \quad \mathbf{r} = x\mathbf{a} + y\mathbf{b}$$

which is the required expression. The vectors $x\mathbf{a}$ and $y\mathbf{b}$ are called *component vectors* of **r** in the directions **a** and **b**, respectively. The scalars x and y may be positive or negative depending on the relative orientations of the vectors. From the manner of construction, it is clear that x and y are unique for a given **a**, **b**, and **r**. The vectors **a** and **b** are called *base vectors* in a plane.

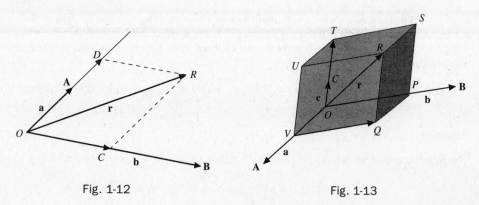

Fig. 1-12   Fig. 1-13

1.10. Given three non-coplanar vectors **a**, **b**, and **c**, find an expression for any vector **r** in three-dimensional space.

Solution

Non-coplanar vectors are vectors that are not parallel to the same plane. Hence, when their initial points coincide, they do not lie in the same plane.

Let **r** be any vector in space having its initial point coincident with the initial points of **a**, **b**, and **c** at O. Through the terminal point of **r**, pass planes parallel respectively to the planes determined by **a** and **b**, **b** and **c**, and **a** and **c**; Refer to Fig. 1-13. Complete the parallelepiped $PQRSTUV$ by extension of the lines of action of **a**, **b**, and **c**, if necessary. From

$$\mathbf{OV} = x(\mathbf{OA}) = x\mathbf{a} \quad \text{where } x \text{ is a scalar}$$
$$\mathbf{OP} = y(\mathbf{OB}) = y\mathbf{b} \quad \text{where } y \text{ is a scalar}$$
$$\mathbf{OT} = x(\mathbf{OC}) = z\mathbf{c} \quad \text{where } z \text{ is a scalar.}$$

But $\mathbf{OR} = \mathbf{OV} + \mathbf{VQ} + \mathbf{QR} = \mathbf{OV} + \mathbf{OP} + \mathbf{OT}$ or $\mathbf{r} = x\mathbf{a} + y\mathbf{b} + z\mathbf{c}$.

From the manner of construction, it is clear that x, y, and z are unique for a given **a**, **b**, **c**, and **r**.

The vectors $x\mathbf{a}$, $y\mathbf{b}$, and $z\mathbf{c}$ are called *component vectors* of **r** in directions **a**, **b**, and **c**, respectively. The vectors **a**, **b**, and **c** are called *base vectors* in three dimensions.

As a special case, if **a**, **b**, and **c** are the unit vectors **i**, **j**, and **k**, which are mutually perpendicular, we see that any vector **r** can be expressed uniquely in terms of **i**, **j**, **k** by the expression $\mathbf{r} = x\mathbf{i} + y\mathbf{j} + z\mathbf{k}$.

Also, if $\mathbf{c} = \mathbf{0}$, then **r** must lie in the plane of **a** and **b**, and so the result of Problem 1.9 is obtained.

1.11. Suppose **a** and **b** are non-collinear. Prove $x\mathbf{a} + y\mathbf{b} = \mathbf{0}$ implies $x = y = 0$.

Solution

Suppose $x \neq 0$. Then $x\mathbf{a} + y\mathbf{b} = \mathbf{0}$ implies $x\mathbf{a} = -y\mathbf{b}$ or $\mathbf{a} = -(y/x)\mathbf{b}$, that is, **a** and **b** must be parallel to the same line (collinear) contrary to hypothesis. Thus, $x = 0$; then $y\mathbf{b} = \mathbf{0}$, from which $y = 0$.

1.12. Suppose $x_1\mathbf{a} + y_1\mathbf{b} = x_2\mathbf{a} + y_2\mathbf{b}$, where **a** and **b** are non-collinear. Prove $x_1 = x_2$ and $y_1 = y_2$.

Solution

Note that $x_1\mathbf{a} + y_1\mathbf{b} = x_2\mathbf{a} + y_2\mathbf{b}$ can be written

$$x_1\mathbf{a} + y_1\mathbf{b} - (x_2\mathbf{a} + y_2\mathbf{b}) = \mathbf{0} \quad \text{or} \quad (x_1 - x_2)\mathbf{a} + (y_1 - y_2)\mathbf{b} = \mathbf{0}.$$

Hence, by Problem 1.11, $x_1 - x_2 = 0$, $y_1 - y_2 = 0$ or $x_1 = x_2$, $y_1 = y_2$.

1.13. Suppose **a**, **b**, and **c** are non-coplanar. Prove $x\mathbf{a} + y\mathbf{b} + z\mathbf{c} = \mathbf{0}$ implies $x = y = z = 0$.

Solution

Suppose $x \neq 0$. Then $x\mathbf{a} + y\mathbf{b} + z\mathbf{c} = \mathbf{0}$ implies $x\mathbf{a} = -y\mathbf{b} - z\mathbf{c}$ or $\mathbf{a} = -(y/x)\mathbf{b} - (z/x)\mathbf{c}$. But $-(y/x)\mathbf{b} - (z/x)\mathbf{c}$ is a vector lying in the plane of **b** and **c** (Problem 1.10); that is, **a** lies in the plane of **b** and **c**, which is clearly a contradiction to the hypothesis that **a**, **b**, and **c** are non-coplanar. Hence, $x = 0$. By similar reasoning, contradictions are obtained upon supposing $y \neq 0$ and $z \neq 0$.

1.14. Suppose $x_1\mathbf{a} + y_1\mathbf{b} + z_1\mathbf{c} = x_2\mathbf{a} + y_2\mathbf{b} + z_2\mathbf{c}$, where **a**, **b**, and **c** are non-coplanar. Prove $x_1 = x_2$, $y_1 = y_2$, and $z_1 = z_2$.

Solution

The equation can be written $(x_1 - x_2)\mathbf{a} + (y_1 - y_2)\mathbf{b} + (z_1 - z_2)\mathbf{c} = \mathbf{0}$. Then, by Problem 1.13,

$$x_1 - x_2 = 0, \; y_1 - y_2 = 0, \; z_1 - z_2 = 0 \quad \text{or} \quad x_1 = x_2, \; y_1 = y_2, \; z_1 = z_2.$$

1.15. Suppose the midpoints of the consecutive sides of a quadrilateral are connected by straight lines. Prove that the resulting quadrilateral is a parallelogram.

Solution

Let $ABCD$ be the given quadrilateral and P, Q, R, S the midpoints of its sides. Refer to Fig. 1-14.
Then, $\mathbf{PQ} = \frac{1}{2}(\mathbf{a} + \mathbf{b})$, $\mathbf{QR} = \frac{1}{2}(\mathbf{b} + \mathbf{c})$, $\mathbf{RS} = \frac{1}{2}(\mathbf{c} + \mathbf{d})$, $\mathbf{SP} = \frac{1}{2}(\mathbf{d} + \mathbf{a})$.
But, $\mathbf{a} + \mathbf{b} + \mathbf{c} + \mathbf{d} = \mathbf{0}$. Then

$$\mathbf{PQ} = \tfrac{1}{2}(\mathbf{a} + \mathbf{b}) = -\tfrac{1}{2}(\mathbf{c} + \mathbf{d}) = \mathbf{SR} \quad \text{and} \quad \mathbf{QR} = \tfrac{1}{2}(\mathbf{b} + \mathbf{c}) = -\tfrac{1}{2}(\mathbf{d} + \mathbf{a}) = \mathbf{PS}$$

Thus, opposite sides are equal and parallel and $PQRS$ is a parallelogram.

Fig. 1-14

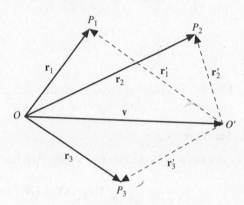

Fig. 1-15

1.16. Let P_1, P_2, and P_3 be points fixed relative to an origin O and let $\mathbf{r}_1$, $\mathbf{r}_2$, and $\mathbf{r}_3$ be position vectors from O to each point. Suppose the vector equation $a_1\mathbf{r}_1 + a_2\mathbf{r}_2 + a_3\mathbf{r}_3 = \mathbf{0}$ holds with respect to origin O. Show that it will hold with respect to any other origin O' if and only if $a_1 + a_2 + a_3 = 0$.

Solution

Let $\mathbf{r}_1'$, $\mathbf{r}_2'$, and $\mathbf{r}_3'$ be the position vectors of P_1, P_2, and P_3 with respect to O' and let $\mathbf{v}$ be the position vector of O' with respect to O. We seek conditions under which the equation $a_1\mathbf{r}_1' + a_2\mathbf{r}_2' + a_3\mathbf{r}_3' = \mathbf{0}$ will hold in the new reference system.

From Fig. 1-15, it is clear that $\mathbf{r}_1 = \mathbf{v} + \mathbf{r}_1'$, $\mathbf{r}_2 = \mathbf{v} + \mathbf{r}_2'$, $\mathbf{r}_3 = \mathbf{v} + \mathbf{r}_3'$ so that $a_1\mathbf{r}_1 + a_2\mathbf{r}_2 + a_3\mathbf{r}_3 = \mathbf{0}$ becomes

$$a_1\mathbf{r}_1 + a_2\mathbf{r}_2 + a_3\mathbf{r}_3 = a_1(\mathbf{v} + \mathbf{r}_1') + a_2(\mathbf{v} + \mathbf{r}_2') + a_3(\mathbf{v} + \mathbf{r}_3')$$
$$= (a_1 + a_2 + a_3)\mathbf{v} + a_1\mathbf{r}_1' + a_2\mathbf{r}_2' + a_3\mathbf{r}_3' = \mathbf{0}$$

The result $a_1\mathbf{r}_1' + a_2\mathbf{r}_2' + a_3\mathbf{r}_3' = \mathbf{0}$ will hold if and only if

$$(a_1 + a_2 + a_3)\mathbf{v} = \mathbf{0}, \quad \text{i.e. } a_1 + a_2 + a_3 = 0.$$

The result can be generalized.

1.17. Prove that the diagonals of a parallelogram bisect each other.

Solution

Let $ABCD$ be the given parallelogram with diagonals intersecting at P as in Fig. 1-16.
Since $\mathbf{BD} + \mathbf{a} = \mathbf{b}$, $\mathbf{BD} = \mathbf{b} - \mathbf{a}$. Then $\mathbf{BP} = x(\mathbf{b} - \mathbf{a})$.
Since $\mathbf{AC} = \mathbf{a} + \mathbf{b}$, $\mathbf{AP} = y(\mathbf{a} + \mathbf{b})$.
But $\mathbf{AB} = \mathbf{AP} + \mathbf{PB} = \mathbf{AP} - \mathbf{BP}$,
that is, $\mathbf{a} = y(\mathbf{a} + \mathbf{b}) - x(\mathbf{b} - \mathbf{a}) = (x + y)\mathbf{a} + (y - x)\mathbf{b}$.

Since **a** and **b** are non-collinear, we have by Problem 1.12, $x + y = 1$ and $y - x = 0$ (i.e., $x = y = \frac{1}{2}$) and P is the mid-point of both diagonals.

Fig. 1-16   Fig. 1-17

1.18. Find the equation of the straight line that passes through two given points A and B having position vectors **a** and **b** with respect to the origin.

Solution

Let **r** be the position vector of a point P on the line through A and B as in Fig. 1-17. Then

$$\mathbf{OA} + \mathbf{AP} = \mathbf{OP} \quad \text{or} \quad \mathbf{a} + \mathbf{AP} = \mathbf{r} \quad \text{(i.e., } \mathbf{AP} = \mathbf{r} - \mathbf{a})$$

and

$$\mathbf{OA} + \mathbf{AB} = \mathbf{OB} \quad \text{or} \quad \mathbf{a} + \mathbf{AB} = \mathbf{b} \quad \text{(i.e., } \mathbf{AB} = \mathbf{b} - \mathbf{a})$$

Since **AP** and **AB** are collinear, $\mathbf{AP} = t\mathbf{AB}$ or $\mathbf{r} - \mathbf{a} = t(\mathbf{b} - \mathbf{a})$. Then the required equation is

$$\mathbf{r} = \mathbf{a} + t(\mathbf{b} - \mathbf{a}) \quad \text{or} \quad \mathbf{r} = (1 - t)\mathbf{a} + t\mathbf{b}$$

If the equation is written $(1 - t)\mathbf{a} + t\mathbf{b} - \mathbf{r} = \mathbf{0}$, the sum of the coefficients of **a**, **b**, and **r** is $1 - t + t - 1 = 0$. Hence, by Problem 18, it is seen that the point P is always on the line joining A and B and does not depend on the choice of origin O, which is, of course, as it should be.

Another Method. Since **AP** and **PB** are collinear, we have for scalars m and n:

$$m\mathbf{AP} = n\mathbf{PB} \quad \text{or} \quad m(\mathbf{r} - \mathbf{a}) = n(\mathbf{b} - \mathbf{r})$$

Solving $\mathbf{r} = (m\mathbf{a} + n\mathbf{b})/(m + n)$, which is called the *symmetric form*.

1.19. Consider points $P(2, 4, 3)$ and $Q(1, -5, 2)$ in 3-space $\mathbf{R}^3$, as in Fig. 1-18.

(a) Find the position vectors $\mathbf{r}_1$ and $\mathbf{r}_2$ for P and Q in terms of the unit vectors **i**, **j**, **k**.
(b) Determine graphically and analytically the resultant of these position vectors.

Solution

(a) $\mathbf{r}_1 = \mathbf{OP} = \mathbf{OC} + \mathbf{CB} + \mathbf{BP} = 2\mathbf{i} + 4\mathbf{j} + 3\mathbf{k}$

   $\mathbf{r}_2 = \mathbf{OQ} = \mathbf{OD} + \mathbf{DE} + \mathbf{EQ} = \mathbf{i} - 5\mathbf{j} + 2\mathbf{k}$

(b) *Graphically*, the resultant of $\mathbf{r}_1$ and $\mathbf{r}_2$ is obtained as the diagonal **OR** of parallelogram *OPRQ*. *Analytically*, the resultant of $\mathbf{r}_1$ and $\mathbf{r}_2$ is given by

$$\mathbf{r}_1 + \mathbf{r}_2 = (2\mathbf{i} + 4\mathbf{j} + 3\mathbf{k}) + (\mathbf{i} - 5\mathbf{j} + 2\mathbf{k}) = 3\mathbf{i} - \mathbf{j} + 5\mathbf{k}$$

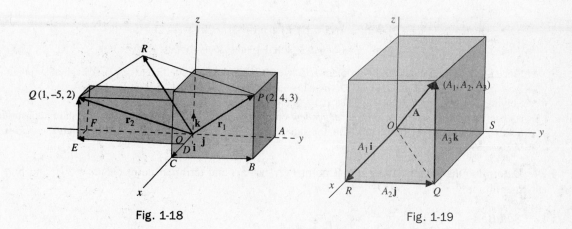

Fig. 1-18   Fig. 1-19

1.20. Prove that the magnitude of the vector $\mathbf{A} = A_1\mathbf{i} + A_2\mathbf{j} + A_3\mathbf{k}$, pictured in Fig. 1-19, is $|\mathbf{A}| = \sqrt{A_1^2 + A_2^2 + A_3^2}$.

Solution

By the Pythagorean theorem,

$$(\overline{OP})^2 = (\overline{OQ})^2 + (\overline{QP})^2$$

where $\overline{OP}$ denotes the magnitude of vector **OP**, and so on. Similarly, $(\overline{OQ})^2 = (\overline{OR})^2 + (\overline{RQ})^2$.

Then $(\overline{OP})^2 = (\overline{OR})^2 + (\overline{RQ})^2 + (\overline{QP})^2$ or $A^2 = A_1^2 + A_2^2 + A_3^2$ (i.e., $\mathbf{A} = \sqrt{A_1^2 + A_2^2 + A_3^2}$).

1.21. Given the radius vectors $\mathbf{r}_1 = 3\mathbf{i} - 2\mathbf{j} + \mathbf{k}$, $\mathbf{r}_2 = 3\mathbf{i} + 4\mathbf{j} + 9\mathbf{k}$, $\mathbf{r}_3 = -\mathbf{i} + 2\mathbf{j} + 2\mathbf{k}$. Find the magnitudes of: (a) $\mathbf{r}_3$, (b) $\mathbf{r}_1 + \mathbf{r}_2 + \mathbf{r}_3$, (c) $\mathbf{r}_1 - \mathbf{r}_2 + 4\mathbf{r}_3$.

Solution

(a) $|\mathbf{r}_3| = |-\mathbf{i} + 2\mathbf{j} + 2\mathbf{k}| = \sqrt{(-1)^2 + (2)^2 + (2)^2} = 3$.

(b) $\mathbf{r}_1 + \mathbf{r}_2 + \mathbf{r}_3 = 3\mathbf{i} + 4\mathbf{j} + 12\mathbf{k}$, hence $|\mathbf{r}_1 + \mathbf{r}_2 + \mathbf{r}_3| = \sqrt{9 + 16 + 144} = \sqrt{169} = 13$.

(c) $\mathbf{r}_1 - \mathbf{r}_2 + 4\mathbf{r}_3 = 2\mathbf{i} + 2\mathbf{j} = \sqrt{4 + 4} = \sqrt{8} = 2\sqrt{2}$.

1.22. Find a unit vector **u** parallel to the resultant **R** of vectors $\mathbf{r}_1 = 2\mathbf{i} + 4\mathbf{j} - 5\mathbf{k}$ and $\mathbf{r}_2 = -\mathbf{i} - 2\mathbf{j} + 3\mathbf{k}$.

Solution

Resultant $\mathbf{R} = \mathbf{r}_1 + \mathbf{r}_2 = (2\mathbf{i} + 4\mathbf{j} - 5\mathbf{k}) + (-\mathbf{i} - 2\mathbf{j} + 3\mathbf{k}) = \mathbf{i} + 2\mathbf{j} - 2\mathbf{k}$. Also,

$$\text{Magnitude of } \mathbf{R} = |\mathbf{R}| = |\mathbf{i} + 2\mathbf{j} - 2\mathbf{k}| = \sqrt{(1)^2 + (2)^2 + (-2)^2} = 3.$$

Then **u** is equal to $\mathbf{R}/|\mathbf{R}|$. That is,

$$\mathbf{u} = \mathbf{R}/|\mathbf{R}| = (\mathbf{i} + 2\mathbf{j} - 2\mathbf{k})/3 = (1/3)\mathbf{i} + (2/3)\mathbf{j} - (2/3)\mathbf{k}$$

Check: $|(1/3)\mathbf{i} + (2/3)\mathbf{j} - (2/3)\mathbf{k}| = \sqrt{(1/3)^2 + (2/3)^2 + (-2/3)^2} = 1$.

1.23. Suppose $\mathbf{r}_1 = 2\mathbf{i} - \mathbf{j} + \mathbf{k}$, $\mathbf{r}_2 = \mathbf{i} - 3\mathbf{j} - 2\mathbf{k}$, $\mathbf{r}_3 = -2\mathbf{i} + \mathbf{j} - 3\mathbf{k}$. Write $\mathbf{r}_4 = \mathbf{i} + 3\mathbf{j} + 2\mathbf{k}$ as a linear combination of $\mathbf{r}_1$, $\mathbf{r}_2$, $\mathbf{r}_3$; that is, find scalars a, b, c such that $\mathbf{r}_4 = a\mathbf{r}_1 + b\mathbf{r}_2 + c\mathbf{r}_3$.

Solution

We require

$$\mathbf{i} + 3\mathbf{j} + 2\mathbf{k} = a(2\mathbf{i} - \mathbf{j} + \mathbf{k}) + b(\mathbf{i} - 3\mathbf{j} - 2\mathbf{k}) + c(-2\mathbf{i} + \mathbf{j} - 3\mathbf{k})$$
$$= (2a + b - 2c)\mathbf{i} + (-a + 3b + c)\mathbf{j} + (a - 2b - 3c)\mathbf{k}$$

Since $\mathbf{i}, \mathbf{j}, \mathbf{k}$ are non-coplanar, by Problem 1.13, we set corresponding coefficients equal to each other obtaining
$2a + b - 2c = 1$, $-a + 3b + c = 3$, $a - 2b - 3c = 2$
Solving, $a = -2$, $b = 1$, $c = -2$. Thus $\mathbf{r}_4 = -2\mathbf{r}_1 + \mathbf{r}_2 - 2\mathbf{r}_3$.

     The vector $\mathbf{r}_4$ is said to be *linearly dependent* on $\mathbf{r}_1$, $\mathbf{r}_2$, and $\mathbf{r}_3$; in other words $\mathbf{r}_1$, $\mathbf{r}_2$, $\mathbf{r}_3$, and $\mathbf{r}_4$ constitute a *linearly dependent* set of vectors. On the other hand, any three (or fewer) of these vectors are *linearly independent*.

1.24. Determine the vector having initial point $P(x_1, y_1, z_1)$ and terminal point $Q(x_2, y_2, z_2)$, and find its magnitude.

Solution

Consider Fig. 1-20. The position vectors of P and Q are, respectively,

$$\mathbf{r}_1 = x_1\mathbf{i} + y_1\mathbf{j} + z_1\mathbf{k} \quad \text{and} \quad \mathbf{r}_2 = x_2\mathbf{i} + y_2\mathbf{j} + z_2\mathbf{k}$$

Then $\mathbf{r}_1 + \mathbf{PQ} = \mathbf{r}_2$ or

$$\mathbf{PQ} = \mathbf{r}_2 - \mathbf{r}_1 = (x_2\mathbf{i} + y_2\mathbf{j} + z_2\mathbf{k}) - (x_1\mathbf{i} + y_1\mathbf{j} + z_1\mathbf{k})$$
$$= (x_2 - x_1)\mathbf{i} + (y_2 - y_1)\mathbf{j} + (z_2 - z_1)\mathbf{k}$$

     Magnitude of $\mathbf{PQ} = \overline{PQ} = \sqrt{(x_2 - x_1)^2 + (y_2 - y_1)^2 + (z_2 - z_1)^2}$. Note that this is the distance between points P and Q.

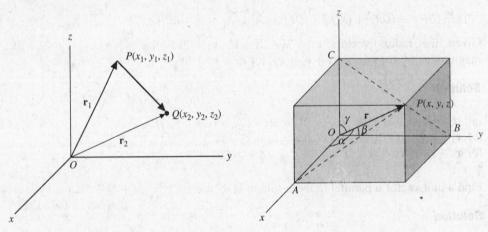

Fig. 1-20   Fig. 1-21

1.25. Determine the angles α, β, and γ that the vector $\mathbf{r} = x\mathbf{i} + y\mathbf{j} + z\mathbf{k}$ makes with the positive directions of the coordinate axes and show that

$$\cos^2 \alpha + \cos^2 \beta + \cos^2 \gamma = 1.$$

Solution

Referring to Fig. 1-21, triangle OAP is a right triangle with right angle at A; then $\cos \alpha = x/|\mathbf{r}|$. Similarly, from right triangles OBP and OCP, $\cos \beta = y/|\mathbf{r}|$ and $\cos \gamma = z/|\mathbf{r}|$, respectively. Also, $|\mathbf{r}| = r = \sqrt{x^2 + y^2 + z^2}$.

Then, $\cos\alpha = x/r$, $\cos\beta = y/r$, and $\cos\gamma = z/r$, from which α, β, and γ can be obtained. From these, it follows that

$$\cos^2\alpha + \cos^2\beta + \cos^2\gamma = \frac{x^2 + y^2 + z^2}{r^2} = 1.$$

The numbers $\cos\alpha$, $\cos\beta$, $\cos\gamma$ are called the *direction cosines* of the vector **OP**.

1.26. Forces **A**, **B**, and **C** acting on an object are given in terms of their components by the vector equations $\mathbf{A} = A_1\mathbf{i} + A_2\mathbf{j} + A_3\mathbf{k}$, $\mathbf{B} = B_1\mathbf{i} + B_2\mathbf{j} + B_3\mathbf{k}$, $\mathbf{C} = C_1\mathbf{i} + C_2\mathbf{j} + C_3\mathbf{k}$. Find the magnitude of the resultant of these forces.

Solution

Resultant force $\mathbf{R} = \mathbf{A} + \mathbf{B} + \mathbf{C} = (A_1 + B_1 + C_1)\mathbf{i} + (A_2 + B_2 + C_2)\mathbf{j} + (A_3 + B_3 + C_3)\mathbf{k}$.

Magnitude of resultant $= \sqrt{(A_1 + B_1 + C_1)^2 + (A_2 + B_2 + C_2)^2 + (A_3 + B_3 + C_3)^2}$.
The result is easily extended to more than three forces.

1.27. Find a set of equations for the straight lines passing through the points $P(x_1, y_1, z_1)$ and $Q(x_2, y_2, z_2)$.

Solution

Let $\mathbf{r}_1$ and $\mathbf{r}_2$ be the position vectors of P and Q, respectively, and $\mathbf{r}$ the position vector of any point R on the line joining P and Q, as pictured in Fig. 1-22

$$\mathbf{r}_1 + \mathbf{PR} = \mathbf{r} \quad \text{or} \quad \mathbf{PR} = \mathbf{r} - \mathbf{r}_1$$
$$\mathbf{r}_1 + \mathbf{PQ} = \mathbf{r}_2 \quad \text{or} \quad \mathbf{PQ} = \mathbf{r}_2 - \mathbf{r}_1$$

But $\mathbf{PR} = t\mathbf{PQ}$ where t is a scalar. Then, $\mathbf{r} - \mathbf{r}_1 = t(\mathbf{r}_2 - \mathbf{r}_1)$ is the required vector equation of the straight line (compare with Problem 1.14).

In rectangular coordinates, we have, since $\mathbf{r} = x\mathbf{i} + y\mathbf{j} + z\mathbf{k}$,

$$(x\mathbf{i} + y\mathbf{j} + z\mathbf{k}) - (x_1\mathbf{i} + y_1\mathbf{j} + z_1\mathbf{k}) = t[(x_2\mathbf{i} + y_2\mathbf{j} + z_2\mathbf{k}) - (x_1\mathbf{i} + y_1\mathbf{j} + z_1\mathbf{k})]$$

or

$$(x - x_1)\mathbf{i} + (y - y_1)\mathbf{j} + (z - z_1)\mathbf{k} = t[(x_2 - x_1)\mathbf{i} + (y_2 - y_1)\mathbf{j} + (z_2 - z_1)\mathbf{k}]$$

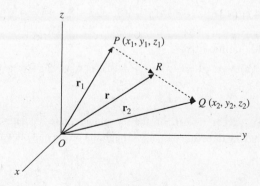

Fig. 1-22

Since $\mathbf{i}, \mathbf{j}, \mathbf{k}$ are non-coplanar vectors, we have by Problem 1.14,

$$x - x_1 = t(x_2 - x_1),\ y - y_1 = t(y_2 - y_1),\ z - z_1 = t(z_2 - z_1)$$

as the parametric equations of the line, t being the parameter. Eliminating t, the equations become

$$\frac{x - x_1}{x_2 - x_1} = \frac{y - y_1}{y_2 - y_1} = \frac{z - z_1}{z_2 - z_1}.$$

1.28. Prove Proposition 1.4: Two or more vectors, $\mathbf{A}_1, \mathbf{A}_2, \ldots, \mathbf{A}_m$, are linearly dependent if and only if one of them is a linear combination of the others.

Solution

Suppose, say, $\mathbf{A}_j$ is a linear combination of the others,

$$\mathbf{A}_j = a_1\mathbf{A}_1 + \cdots + a_{j-1}\mathbf{A}_{j-1} + a_{j+1}\mathbf{A}_{j+1} + \cdots + a_m\mathbf{A}_m$$

Then, by adding $-\mathbf{A}_j$ to both sides, we obtain

$$a_1\mathbf{A}_1 + \cdots + a_{j-1}\mathbf{A}_{j-1} - \mathbf{A}_j + a_{j+1}\mathbf{A}_{j+1} + \cdots + a_m\mathbf{A}_m = \mathbf{0}$$

where the coefficient of $\mathbf{A}_j$ is not 0. Thus the vectors are linearly dependent.

Conversely, suppose the vectors are linearly dependent, say

$$b_1\mathbf{A}_1 + \cdots + b_j\mathbf{A}_j + \cdots + b_m\mathbf{A}_m = \mathbf{0} \quad \text{where } b_j \neq 0$$

Then we can solve for $\mathbf{A}_j$ obtaining

$$\mathbf{A}_j = (b_1/b_j)\mathbf{A}_1 + \cdots + (b_{j-1}/b_j)\mathbf{A}_{j-1} + (b_{j+1}/b_j)\mathbf{A}_{j+1} + \cdots + (b_m/b_j)\mathbf{A}_m$$

Thus $\mathbf{A}_j$ is a linear combination of the others.

1.29. Consider the scalar field φ defined by $\varphi(x, y, z) = 3x^2z^2 - xy^3 - 15$. Find φ at the points (a) $(0, 0, 0)$, (b) $(1, -2, 2)$, (c) $(-1, -2, -3)$.

Solution

(a) $\varphi(0, 0, 0) = 3(0)^2(0)^2 - (0)(0)^3 - 15 = 0 - 0 - 15 = -15$.

(b) $\varphi(1, -2, 2) = 3(1)^2(2)^2 - (1)(-2)^3 - 15 = 12 + 8 - 15 = 5$.

(c) $\varphi(-1, -2, -3) = 3(-1)^2(-3)^2 - (-1)(-2)^3 - 15 = 27 - 8 + 15 = 4$.

1.30. Describe the vector fields defined by:

(a) $\mathbf{V}(x, y) = x\mathbf{i} + y\mathbf{j}$, (b) $\mathbf{V}(x, y) = -x\mathbf{i} - y\mathbf{j}$, (c) $\mathbf{V}(x, y, z) = x\mathbf{i} + y\mathbf{j} + z\mathbf{k}$

Solution

(a) At each point (x, y), except $(0, 0)$, of the xy plane, there is defined a unique vector $x\mathbf{i} + y\mathbf{j}$ of magnitude $\sqrt{x^2 + y^2}$ having direction passing through the origin and outward from it. To simplify graphing procedures, note that all vectors associated with points on the circles $x^2 + y^2 = a^2$, $a > 0$ have magnitude a. The field therefore appears in Fig. 1-23(a) where an appropriate scale is used.

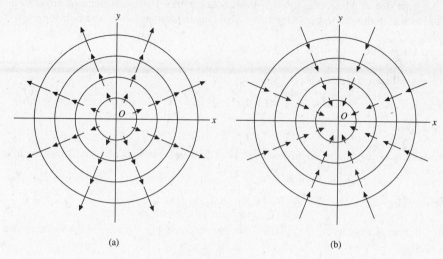

Fig. 1-23

(b) Here each vector is equal to but opposite in direction to the corresponding one in Part (a). The field therefore appears in Fig. 1-23(b).

In Fig. 1-23(a), the field has the appearance of a fluid emerging from a point source O and flowing in the directions indicated. For this reason, the field is called a *source field* and O is a *source*.

In Fig. 1-23(b), the field seems to be flowing toward O, and the field is therefore called a *sink field* and O is a *sink*.

In three dimensions, the corresponding interpretation is that a fluid is emerging radially from (or proceeding radially toward) a line source (or line sink).

The vector field is called two-dimensional since it is independent of z.

(c) Since the magnitude of each vector is $\sqrt{x^2 + y^2 + z^2}$, all points on the sphere $x^2 + y^2 + z^2 = a^2$, $a > 0$ have vectors of magnitude a associated with them. The field therefore takes on the appearance of that of a fluid emerging from source O and proceeding in all directions in space. This is a *three-dimensional source field*.

SUPPLEMENTARY PROBLEMS

1.31. Determine which of the following are scalar and which are vectors:

(a) Kinetic energy, (b) electric field intensity, (c) entropy, (d) work, (e) centrifugal force, (f) temperature, (g) charge, (h) shearing stress, (i) frequency.

1.32. An airplane travels 200 miles due west, and then 150 miles 60° north of west. Determine the resultant displacement.

1.33. Find the resultant of the following displacements: **A**: 20 miles 30° south of east; **B**: 50 miles due west; **C**: 40 miles 30° northeast; **D**: 30 miles 60° south of west.

1.34. Suppose ABCDEF are the vertices of a regular hexagon. Find the resultant of the forces represented by the vectors **AB**, **AC**, **AD**, **AE**, and **AF**.

1.35. Consider vectors **A** and **B**. Show that: (a) $|\mathbf{A} + \mathbf{B}| \leq |\mathbf{A}| + |\mathbf{B}|$; (b) $|\mathbf{A} - \mathbf{B}| \geq |\mathbf{A}| - |\mathbf{B}|$.

1.36. Show that: $|\mathbf{A} + \mathbf{B} + \mathbf{C}| \leq |\mathbf{A}| + |\mathbf{B}| + |\mathbf{C}|$.

1.37. Two towns, A and B, are situated directly opposite each other on the banks of a river whose width is 8 miles and which flows at a speed of 4 mi/hr. A man located at A wishes to reach town C which is 6 miles upstream from and on the same side of the river as town B. If his boat can travel at a maximum speed of 10 mi/hr and if he wishes to reach C in the shortest possible time, what course must he follow and how long will the trip take?

1.38. Simplify: $2\mathbf{A} + \mathbf{B} + 3\mathbf{C} - \{\mathbf{A} - 2\mathbf{B} - 2(2\mathbf{A} - 3\mathbf{B} - \mathbf{C})\}$.

1.39. Consider non-collinear vectors $\mathbf{a}$ and $\mathbf{b}$. Suppose

$$\mathbf{A} = (x + 4y)\mathbf{a} + (2x + y + 1)\mathbf{b} \quad \text{and} \quad \mathbf{B} = (y - 2x + 2)\mathbf{a} + (2x - 3y - 1)\mathbf{b}$$

Find x and y such that $3\mathbf{A} = 2\mathbf{B}$.

1.40. The base vectors $\mathbf{a}_1$, $\mathbf{a}_2$, and $\mathbf{a}_3$ are given in terms of the base vectors $\mathbf{b}_1$, $\mathbf{b}_2$, and $\mathbf{b}_3$ by the relations

$$\mathbf{a}_1 = 2\mathbf{b}_1 + 3\mathbf{b}_2 - \mathbf{b}_3, \quad \mathbf{a}_2 = \mathbf{b}_1 - 2\mathbf{b}_2 + 2\mathbf{b}_3, \quad \mathbf{a}_3 = -2\mathbf{b}_1 + \mathbf{b}_2 - 2\mathbf{b}_3$$

Suppose $\mathbf{F} = 3\mathbf{b}_1 - \mathbf{b}_2 + 2\mathbf{b}_3$. Express $\mathbf{F}$ in terms of $\mathbf{a}_1$, $\mathbf{a}_2$, and $\mathbf{a}_3$.

1.41. An object P is acted upon by three coplanar forces as shown in Fig. 1-24. Find the force needed to prevent P from moving.

Fig. 1-24	Fig. 1-25

1.42. A 100 lb weight is suspended from the center of a rope as shown in Fig. 1-25. Determine the tension $\mathbf{T}$ in the rope.

1.43. Suppose $\mathbf{a}$, $\mathbf{b}$, and $\mathbf{c}$ are non-coplanar vectors. Determine whether the following vectors are linearly independent or linearly dependent:

$$\mathbf{r}_1 = 2\mathbf{a} - 3\mathbf{b} + \mathbf{c}, \quad \mathbf{r}_2 = 3\mathbf{a} - 5\mathbf{b} + 2\mathbf{c}, \quad \mathbf{r}_3 = 4\mathbf{a} - 5\mathbf{b} + \mathbf{c}.$$

1.44. (a) If O is any point within triangle ABC and P, Q, and R are midpoints of the sides AB, BC, and CA, respectively, prove that $\mathbf{OA} + \mathbf{OB} + \mathbf{OC} = \mathbf{OP} + \mathbf{OQ} + \mathbf{OR}$.

(b) Does the result hold if O is any point outside the triangle? Prove your result.

1.45. In Fig. 1-26, $ABCD$ is a parallelogram with P and Q the midpoints of sides $\mathbf{BC}$ and $\mathbf{CD}$, respectively. Prove that $\mathbf{AP}$ and $\mathbf{AQ}$ trisect diagonal $\mathbf{BD}$ at points E and F.

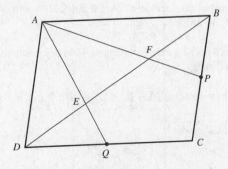

Fig. 1-26

1.46. Prove that the line joining the midpoints of two sides of a triangle is parallel to the third side and has one half of its magnitude.

1.47. Prove that the medians of a triangle meet in a common point, which is a point of trisection of the medians.

1.48. Prove that the angle bisectors of a triangle meet in a common point.

1.49. Let the position vectors of points P and Q relative to the origin O is given by vectors $\mathbf{p}$ and $\mathbf{q}$, respectively. Suppose R is a point which divides PQ into segments that are in the ratio $m:n$. Show that the position vector of R is given by $\mathbf{r} = (m\mathbf{p} + n\mathbf{q})/(m + n)$ and that this is independent of the origin.

1.50. A quadrilateral $ABCD$ has masses of 1, 2, 3, and 4 units located, respectively, at its vertices $A(-1, -2, 2)$, $B(3, 2, -1)$, $C(1, -2, 4)$, and $D(3, 1, 2)$. Find the coordinates of the centroid.

1.51. Show that the equation of a plane which passes through three given points A, B, and C not in the same straight line and having position vectors $\mathbf{a}$, $\mathbf{b}$, and $\mathbf{c}$ relative to an origin O, can be written

$$\mathbf{r} = \frac{m\mathbf{a} + n\mathbf{b} + p\mathbf{c}}{m + n + p}$$

where m, n, p are scalars. Verify that the equation is independent of the origin.

1.52. The position vectors of points P and Q are given by $\mathbf{r}_1 = 2\mathbf{i} + 3\mathbf{j} - \mathbf{k}$, $\mathbf{r}_2 = 4\mathbf{i} - 3\mathbf{j} + 2\mathbf{k}$. Determine $\mathbf{PQ}$ in terms of $\mathbf{i}$, $\mathbf{j}$, and $\mathbf{k}$, and find its magnitude.

1.53. Suppose $\mathbf{A} = 3\mathbf{i} - \mathbf{j} - 4\mathbf{k}$, $\mathbf{B} = -2\mathbf{i} + 4\mathbf{j} - 3\mathbf{k}$, $\mathbf{C} = \mathbf{i} + 2\mathbf{j} - \mathbf{k}$. Find
(a) $2\mathbf{A} - \mathbf{B} + 3\mathbf{C}$, (b) $|\mathbf{A} + \mathbf{B} + \mathbf{C}|$, (c) $|3\mathbf{A} - 2\mathbf{B} + 4\mathbf{C}|$, (d) a unit vector parallel to $3\mathbf{A} - 2\mathbf{B} + 4\mathbf{C}$.

1.54 The following forces act on a particle P: $\mathbf{F}_1 = 2\mathbf{i} + 3\mathbf{j} - 5\mathbf{k}$, $\mathbf{F}_2 = -5\mathbf{i} + \mathbf{j} + 3\mathbf{k}$, $\mathbf{F}_3 = \mathbf{i} - 2\mathbf{j} + 4\mathbf{k}$, $\mathbf{F}_4 = 4\mathbf{i} - 3\mathbf{j} - 2\mathbf{k}$, measured in pounds. Find (a) the resultant of the forces, (b) the magnitude of the resultant.

1.55. In each case, determine whether the vectors are linearly independent or linearly dependent:
(a) $\mathbf{A} = 2\mathbf{i} + \mathbf{j} - 3\mathbf{k}$, $\mathbf{B} = \mathbf{i} - 4\mathbf{k}$, $\mathbf{C} = 4\mathbf{i} + 3\mathbf{j} - \mathbf{k}$, (b) $\mathbf{A} = \mathbf{i} - 3\mathbf{j} + 2\mathbf{k}$, $\mathbf{B} = 2\mathbf{i} - 4\mathbf{j} - \mathbf{k}$, $\mathbf{C} = 3\mathbf{i} + 2\mathbf{j} - \mathbf{k}$.

1.56. Prove that any four vectors in three dimensions must be linearly dependent.

1.57. Show that a necessary and sufficient condition that the vectors $\mathbf{A} = A_1\mathbf{i} + A_2\mathbf{j} + A_3\mathbf{k}$, $\mathbf{B} = B_1\mathbf{i} + B_2\mathbf{j} + B_3\mathbf{k}$, $\mathbf{C} = C_1\mathbf{i} + C_2\mathbf{j} + C_3\mathbf{k}$ be linearly independent is that the determinant

$$\begin{vmatrix} A_1 & A_2 & A_3 \\ B_1 & B_2 & B_3 \\ C_1 & C_2 & C_3 \end{vmatrix} \text{ be different from zero.}$$

1.58. (a) Prove that the vectors $\mathbf{A} = 3\mathbf{i} + \mathbf{j} - 2\mathbf{k}$, $\mathbf{B} = -\mathbf{i} + 3\mathbf{j} + 4\mathbf{k}$, $\mathbf{C} = 4\mathbf{i} - 2\mathbf{j} - 6\mathbf{k}$ can form the sides of a triangle.
(b) Find the lengths of the medians of the triangle.

1.59. Given the scalar field defined by $\phi(x, y, z) = 4yx^3 + 3xyz - z^2 + 2$. Find (a) $\phi(1, -1, -2)$, (b) $\phi(0, -3, 1)$.

ANSWERS TO SUPPLEMENTARY PROBLEMS

1.31. (a) s, (b) v, (c) s, (d) s, (e) v, (f) s, (g) s, (h) v, (i) s.

1.32. Magnitude 304.1 ($50\sqrt{37}$), direction $25°17'$ north of east (arcsin $3\sqrt{111}/74$).

1.33. Magnitude: 20.9 mi, direction $21°39'$ south of west.

1.34. $3\mathbf{AD}$.

1.37. Straight line course upstream making an angle $34°28'$ with the shore line. 1 hr 25 min.

1.38. $5\mathbf{A} - 3\mathbf{B} + \mathbf{C}$.

1.39. $x = 2$, $y = -1$.

1.40. $2\mathbf{a}_1 + 5\mathbf{a}_2 + 3\mathbf{a}_3$.

1.41. 323 lb directly opposite 150 lb force.

1.42. 100 lb

1.43. Linearly dependent since $\mathbf{r}_3 = 5\mathbf{r}_1 - 2\mathbf{r}_2$.

1.44. Yes.

1.50. $(2, 0, 2)$.

1.52. $2\mathbf{i} - 6\mathbf{j} + 3\mathbf{k}$, 7.

1.53. (a) $11\mathbf{i} - 8\mathbf{k}$, (b) $\sqrt{93}$,

1.54. (a) $2\mathbf{i} - \mathbf{j}$, (b) $\sqrt{5}$.

       (c) $\sqrt{398}$, $(3\mathbf{A} - 2\mathbf{B} + 4\mathbf{C})/\sqrt{398}$.

1.55. (a) Linearly dependent, (b) Linearly independent.

1.58. (b) $\sqrt{6}$, $(1/2)\sqrt{114}$, $(1/2)\sqrt{150}$.

1.59. (a) 36, (b) -11.

CHAPTER 2

The DOT and CROSS Product

2.1 Introduction

Operations of vector addition and scalar multiplication were defined for our vectors and scalars in Chapter 1. Here, we define two new operations of multiplication for our vectors. One of the operations, the DOT product, yields a scalar, while the other operation, the CROSS product yields a vector. We then combine these operations to define certain triple products.

2.2 Dot or Scalar Product

The dot or scalar product of two vectors $\mathbf{A}$ and $\mathbf{B}$, denoted by $\mathbf{A} \cdot \mathbf{B}$ (read: $\mathbf{A}$ dot $\mathbf{B}$), is defined as the product of the magnitudes of $\mathbf{A}$ and $\mathbf{B}$ and the cosine of the angle θ between them. In symbols,

$$\mathbf{A} \cdot \mathbf{B} = |\mathbf{A}||\mathbf{B}| \cos \theta, \qquad 0 \leq \theta \leq \pi$$

We emphasize that $\mathbf{A} \cdot \mathbf{B}$ is a scalar and not a vector.

The following proposition applies.

PROPOSITION 2.1: Suppose $\mathbf{A}$, $\mathbf{B}$, and $\mathbf{C}$ are vectors and m is a scalar. Then the following laws hold:

    (i) $\mathbf{A} \cdot \mathbf{B} = \mathbf{B} \cdot \mathbf{A}$ Commutative Law for Dot Products

    (ii) $\mathbf{A} \cdot (\mathbf{B} + \mathbf{C}) = \mathbf{A} \cdot \mathbf{B} + \mathbf{A} \cdot \mathbf{C}$ Distributive Law

    (iii) $m(\mathbf{A} \cdot \mathbf{B}) = (m\mathbf{A}) \cdot \mathbf{B} = \mathbf{A} \cdot (m\mathbf{B}) = (\mathbf{A} \cdot \mathbf{B})m$

    (iv) $\mathbf{i} \cdot \mathbf{i} = \mathbf{j} \cdot \mathbf{j} = \mathbf{k} \cdot \mathbf{k} = 1, \qquad \mathbf{i} \cdot \mathbf{j} = \mathbf{j} \cdot \mathbf{k} = \mathbf{k} \cdot \mathbf{i} = 0$

    (v) If $\mathbf{A} \cdot \mathbf{B} = 0$ and $\mathbf{A}$ and $\mathbf{B}$ are not null vectors, then $\mathbf{A}$ and $\mathbf{B}$ are perpendicular.

There is a simple formula for $\mathbf{A} \cdot \mathbf{B}$ when the unit vectors $\mathbf{i}$, $\mathbf{j}$, $\mathbf{k}$ are used.

PROPOSITION 2.2: Given $\mathbf{A} = A_1\mathbf{i} + A_2\mathbf{j} + A_3\mathbf{k}$ and $\mathbf{B} = B_1\mathbf{i} + B_2\mathbf{j} + B_3\mathbf{k}$. Then

$$\mathbf{A} \cdot \mathbf{B} = A_1B_1 + A_2B_2 + A_3B_3$$

COROLLARY 2.3: Suppose $\mathbf{A} = A_1\mathbf{i} + A_2\mathbf{j} + A_3\mathbf{k}$. Then $\mathbf{A} \cdot \mathbf{A} = A_1^2 + A_2^2 + A_3^2$.

EXAMPLE 2.1 Given $\mathbf{A} = 4\mathbf{i} + 2\mathbf{j} - 3\mathbf{k}$, $\mathbf{B} = 5\mathbf{i} - \mathbf{j} - 2\mathbf{k}$, $\mathbf{C} = 3\mathbf{i} + \mathbf{j} + 7\mathbf{k}$. Then:

$$\mathbf{A} \cdot \mathbf{B} = (4)(5) + (2)(-1) + (-3)(-2) = 20 - 2 + 6 = 24, \quad \mathbf{A} \cdot \mathbf{C} = 12 + 2 - 21 = -7,$$

$$\mathbf{B} \cdot \mathbf{C} = 15 - 1 - 14 = 0, \quad \mathbf{A} \cdot \mathbf{A} = 4^2 + 2^2 + (-3)^2 = 16 + 4 + 9 = 29$$

Thus vectors $\mathbf{B}$ and $\mathbf{C}$ are perpendicular.

2.3   Cross Product

The cross product of vectors **A** and **B** is a vector $\mathbf{C} = \mathbf{A} \times \mathbf{B}$ (read: **A** cross **B**) defined as follows. The magnitude of $\mathbf{C} = \mathbf{A} \times \mathbf{B}$ is equal to the product of the magnitudes of **A** and **B** and the sine of the angle θ between them. The direction of $\mathbf{C} = \mathbf{A} \times \mathbf{B}$ is perpendicular to the plane of **A** and **B** so that **A**, **B**, and **C** form a right-handed system. In symbols,

$$\mathbf{A} \times \mathbf{B} = |\mathbf{A}||\mathbf{B}| \sin \theta \, \mathbf{u} \qquad 0 \le \theta \le \pi$$

where **u** is a unit vector indicating the direction of $\mathbf{A} \times \mathbf{B}$. [Thus **A**, **B**, and **u** form a right-handed system.] If $\mathbf{A} = \mathbf{B}$, or if **A** is parallel to **B**, then $\sin \theta = 0$ and we define $\mathbf{A} \times \mathbf{B} = \mathbf{0}$.

The following proposition applies.

PROPOSITION 2.4:   Suppose **A**, **B**, and **C** are vectors and m is a scalar. Then the following laws hold:

     (i) $\mathbf{A} \times \mathbf{B} = -(\mathbf{B} \times \mathbf{A})$ Commutative Law for Cross Products Fails

     (ii) $\mathbf{A} \times (\mathbf{B} + \mathbf{C}) = \mathbf{A} \times \mathbf{B} + \mathbf{A} \times \mathbf{C}$ Distributive Law

     (iii) $m(\mathbf{A} \times \mathbf{B}) = (m\mathbf{A}) \times \mathbf{B} = \mathbf{A} \times (m\mathbf{B}) = (\mathbf{A} \times \mathbf{B})m$

     (iv) $\mathbf{i} \times \mathbf{i} = \mathbf{j} \times \mathbf{j} = \mathbf{k} \times \mathbf{k} = 0, \quad \mathbf{i} \times \mathbf{j} = \mathbf{k}, \, \mathbf{j} \times \mathbf{k} = \mathbf{i}, \, \mathbf{k} \times \mathbf{i} = \mathbf{j}$

     (v)   If $\mathbf{A} \times \mathbf{B} = 0$ and **A** and **B** are not null vectors, then **A** and **B** are parallel.

     (vi)   The magnitude of $\mathbf{A} \times \mathbf{B}$ is the same as the area of a parallelogram with sides **A** and **B**.

There is a simple formula for $\mathbf{A} \times \mathbf{B}$ when the unit vectors **i**, **j**, **k** are used.

PROPOSITION 2.5:   Given $\mathbf{A} = A_1\mathbf{i} + A_2\mathbf{j} + A_3\mathbf{k}$ and $\mathbf{B} = B_1\mathbf{i} + B_2\mathbf{j} + B_3\mathbf{k}$. Then

$$\mathbf{A} \times \mathbf{B} = \begin{vmatrix} i & j & k \\ A_1 & A_2 & A_3 \\ B_1 & B_2 & B_3 \end{vmatrix} = \begin{vmatrix} A_2 & A_3 \\ B_2 & B_3 \end{vmatrix}\mathbf{i} - \begin{vmatrix} A_1 & A_3 \\ B_1 & B_3 \end{vmatrix}\mathbf{j} + \begin{vmatrix} A_1 & A_2 \\ B_1 & B_2 \end{vmatrix}\mathbf{k}$$

EXAMPLE 2.2   Given: $\mathbf{A} = 4\mathbf{i} + 2\mathbf{j} - 3\mathbf{k}$ and $\mathbf{B} = 3\mathbf{i} + 5\mathbf{j} + 2\mathbf{k}$. Then

$$\mathbf{A} \times \mathbf{B} = \begin{vmatrix} i & j & k \\ 4 & 2 & -3 \\ 3 & 5 & 2 \end{vmatrix} = 19\mathbf{i} - 17\mathbf{j} + 14\mathbf{k}$$

2.4   Triple Products

Dot and cross multiplication of three vectors **A**, **B**, and **C** may produce meaningful products, called *triple products*, of the form $(\mathbf{A} \cdot \mathbf{B})\mathbf{C}$, $\mathbf{A} \cdot (\mathbf{B} \times \mathbf{C})$, and $\mathbf{A} \times (\mathbf{B} \times \mathbf{C})$.

The following proposition applies.

PROPOSITION 2.6:   Suppose **A**, **B**, and **C** are vectors and m is a scalar. Then the following laws hold:

     (i)   In general, $(\mathbf{A} \cdot \mathbf{B})\mathbf{C} \ne \mathbf{A}(\mathbf{B} \cdot \mathbf{C})$.

     (ii) $\mathbf{A} \cdot (\mathbf{B} \times \mathbf{C}) = \mathbf{B} \cdot (\mathbf{C} \times \mathbf{A}) = \mathbf{C} \cdot (\mathbf{A} \times \mathbf{B})$ = volume of a parallelepiped having **A**, **B**, and **C** as edges, or the negative of this volume, according as **A**, **B**, and **C** do or do not form a right-handed system.

     (iii)   In general, $\mathbf{A} \times (\mathbf{B} \times \mathbf{C}) \ne (\mathbf{A} \times \mathbf{B}) \times \mathbf{C}$
             (Associative Law for Cross Products Fails)

     (iv) $\mathbf{A} \times (\mathbf{B} \times \mathbf{C}) = (\mathbf{A} \cdot \mathbf{C})\mathbf{B} - (\mathbf{A} \cdot \mathbf{B})\mathbf{C}$

          $(\mathbf{A} \times \mathbf{B}) \times \mathbf{C} = (\mathbf{A} \cdot \mathbf{C})\mathbf{B} - (\mathbf{B} \cdot \mathbf{C})\mathbf{A}$

There is a simple formula for $\mathbf{A} \cdot (\mathbf{B} \times \mathbf{C})$ when the unit vectors **i**, **j**, **k** are used.

PROPOSITION 2.7: Given $\mathbf{A} = A_1\mathbf{i} + A_2\mathbf{j} + A_3\mathbf{k}$, $\mathbf{B} = B_1\mathbf{i} + B_2\mathbf{j} + B_3\mathbf{k}$, $\mathbf{C} = C_1\mathbf{i} + C_2\mathbf{j} + C_3\mathbf{k}$. Then

$$\mathbf{A} \cdot (\mathbf{B} \times \mathbf{C}) = \begin{vmatrix} A_1 & A_2 & A_3 \\ B_1 & B_2 & B_3 \\ C_1 & C_2 & C_3 \end{vmatrix}$$

EXAMPLE 2.3 Given $\mathbf{A} = 4\mathbf{i} + 2\mathbf{j} - 3\mathbf{k}$, $\mathbf{B} = 5\mathbf{i} + \mathbf{j} - 2\mathbf{k}$, $\mathbf{C} = 3\mathbf{i} - \mathbf{j} + 2\mathbf{k}$. Then:

$$\mathbf{A} \cdot (\mathbf{B} \times \mathbf{C}) = \begin{vmatrix} 4 & 2 & -3 \\ 5 & 1 & -2 \\ 3 & -1 & 2 \end{vmatrix} = 8 - 12 + 15 + 9 - 8 - 20 = -8.$$

2.5 Reciprocal Sets of Vectors

The sets $\mathbf{a}$, $\mathbf{b}$, $\mathbf{c}$ and $\mathbf{a'}$, $\mathbf{b'}$, $\mathbf{c'}$ are called *reciprocal sets* or *reciprocal systems* of vectors if:

$$\mathbf{a} \cdot \mathbf{a'} = \mathbf{b} \cdot \mathbf{b'} = \mathbf{c} \cdot \mathbf{c'} = 1$$
$$\mathbf{a'} \cdot \mathbf{b} = \mathbf{a'} \cdot \mathbf{c} = \mathbf{b'} \cdot \mathbf{a} = \mathbf{b'} \cdot \mathbf{c} = \mathbf{c'} \cdot \mathbf{a} = \mathbf{c'} \cdot \mathbf{b} = 0$$

That is, each vector is orthogonal to the reciprocal of the other two vectors in the system.

PROPOSITION 2.8: The sets $\mathbf{a}$, $\mathbf{b}$, $\mathbf{c}$ and $\mathbf{a'}$, $\mathbf{b'}$, $\mathbf{c'}$ are *reciprocal sets* of vectors if and only if

$$\mathbf{a'} = \frac{\mathbf{b} \times \mathbf{c}}{\mathbf{a} \cdot \mathbf{b} \times \mathbf{c}}, \quad \mathbf{b'} = \frac{\mathbf{c} \times \mathbf{a}}{\mathbf{a} \cdot \mathbf{b} \times \mathbf{c}}, \quad \mathbf{c'} = \frac{\mathbf{a} \times \mathbf{b}}{\mathbf{a} \cdot \mathbf{b} \times \mathbf{c}}$$

where $\mathbf{a} \cdot \mathbf{b} \times \mathbf{c} \neq 0$.

SOLVED PROBLEMS

Dot or Scalar Product

2.1. Prove Proposition 2.1(i): $\mathbf{A} \cdot \mathbf{B} = \mathbf{B} \cdot \mathbf{A}$.

Solution

$$\mathbf{A} \cdot \mathbf{B} = |A||B| \cos \theta = |B||A| \cos \theta = \mathbf{B} \cdot \mathbf{A}.$$

Thus the commutative law for dot products is valid.

2.2. Prove that the projection of $\mathbf{A}$ on $\mathbf{B}$ is equal to $\mathbf{A} \cdot \mathbf{b}$ where $\mathbf{b}$ is a unit vector in the direction of $\mathbf{B}$.

Solution

Through the initial and terminal points of $\mathbf{A}$ pass planes perpendicular to $\mathbf{B}$ at G and H as in Fig. 2-1. Thus

$$\text{Projection of } \mathbf{A} \text{ on } \mathbf{B} = \overline{GH} = \overline{EF} = A \cos \theta = \mathbf{A} \cdot \mathbf{b}$$

Fig. 2-1

Fig. 2-2

2.3. Prove Proposition 2.1(ii): $\mathbf{A} \cdot (\mathbf{B} + \mathbf{C}) = \mathbf{A} \cdot \mathbf{B} + \mathbf{A} \cdot \mathbf{C}$.

Solution

Let $\mathbf{a}$ be a unit vector in the direction of $\mathbf{A}$. Then, as pictured in Fig. 2-2

$$\text{Proj}(\mathbf{B} + \mathbf{C}) \text{ on } \mathbf{A} = \text{Proj}(\mathbf{B}) \text{ on } \mathbf{A} + \text{Proj}(\mathbf{C}) \text{ on } \mathbf{A} \text{ and so } (\mathbf{B} + \mathbf{C}) \cdot \mathbf{a} = \mathbf{B} \cdot \mathbf{a} + \mathbf{C} \cdot \mathbf{a}$$

Multiplying by A,

$$(\mathbf{B} + \mathbf{C}) \cdot A\mathbf{a} = \mathbf{B} \cdot A\mathbf{a} + \mathbf{C} \cdot A\mathbf{a} \quad \text{and} \quad (\mathbf{B} + \mathbf{C}) \cdot \mathbf{A} = \mathbf{B} \cdot \mathbf{A} + \mathbf{C} \cdot \mathbf{A}$$

Then by the commutative law for dot products,

$$\mathbf{A} \cdot (\mathbf{B} + \mathbf{C}) = \mathbf{A} \cdot \mathbf{B} + \mathbf{A} \cdot \mathbf{C}$$

Thus the distributive law is valid.

2.4. Prove that $(\mathbf{A} + \mathbf{B}) \cdot (\mathbf{C} + \mathbf{D}) = \mathbf{A} \cdot \mathbf{C} + \mathbf{A} \cdot \mathbf{D} + \mathbf{B} \cdot \mathbf{C} + \mathbf{B} \cdot \mathbf{D}$.

Solution

By Problem 2.3, $(\mathbf{A} + \mathbf{B}) \cdot (\mathbf{C} + \mathbf{D}) = \mathbf{A} \cdot (\mathbf{C} + \mathbf{D}) + \mathbf{B} \cdot (\mathbf{C} + \mathbf{D}) = \mathbf{A} \cdot \mathbf{C} + \mathbf{A} \cdot \mathbf{D} + \mathbf{B} \cdot \mathbf{C} + \mathbf{B} \cdot \mathbf{D}$.
The ordinary laws of algebra are valid for dot products.

2.5. Evaluate: (a) $\mathbf{i} \cdot \mathbf{i}$,   (b) $\mathbf{i} \cdot \mathbf{k}$,   (c) $\mathbf{k} \cdot \mathbf{j}$,   (d) $\mathbf{j} \cdot (2\mathbf{j} - 3\mathbf{j} + \mathbf{k})$,   (e) $(2\mathbf{i} - \mathbf{j}) \cdot (3\mathbf{i} + \mathbf{k})$.

Solution

(a) $\mathbf{i} \cdot \mathbf{i} = |\mathbf{i}||\mathbf{i}| \cos 0° = (1)(1)(1) = 1$
(b) $\mathbf{i} \cdot \mathbf{k} = |\mathbf{i}||\mathbf{k}| \cos 90° = (1)(1)(0) = 0$
(c) $\mathbf{k} \cdot \mathbf{j} = |\mathbf{k}||\mathbf{j}| \cos 90° = (1)(1)(0) = 0$
(d) $\mathbf{j} \cdot (2\mathbf{i} - 3\mathbf{j} + \mathbf{k}) = 2\mathbf{j} \cdot \mathbf{i} - 3\mathbf{j} \cdot \mathbf{j} + \mathbf{j} \cdot \mathbf{k} = 0 - 3 + 0 = -3$
(e) $(2\mathbf{i} - \mathbf{j}) \cdot (3\mathbf{i} + \mathbf{k}) = 2\mathbf{i} \cdot (3\mathbf{i} + \mathbf{k}) - \mathbf{j} \cdot (3\mathbf{i} + \mathbf{k}) = 6\mathbf{i} \cdot \mathbf{i} + 2\mathbf{i} \cdot \mathbf{k} - 3\mathbf{j} \cdot \mathbf{i} - \mathbf{j} \cdot \mathbf{k} = 6 + 0 - 0 - 0 = 6$

2.6. Suppose $\mathbf{A} = A_1\mathbf{i} + A_2\mathbf{j} + A_3\mathbf{k}$ and $\mathbf{B} = B_1\mathbf{i} + B_2\mathbf{j} + B_3\mathbf{k}$. Prove that $\mathbf{A} \cdot \mathbf{B} = A_1B_1 + A_2B_2 + A_3B_3$.

Solution

Since $\mathbf{i} \cdot \mathbf{i} = \mathbf{j} \cdot \mathbf{j} = \mathbf{k} \cdot \mathbf{k} = 1$, and all other dot products are zero, we have:

$$\begin{aligned}
\mathbf{A} \cdot \mathbf{B} &= (A_1\mathbf{i} + A_2\mathbf{j} + A_3\mathbf{k}) \cdot (B_1\mathbf{i} + B_2\mathbf{j} + B_3\mathbf{k}) \\
&= A_1\mathbf{i} \cdot (B_1\mathbf{i} + B_2\mathbf{j} + B_3\mathbf{k}) + A_2\mathbf{j} \cdot (B_1\mathbf{i} + B_2\mathbf{j} + B_3\mathbf{k}) + A_3\mathbf{k} \cdot (B_1\mathbf{i} + B_2\mathbf{j} + B_3\mathbf{k}) \\
&= A_1B_1\mathbf{i} \cdot \mathbf{i} + A_1B_2\mathbf{i} \cdot \mathbf{j} + A_1B_3\mathbf{i} \cdot \mathbf{k} + A_2B_1\mathbf{j} \cdot \mathbf{i} + A_2B_2\mathbf{j} \cdot \mathbf{j} + A_2B_3\mathbf{j} \cdot \mathbf{k} \\
&\quad + A_3B_1\mathbf{k} \cdot \mathbf{i} + A_3B_2\mathbf{k} \cdot \mathbf{j} + A_3B_3\mathbf{k} \cdot \mathbf{k} \\
&= A_1B_1 + A_2B_2 + A_3B_3
\end{aligned}$$

2.7. Let $\mathbf{A} = A_1\mathbf{i} + A_2\mathbf{j} + A_3\mathbf{k}$. Show that $A = \sqrt{\mathbf{A} \cdot \mathbf{A}} = \sqrt{A_1^2 + A_2^2 + A_3^2}$.

Solution

$\mathbf{A} \cdot \mathbf{A} = (A)(A) \cos 0° = A^2$. Then $A = \sqrt{\mathbf{A} \cdot \mathbf{A}}$.

By Problem 2.6 and taking $\mathbf{B} = \mathbf{A}$, we have

$$\mathbf{A} \cdot \mathbf{A} = (A_1\mathbf{i} + A_2\mathbf{j} + A_3\mathbf{k}) \cdot (A_1\mathbf{i} + A_2\mathbf{j} + A_3\mathbf{k})$$

$$= (A_1)(A_1) + (A_2)(A_2) + (A_3)(A_3) = A_1^2 + A_2^2 + A_3^2$$

Then $A = \sqrt{\mathbf{A} \cdot \mathbf{A}} = \sqrt{A_1^2 + A_2^2 + A_3^2}$ is the magnitude of $\mathbf{A}$. Sometimes $\mathbf{A} \cdot \mathbf{A}$ is written $\mathbf{A}^2$.

2.8. Suppose $\mathbf{A} \cdot \mathbf{B} = 0$ and A and B are not zero. Show that $\mathbf{A}$ is perpendicular to $\mathbf{B}$.

Solution

If $\mathbf{A} \cdot \mathbf{B} = AB \cos \theta = 0$, then $\cos \theta = 0$ or $\theta = 90°$. Conversely, if $\theta = 90°$, $\mathbf{A} \cdot \mathbf{B} = 0$.

2.9. Find the angle between $\mathbf{A} = 2\mathbf{i} + 2\mathbf{j} - \mathbf{k}$ and $\mathbf{B} = 7\mathbf{i} + 24\mathbf{k}$.

Solution

We have $\mathbf{A} \cdot \mathbf{B} = |\mathbf{A}||\mathbf{B}| \cos \theta$.

$$|A| = \sqrt{(2)^2 + (2)^2 + (-1)^2} = 3 \quad \text{and} \quad |B| = \sqrt{(7)^2 + (0)^2 + (24)^2} = 25$$
$$\mathbf{A} \cdot \mathbf{B} = (2)(7) + (2)(0) + (-1)(24) = -10$$

Therefore,

$$\cos \theta = \frac{A \cdot B}{|A||B|} = \frac{-10}{(3)(25)} = \frac{-2}{15} = -0.1333 \text{ and } \theta = 98° \text{ (approximately)}.$$

2.10. Determine the value of α so that $\mathbf{A} = 2\mathbf{i} + \alpha\mathbf{j} + \mathbf{k}$ and $\mathbf{B} = \mathbf{i} + 3\mathbf{j} - 8\mathbf{k}$ are perpendicular.

Solution

By Proposition 2.1(v), $\mathbf{A}$ and $\mathbf{B}$ are perpendicular when $\mathbf{A} \cdot \mathbf{B} = 0$. Thus,

$$\mathbf{A} \cdot \mathbf{B} = (2)(1) + (\alpha)(3) + (1)(-8) = 2 + 3\alpha - 8 = 0$$

and if $\alpha = 2$.

2.11. Show that the vectors $\mathbf{A} = -\mathbf{i} + \mathbf{j}$, $\mathbf{B} = -\mathbf{i} - \mathbf{j} - 2\mathbf{k}$, $\mathbf{C} = 2\mathbf{j} + 2\mathbf{k}$ form a right triangle.

Solution

First we show that the vectors form a triangle. From Fig. 2-3, we see that the vectors form a triangle if:

(a)   one of the vectors, say (3), is the sum of (1) and (2) or

(b)   the sum of the vectors $(1) + (2) + (3)$ is zero

according as (a) two vectors have a common terminal point, or (b) none of the vectors have a common terminal point. By trial, we find $\mathbf{A} = \mathbf{B} + \mathbf{C}$ so the vectors do form a triangle.

Since $\mathbf{A} \cdot \mathbf{B} = (-1)(-1) + (1)(-1) + (0)(-2) = 0$, it follows that $\mathbf{A}$ and $\mathbf{B}$ are perpendicular and the triangle is a right triangle.

Fig. 2-3

2.12. Find the angles that the vector $\mathbf{A} = 4\mathbf{i} - 8\mathbf{j} + \mathbf{k}$ makes with the coordinate axes.

Solution

Let α, β, γ be the angles that $\mathbf{A}$ makes with the positive x, y, z axes, respectively.

$$\mathbf{A} \cdot \mathbf{i} = |A|(1) \cos \alpha = \sqrt{(4)^2 + (-8)^2 + (1)^2} \cos \alpha = 9 \cos \alpha$$

$$\mathbf{A} \cdot \mathbf{i} = (4\mathbf{i} - 8\mathbf{j} + \mathbf{k}) \cdot \mathbf{i} = 4$$

Then $\cos \alpha = 4/9 = 0.4444$ and $\alpha = 63.6°$ approximately. Similarly,

$$\cos \beta = -8/9, \ \beta = 152.7° \quad \text{and} \quad \cos \gamma = 1/9, \ \gamma = 83.6°$$

The cosines of α, β, γ are called the *direction cosines* of the vector $\mathbf{A}$.

2.13. Find the projection of the vector $\mathbf{A} = \mathbf{i} - 2\mathbf{j} + 3\mathbf{k}$ on the vector $\mathbf{B} = \mathbf{i} + 2\mathbf{j} + 2\mathbf{k}$.

Solution

We use the result of Problem 2.2. A unit vector in the direction of $\mathbf{B}$ is

$$\mathbf{b} = \mathbf{B}/|B| = (\mathbf{i} + 2\mathbf{j} + 2\mathbf{k})/\sqrt{1 + 4 + 4} = \mathbf{i}/3 + 2\mathbf{j}/3 + 2\mathbf{k}/3$$

The projection of $\mathbf{A}$ on vector $\mathbf{B}$ is

$$\mathbf{A} \cdot \mathbf{b} = (\mathbf{i} - 2\mathbf{j} + 3\mathbf{k}) \cdot (\mathbf{i}/3 + 2\mathbf{j}/3 + 2\mathbf{k}/3) = (1)(1/3) + (-2)(2/3) + (3)(2/3) = 1.$$

2.14. Without making use of the cross product, determine a unit vector perpendicular to the plane of $\mathbf{A} = 2\mathbf{i} - 6\mathbf{j} - 3\mathbf{k}$ and $\mathbf{B} = 4\mathbf{i} + 3\mathbf{j} - \mathbf{k}$.

Solution

Let vector $\mathbf{C} = c_1\mathbf{i} + c_2\mathbf{j} + c_3\mathbf{k}$ be perpendicular to the plane of $\mathbf{A}$ and $\mathbf{B}$. Then $\mathbf{C}$ is perpendicular to $\mathbf{A}$ and also to $\mathbf{B}$. Hence,

$$\mathbf{C} \cdot \mathbf{A} = 2c_1 - 6c_2 - 3c_3 = 0 \quad \text{or} \quad (1) \; 2c_1 - 6c_2 = 3c_3$$
$$\mathbf{C} \cdot \mathbf{B} = 4c_1 + 3c_2 - c_3 = 0 \quad \text{or} \quad (2) \; 4c_1 + 3c_2 = c_3$$

Solving (1) and (2) simultaneously: $c_1 = \frac{1}{2}c_3$, $c_2 = -\frac{1}{3}c_3$, $\mathbf{C} = c_3\left(\frac{1}{2}\mathbf{i} - \frac{1}{3}\mathbf{j} + \mathbf{k}\right)$.

Then a unit vector in the direction of $\mathbf{C}$ is $\dfrac{\mathbf{C}}{|C|} = \dfrac{c_3\left(\frac{1}{2}\mathbf{i} - \frac{1}{3}\mathbf{j} + \mathbf{k}\right)}{\sqrt{c_3^2\left[\left(\frac{1}{2}\right)^2 + \left(-\frac{1}{3}\right)^2 + (1)^2\right]}} = \pm\left(\frac{3}{7}\mathbf{i} - \frac{2}{7}\mathbf{j} + \frac{6}{7}\mathbf{k}\right)$.

2.15. Prove the law of cosines for plane triangles.

Solution

From Fig. 2-4,

$$\mathbf{B} + \mathbf{C} = \mathbf{A} \quad \text{or} \quad \mathbf{C} = \mathbf{A} - \mathbf{B}$$

Then

$$\mathbf{C} \cdot \mathbf{C} = (\mathbf{A} - \mathbf{B}) \cdot (\mathbf{A} - \mathbf{B}) = \mathbf{A} \cdot \mathbf{A} + \mathbf{B} \cdot \mathbf{B} - 2\mathbf{A} \cdot \mathbf{B}$$

and

$$C^2 = A^2 + B^2 - 2AB\cos\theta$$

Fig. 2-4   Fig. 2-5

2.16. Prove the diagonals of a rhombus are perpendicular. (Refer to Fig. 2-5.)

Solution

$$OQ = OP + PQ = \mathbf{A} + \mathbf{B}$$

$$OR + RP = OP \quad \text{or} \quad \mathbf{B} + RP = \mathbf{A} \quad \text{and} \quad RP = \mathbf{A} - \mathbf{B}$$

Then, since $|\mathbf{A}| = |\mathbf{B}|$,

$$OQ \cdot RP = (\mathbf{A} + \mathbf{B}) \cdot (\mathbf{A} - \mathbf{B}) = |\mathbf{A}|^2 - |\mathbf{B}|^2 = 0$$

Thus OQ is perpendicular to RP.

2.17. Let $\mathbf{A} = A_1\mathbf{i} + A_2\mathbf{j} + A_3\mathbf{k}$ be any vector. Prove that $\mathbf{A} = (\mathbf{A} \cdot \mathbf{i})\mathbf{i} + (\mathbf{A} \cdot \mathbf{j})\mathbf{j} + (\mathbf{A} \cdot \mathbf{k})\mathbf{k}$.

Solution

Since $\mathbf{A} = A_1\mathbf{i} + A_2\mathbf{j} + A_3\mathbf{k}$,

$$\mathbf{A} \cdot \mathbf{i} = A_1\mathbf{i} \cdot \mathbf{i} + A_2\mathbf{j} \cdot \mathbf{i} + A_3\mathbf{k} \cdot \mathbf{i} = A_1$$

Similarly, $\mathbf{A} \cdot \mathbf{j} = A_2$ and $\mathbf{A} \cdot \mathbf{k} = A_3$. Then

$$\mathbf{A} = A_1\mathbf{i} + A_2\mathbf{j} + A_3\mathbf{k} = (\mathbf{A} \cdot \mathbf{i})\mathbf{i} + (\mathbf{A} \cdot \mathbf{j})\mathbf{j} - (\mathbf{A} \cdot \mathbf{k})\mathbf{k}$$

2.18. Find the work done in moving an object along a vector $\mathbf{r} = 3\mathbf{i} + \mathbf{j} - 5\mathbf{k}$ if the applied force is $\mathbf{F} = 2\mathbf{i} - \mathbf{j} - \mathbf{k}$.

Solution

Consider Fig. 2-6.

$$\text{Work done} = (\text{magnitude of force in direction of motion})(\text{distance moved})$$
$$= (\mathbf{F} \cos \theta)(\mathbf{r}) = \mathbf{F} \cdot \mathbf{r} = (2\mathbf{i} - \mathbf{j} - \mathbf{k}) \cdot (3\mathbf{i} + \mathbf{j} - 5\mathbf{k})$$
$$= 6 - 1 + 5 = 10$$

Fig. 2-6   Fig. 2-7

2.19. Find an equation of the plane perpendicular to the vector $\mathbf{A} = 2\mathbf{i} - 3\mathbf{j} + 6\mathbf{k}$ and passing through the terminal point of the vector $\mathbf{B} = \mathbf{i} + 2\mathbf{j} + 3\mathbf{k}$. [See Fig. 2-7.]

Solution

Since $\mathbf{PQ} = \mathbf{B} - \mathbf{r}$ is perpendicular to $\mathbf{A}$, we have $(\mathbf{B} - \mathbf{r}) \cdot \mathbf{A} = 0$ or $\mathbf{r} \cdot \mathbf{A} = \mathbf{B} \cdot \mathbf{A}$ is the required equation of the plane in vector form. In rectangular form this becomes

$$(x\mathbf{i} + y\mathbf{j} + z\mathbf{k}) \cdot (2\mathbf{i} - 3\mathbf{j} + 6\mathbf{k}) = (\mathbf{i} + 2\mathbf{j} + 3\mathbf{k}) \cdot (2\mathbf{i} - 3\mathbf{j} + 6\mathbf{k})$$

or

$$2x - 3y + 6z = 2 - 6 + 18 = 14$$

2.20. Find the distance from the origin to the plane in Problem 2.19.

Solution

The distance from the origin to the plane is the projection of $\mathbf{B}$ on $\mathbf{A}$. A unit vector in the direction of A is

$$\mathbf{a} = \mathbf{A}/|A| = \frac{2\mathbf{i} - 3\mathbf{j} + 6\mathbf{k}}{\sqrt{(2)^2 + (-3)^2 + (6)^2}} = \frac{2}{7}\mathbf{i} - \frac{3}{7}\mathbf{j} + \frac{6}{7}\mathbf{k}$$

Then the projection of $\mathbf{B}$ on $\mathbf{A}$ is equal to

$$\mathbf{B} \cdot \mathbf{a} = (\mathbf{i} + 2\mathbf{j} + 3\mathbf{k})\left(\frac{2}{7}\mathbf{i} - \frac{3}{7}\mathbf{j} + \frac{6}{7}\mathbf{k}\right) = (1)\frac{2}{7} - (2)\frac{3}{7} + (3)\frac{6}{7} = 2.$$

Cross or Vector Product

2.21. Prove $\mathbf{A} \times \mathbf{B} = -(\mathbf{A} \times \mathbf{B})$.

Solution

$\mathbf{A} \times \mathbf{B} = \mathbf{C}$ has magnitude $AB \sin \theta$ and direction such that $\mathbf{A}, \mathbf{B}, \mathbf{C}$ form a right-handed system as in Fig. 2-8(a).
$\mathbf{B} \times \mathbf{A} = \mathbf{D}$ has magnitude $BA \sin \theta$ and direction such that $\mathbf{B}, \mathbf{A}, \mathbf{D}$ form a right-handed system as in Fig. 2-8(b).
Then $\mathbf{D}$ has the same magnitude as $\mathbf{C}$ but in the opposite direction, i.e. $\mathbf{C} = -\mathbf{D}$. Thus $\mathbf{A} \times \mathbf{B} = -(\mathbf{A} \times \mathbf{B})$.
Accordingly, the commutative law for cross products is not valid.

(a)   (b)

Fig. 2-8

2.22. Suppose $\mathbf{A} \times \mathbf{B} = 0$ and $\mathbf{A}$ and $\mathbf{B}$ are not zero. Show that $\mathbf{A}$ is parallel to $\mathbf{B}$.

Solution

Since $\mathbf{A} \times \mathbf{B} = AB \sin \theta \, \mathbf{u} = \mathbf{0}$, we have $\sin \theta = 0$ and hence $\theta = 0°$ or $180°$.

2.23. Show that $|\mathbf{A} \times \mathbf{B}|^2 + |\mathbf{A} \cdot \mathbf{B}|^2 = |\mathbf{A}|^2|\mathbf{B}|^2$.

Solution

$$|\mathbf{A} \times \mathbf{B}|^2 + |\mathbf{A} \cdot \mathbf{B}|^2 = |AB \sin\theta\, \mathbf{u}|^2 + |AB \cos\theta|^2$$
$$= A^2B^2 \sin^2\theta + A^2B^2 \cos^2\theta$$
$$= A^2B^2 = |\mathbf{A}|^2|\mathbf{B}|^2$$

2.24. Evaluate: (a) $2\mathbf{j} \times 3\mathbf{k}$ (b) $2\mathbf{j} \times -\mathbf{k}$ (c) $-3\mathbf{i} \times -2\mathbf{k}$, $2\mathbf{j} \times 3\mathbf{i} - \mathbf{k}$

Solution

(a) $(2\mathbf{j}) \times (3\mathbf{k}) = 6(\mathbf{j} \times \mathbf{k}) = 6\mathbf{i}$
(b) $(2\mathbf{j}) \times (-\mathbf{k}) = -2(\mathbf{j} \times \mathbf{k}) = -2\mathbf{i}$
(c) $(-3\mathbf{i}) \times (-2\mathbf{k}) = 6(\mathbf{i} \times \mathbf{k}) = -6\mathbf{j}$
(d) $2\mathbf{j} \times 3\mathbf{i} - \mathbf{k} = 6(\mathbf{j} \times \mathbf{i}) - \mathbf{k} = -6\mathbf{k} - \mathbf{k} = -7\mathbf{k}$.

2.25. Prove that $\mathbf{A} \times (\mathbf{B} + \mathbf{C}) = \mathbf{A} \times \mathbf{B} + \mathbf{A} \times \mathbf{C}$ for the case where $\mathbf{A}$ is perpendicular to both $\mathbf{B}$ and $\mathbf{C}$. [See Fig. 2-9.]

Solution

Since $\mathbf{A}$ is perpendicular to $\mathbf{B}$, $\mathbf{A} \times \mathbf{B}$ is a vector perpendicular to the plane of $\mathbf{A}$ and $\mathbf{B}$ and having magnitude $AB \sin 90° = AB$ or magnitude of $A\mathbf{B}$. This is equivalent to multiplying vector $\mathbf{B}$ by A and rotating the resultant vector through 90° to the position shown in Fig. 2-9.

Similarly, $\mathbf{A} \times \mathbf{C}$ is the vector obtained by multiplying $\mathbf{C}$ by A and rotating the resultant vector through 90° to the position shown.

In like manner, $\mathbf{A} \times (\mathbf{B} + \mathbf{C})$ is the vector obtained by multiplying $\mathbf{B} + \mathbf{C}$ by A and rotating the resultant vector through 90° to the position shown.

Since $\mathbf{A} \times (\mathbf{B} + \mathbf{C})$ is the diagonal of the parallelogram with $\mathbf{A} \times \mathbf{B}$ and $\mathbf{A} \times \mathbf{C}$ as sides, we have $\mathbf{A} \times (\mathbf{B} + \mathbf{C}) = \mathbf{A} \times \mathbf{B} + \mathbf{A} \times \mathbf{C}$.

Fig. 2-9   Fig. 2-10

2.26. Prove that $\mathbf{A} \times (\mathbf{B} + \mathbf{C}) = \mathbf{A} \times \mathbf{B} + \mathbf{A} \times \mathbf{C}$ for the general case where $\mathbf{A}$, $\mathbf{B}$, and $\mathbf{C}$ are non-coplanar. [See Fig. 2-10.]

Solution

Resolve $\mathbf{B}$ into two component vectors, one perpendicular to $\mathbf{A}$ and the other parallel to $\mathbf{A}$, and denote them by $\mathbf{B}_\perp$ and $\mathbf{B}_\parallel$, respectively. Then $\mathbf{B} = \mathbf{B}_\perp + \mathbf{B}_\parallel$.

If θ is the angle between $\mathbf{A}$ and $\mathbf{B}$, then $\mathbf{B}_\perp = B \sin\theta$. Thus the magnitude of $\mathbf{A} \times \mathbf{B}_\perp$ is $AB \sin\theta$, the same as the magnitude of $\mathbf{A} \times \mathbf{B}$. Also, the direction of $\mathbf{A} \times \mathbf{B}_\perp$ is the same as the direction of $\mathbf{A} \times \mathbf{B}$. Hence $\mathbf{A} \times \mathbf{B}_\perp = \mathbf{A} \times \mathbf{B}$.

Similarly, if $\mathbf{C}$ is resolved into two component vectors $\mathbf{C}_\parallel$ and $\mathbf{C}_\perp$, parallel and perpendicular respectively to $\mathbf{A}$, then $\mathbf{A} \times \mathbf{C}_\perp = \mathbf{A} \times \mathbf{C}$.

Also, since $\mathbf{B} + \mathbf{C} = \mathbf{B}_\perp + \mathbf{B}_\parallel + \mathbf{C}_\perp + \mathbf{C}_\parallel = (\mathbf{B}_\perp + \mathbf{C}_\perp) + (\mathbf{B}_\parallel + \mathbf{C}_\parallel)$ it follows that

$$\mathbf{A} \times (\mathbf{B}_\perp + \mathbf{C}_\perp) = \mathbf{A} \times (\mathbf{B} + \mathbf{C}).$$

Now $\mathbf{B}_\perp$ and $\mathbf{C}_\perp$ are vectors perpendicular to $\mathbf{A}$ and so by Problem 2.25,

$$\mathbf{A} \times (\mathbf{B}_\perp + \mathbf{C}_\perp) = \mathbf{A} \times \mathbf{B}_\perp + \mathbf{A} \times \mathbf{C}_\perp$$

Then

$$\mathbf{A} \times (\mathbf{B} + \mathbf{C}) = \mathbf{A} \times \mathbf{B} + \mathbf{A} \times \mathbf{C}$$

and the distributive law holds. Multiplying by -1, using Problem 2.21, this becomes $(\mathbf{B} + \mathbf{C}) \times \mathbf{A} = \mathbf{B} \times \mathbf{A} + \mathbf{C} \times \mathbf{A}$. Note that the order of factors in cross products is important. The usual laws of algebra apply only if proper order is maintained.

2.27. Suppose $\mathbf{A} = A_1\mathbf{i} + A_2\mathbf{j} + A_3\mathbf{k}$ and $\mathbf{B} = B_1\mathbf{i} + B_2\mathbf{j} + B_3\mathbf{k}$. Prove $\mathbf{A} \times \mathbf{B} = \begin{vmatrix} \mathbf{i} & \mathbf{j} & \mathbf{k} \\ A_1 & A_2 & A_3 \\ B_1 & B_2 & B_3 \end{vmatrix}$.

Solution

$$\begin{aligned}
\mathbf{A} \times \mathbf{B} &= (A_1\mathbf{i} + A_2\mathbf{j} + A_3\mathbf{k}) \times (B_1\mathbf{i} + B_2\mathbf{j} + B_3\mathbf{k}) \\
&= A_1\mathbf{i} \times (B_1\mathbf{i} + B_2\mathbf{j} + B_3\mathbf{k}) + A_2\mathbf{j} \times (B_1\mathbf{i} + B_2\mathbf{j} + B_3\mathbf{k}) + A_3\mathbf{k} \times (B_1\mathbf{i} + B_2\mathbf{j} + B_3\mathbf{k}) \\
&= A_1B_1\mathbf{i} \times \mathbf{i} + A_1B_2\mathbf{i} \times \mathbf{j} + A_1B_3\mathbf{i} \times \mathbf{k} + A_2B_1\mathbf{j} \times \mathbf{i} + A_2B_2\mathbf{j} \times \mathbf{j} + A_2B_3\mathbf{j} \times \mathbf{k} \\
&\quad + A_3B_1\mathbf{k} \times \mathbf{i} + A_3B_2\mathbf{k} \times \mathbf{j} + A_3B_3\mathbf{k} \times \mathbf{k} \\
&= (A_2B_3 - A_3B_2)\mathbf{i} + (A_3B_1 - A_1B_3)\mathbf{j} + (A_1B_2 - A_2B_1)\mathbf{k} = \begin{vmatrix} \mathbf{i} & \mathbf{j} & \mathbf{k} \\ A_1 & A_2 & A_3 \\ B_1 & B_2 & B_3 \end{vmatrix}.
\end{aligned}$$

2.28. Suppose $\mathbf{A} = \mathbf{j} + 2\mathbf{k}$ and $\mathbf{B} = \mathbf{i} + 2\mathbf{j} + 3\mathbf{k}$. Find: (a) $\mathbf{A} \times \mathbf{B}$, (b) $\mathbf{B} \times \mathbf{A}$, (c) $(\mathbf{A} + \mathbf{B}) \times (\mathbf{A} - \mathbf{B})$.

Solution

(a) $\mathbf{A} \times \mathbf{B} = (\mathbf{j} + 2\mathbf{k}) \times (\mathbf{i} + 2\mathbf{j} + 3\mathbf{k}) = \begin{vmatrix} \mathbf{i} & \mathbf{j} & \mathbf{k} \\ 0 & 1 & 2 \\ 1 & 2 & 3 \end{vmatrix}$

$$= \begin{vmatrix} 1 & 2 \\ 2 & 3 \end{vmatrix}\mathbf{i} - \begin{vmatrix} 0 & 2 \\ 1 & 3 \end{vmatrix}\mathbf{j} + \begin{vmatrix} 0 & 1 \\ 1 & 2 \end{vmatrix}\mathbf{k} = -\mathbf{i} + 2\mathbf{j} - \mathbf{k}.$$

(b) $\mathbf{B} \times \mathbf{A} = (\mathbf{i} + 2\mathbf{j} + 3\mathbf{k}) \times (\mathbf{j} + 2\mathbf{k}) = \begin{vmatrix} \mathbf{i} & \mathbf{j} & \mathbf{k} \\ 1 & 2 & 3 \\ 0 & 1 & 2 \end{vmatrix}$

$$= \begin{vmatrix} 2 & 3 \\ 1 & 2 \end{vmatrix}\mathbf{i} - \begin{vmatrix} 1 & 3 \\ 0 & 2 \end{vmatrix}\mathbf{j} + \begin{vmatrix} 1 & 2 \\ 0 & 1 \end{vmatrix}\mathbf{k} = \mathbf{i} - 2\mathbf{j} + \mathbf{k}.$$

Comparing with (a), we have $\mathbf{A} \times \mathbf{B} = -(\mathbf{B} \times \mathbf{A})$. Note this is equivalent to the theorem: If two rows of a determinant are interchanged, the determinant changes sign.

(c) $\mathbf{A} + \mathbf{B} = \mathbf{i} + 3\mathbf{j} + 5\mathbf{k}$ and $\mathbf{A} - \mathbf{B} = -\mathbf{i} - \mathbf{j} - \mathbf{k}$. Then

$$(\mathbf{A} + \mathbf{B}) \times (\mathbf{A} - \mathbf{B}) = \begin{vmatrix} \mathbf{i} & \mathbf{j} & \mathbf{k} \\ 1 & 3 & 5 \\ -1 & -1 & -1 \end{vmatrix} = \begin{vmatrix} 3 & 5 \\ -1 & -1 \end{vmatrix}\mathbf{i} - \begin{vmatrix} 1 & 5 \\ -1 & -1 \end{vmatrix}\mathbf{j} + \begin{vmatrix} 1 & 3 \\ -1 & -1 \end{vmatrix}\mathbf{k}$$

$$= 2\mathbf{i} - 4\mathbf{j} + 2\mathbf{k}.$$

2.29. Suppose $\mathbf{A} = -\mathbf{i} + \mathbf{j} + \mathbf{k}$, $\mathbf{B} = \mathbf{i} - \mathbf{j} + \mathbf{k}$, $\mathbf{C} = \mathbf{i} + \mathbf{j} - \mathbf{k}$. Find: (a) $(\mathbf{A} \times \mathbf{B}) \times \mathbf{C}$, (b) $\mathbf{A} \times (\mathbf{B} \times \mathbf{C})$.

Solution

(a) $\quad \mathbf{A} \times \mathbf{B} = \begin{vmatrix} \mathbf{i} & \mathbf{j} & \mathbf{k} \\ -1 & 1 & 1 \\ 1 & -1 & 1 \end{vmatrix} = 2\mathbf{i} + 2\mathbf{j}$

Then $(\mathbf{A} \times \mathbf{B}) \times \mathbf{C} = (2\mathbf{i} + 2\mathbf{j}) \times (\mathbf{i} + \mathbf{j} - \mathbf{k}) = \begin{vmatrix} \mathbf{i} & \mathbf{j} & \mathbf{k} \\ 2 & 2 & 0 \\ 1 & 1 & -1 \end{vmatrix} = -2\mathbf{i} + 2\mathbf{j}$

(b) $\quad \mathbf{B} \times \mathbf{C} = (\mathbf{i} - \mathbf{j} + \mathbf{k}) \times (\mathbf{i} + \mathbf{j} - \mathbf{k}) = \begin{vmatrix} \mathbf{i} & \mathbf{j} & \mathbf{k} \\ 1 & -1 & 1 \\ 1 & 1 & -1 \end{vmatrix} = 2\mathbf{j} + 2\mathbf{k}.$

Then $\mathbf{A} \times (\mathbf{B} \times \mathbf{C}) = (-\mathbf{i} + \mathbf{j} + \mathbf{k}) \times (2\mathbf{j} + 2\mathbf{k}) = \begin{vmatrix} \mathbf{i} & \mathbf{j} & \mathbf{k} \\ -1 & 1 & 1 \\ 0 & 2 & 2 \end{vmatrix} = 2\mathbf{j} - 2\mathbf{k}.$

Thus $(\mathbf{A} \times \mathbf{B}) \times \mathbf{C} \neq \mathbf{A} \times (\mathbf{B} \times \mathbf{C})$. This shows the need for parentheses in $\mathbf{A} \times \mathbf{B} \times \mathbf{C}$ to avoid ambiguity.

2.30. Prove: (a) The area of a parallelogram with touching sides $\mathbf{A}$ and $\mathbf{B}$, as in Fig. 2-11, is $|\mathbf{A} \times \mathbf{B}|$. (b) The area of a triangle with sides $\mathbf{A}$ and $\mathbf{B}$ is $\frac{1}{2}|\mathbf{A} \times \mathbf{B}|$.

Solution

(a) Area of parallelogram $= h|\mathbf{B}| = |\mathbf{A}| \sin \theta \, |\mathbf{B}| = |\mathbf{A} \times \mathbf{B}|$.

(b) Area of triangle $= \frac{1}{2}$ area of parallelogram $= \frac{1}{2}|\mathbf{A} \times \mathbf{B}|$.

Fig. 2-11

Fig. 2-12

Fig. 2-13

2.31. Prove the law of sines for plane triangles.

Solution

Let $\mathbf{a}$, $\mathbf{b}$, $\mathbf{c}$ represent the sides of a triangle ABC as in Fig. 2-12. Then, $\mathbf{a} + \mathbf{b} + \mathbf{c} = \mathbf{0}$. Multiplying by $\mathbf{a}\times$, $\mathbf{b}\times$, and $\mathbf{c}\times$ in succession, we find

$$\mathbf{a} \times \mathbf{b} = \mathbf{b} \times \mathbf{c} = \mathbf{c} \times \mathbf{a}$$

that is

$$ab \sin C = bc \sin A = ca \sin B$$

or

$$\frac{\sin A}{a} = \frac{\sin B}{b} = \frac{\sin C}{c}.$$

2.32. Consider a tetrahedron, as in Fig. 2-13, with faces F_1, F_2, F_3, F_4. Let $\mathbf{V}_1, \mathbf{V}_2, \mathbf{V}_3, \mathbf{V}_4$ be vectors whose magnitudes are equal to the areas of F_1, F_2, F_3, F_4, respectively, and whose directions are perpendicular to these faces in the outward direction. Show that $\mathbf{V}_1 + \mathbf{V}_2 + \mathbf{V}_3 + \mathbf{V}_4 = \mathbf{0}$.

Solution

By Problem 2.30, the area of a triangular face determined by $\mathbf{R}$ and $\mathbf{S}$ is $\frac{1}{2}|\mathbf{R} \times \mathbf{S}|$.

The vectors associated with each of the faces of the tetrahedron are

$$\mathbf{V}_1 = \tfrac{1}{2}\mathbf{A} \times \mathbf{B}, \quad \mathbf{V}_2 = \tfrac{1}{2}\mathbf{B} \times \mathbf{C}, \quad \mathbf{V}_3 = \tfrac{1}{2}\mathbf{C} \times \mathbf{A}, \quad \mathbf{V}_4 = \tfrac{1}{2}(\mathbf{C} - \mathbf{A}) \times (\mathbf{B} - \mathbf{A})$$

Then

$$\mathbf{V}_1 + \mathbf{V}_2 + \mathbf{V}_3 + \mathbf{V}_4 = \tfrac{1}{2}[\mathbf{A} \times \mathbf{B} + \mathbf{B} \times \mathbf{C} + \mathbf{C} \times \mathbf{A} + (\mathbf{C} - \mathbf{A}) \times (\mathbf{B} - \mathbf{A})]$$

$$= \tfrac{1}{2}[\mathbf{A} \times \mathbf{B} + \mathbf{B} \times \mathbf{C} + \mathbf{C} \times \mathbf{A} + \mathbf{C} \times \mathbf{B} - \mathbf{C} \times \mathbf{A} - \mathbf{A} \times \mathbf{B} + \mathbf{A} \times \mathbf{A}] = 0.$$

This result can be generalized to closed polyhedra and in the limiting case to any closed surface.

Because of the application presented here, it is sometimes convenient to assign a direction to area and we speak of the *vector area*.

2.33. Find the area of the triangle having vertices at $P(1, 3, 2)$, $Q(2, -1, 1)$, $R(-1, 2, 3)$.

Solution

$$\mathbf{PQ} = (2 - 1)\mathbf{i} + (-1 - 3)\mathbf{j} + (1 - 2)\mathbf{k} = \mathbf{i} - 4\mathbf{j} - \mathbf{k}$$
$$\mathbf{PR} = (-1 - 1)\mathbf{i} + (2 - 3)\mathbf{j} + (3 - 2)\mathbf{k} = -2\mathbf{i} - \mathbf{j} + \mathbf{k}$$

From Problem 2.30,

$$\text{area of triangle} = \frac{1}{2}\left|\mathbf{PQ} \times \mathbf{PR}\right| = \frac{1}{2}\left|(\mathbf{i} - 4\mathbf{j} - \mathbf{k}) \times (-2\mathbf{i} - \mathbf{j} + \mathbf{k})\right|$$

$$= \frac{1}{2}\begin{vmatrix} \mathbf{i} & \mathbf{j} & \mathbf{k} \\ 1 & -4 & -1 \\ -2 & -1 & 1 \end{vmatrix} = \frac{1}{2}\left|-5\mathbf{i} + \mathbf{j} - 9\mathbf{k}\right| = \frac{1}{2}\sqrt{(-5)^2 + (1)^2 + (-9)^2} = \frac{1}{2}\sqrt{107}.$$

2.34. Determine a unit vector perpendicular to the plane of $\mathbf{A} = 2\mathbf{i} - 6\mathbf{j} - 3\mathbf{k}$ and $\mathbf{B} = 4\mathbf{i} + 3\mathbf{j} - \mathbf{k}$.

Solution

$\mathbf{A} \times \mathbf{B}$ is a vector perpendicular to the plane of $\mathbf{A}$ and $\mathbf{B}$.

$$\mathbf{A} \times \mathbf{B} = \begin{vmatrix} \mathbf{i} & \mathbf{j} & \mathbf{k} \\ 2 & -6 & -3 \\ 4 & 3 & -1 \end{vmatrix} = 15\mathbf{i} - 10\mathbf{j} + 30\mathbf{k}$$

A unit vector parallel to $\mathbf{A} \times \mathbf{B}$ is $\dfrac{\mathbf{A} \times \mathbf{B}}{|\mathbf{A} \times \mathbf{B}|} = \dfrac{15\mathbf{i} - 10\mathbf{j} + 30\mathbf{k}}{\sqrt{(15)^2 + (-10)^2 + (30)^2}} = \dfrac{3}{7}\mathbf{i} - \dfrac{2}{7}\mathbf{j} + \dfrac{6}{7}\mathbf{k}.$

Another unit vector, opposite in direction, is $(-3\mathbf{i} + 2\mathbf{j} - 6\mathbf{k})/7$. Compare with Problem 2.14.

2.35. Find an expression for the moment of a force $\mathbf{F}$ about a point P as in Fig. 2-14.

Solution

The moment $\mathbf{M}$ of $\mathbf{F}$ about P is in magnitude equal to P to the line of action of $\mathbf{F}$. Then, if $\mathbf{r}$ is the vector from P to the initial point Q of $\mathbf{F}$,

$$M = F(r \sin \theta) = rF \sin \theta = |\mathbf{r} \times \mathbf{F}|$$

If we think of a right-threaded screw at P perpendicular to the plane of $\mathbf{r}$ and $\mathbf{F}$, then when the force $\mathbf{F}$ acts, the screw will move in the direction of $\mathbf{r} \times \mathbf{F}$. Because of this, it is convenient to define the moment as the vector $\mathbf{M} = \mathbf{r} \times \mathbf{F}$.

Fig. 2-14   Fig. 2-15

2.36. As in Fig. 2-15, a rigid body rotates about an axis through point O with angular speed ω. Prove that the linear velocity $\mathbf{v}$ of a point P of the body with position vector $\mathbf{r}$ is given by $\mathbf{v} = \boldsymbol{\omega} \times \mathbf{r}$, where $\boldsymbol{\omega}$ is the vector with magnitude ω whose direction is that in which a right-handed screw would advance under the given rotation.

Solution

Since P travels in a circle of radius $r \sin \theta$, the magnitude of the linear velocity $\mathbf{v}$ is $\omega(r \sin \theta) = |\boldsymbol{\omega} \times \mathbf{r}|$. Also, $\mathbf{v}$ must be perpendicular to both $\boldsymbol{\omega}$ and $\mathbf{r}$ and is such that $\mathbf{r}$, $\boldsymbol{\omega}$, and $\mathbf{v}$ form a right-handed system.

Then $\mathbf{v}$ agrees both in magnitude and direction with $\boldsymbol{\omega} \times \mathbf{r}$; hence $\mathbf{v} = \boldsymbol{\omega} \times \mathbf{r}$. The vector $\boldsymbol{\omega}$ is called the *angular velocity*.

Triple Products

2.37. Suppose $\mathbf{A} = A_1\mathbf{i} + A_2\mathbf{j} + A_3\mathbf{k}$, $\mathbf{B} = B_1\mathbf{i} + B_2\mathbf{j} + B_3\mathbf{k}$, $\mathbf{C} = C_1\mathbf{i} + C_2\mathbf{j} + C_3\mathbf{k}$.
Show that

$$\mathbf{A} \cdot (\mathbf{B} \times \mathbf{C}) = \begin{vmatrix} A_1 & A_2 & A_3 \\ B_1 & B_2 & B_3 \\ C_1 & C_2 & C_3 \end{vmatrix}$$

Solution

$$\mathbf{A} \cdot (\mathbf{B} \times \mathbf{C}) = \mathbf{A} \cdot \begin{vmatrix} \mathbf{i} & \mathbf{j} & \mathbf{k} \\ B_1 & B_2 & B_3 \\ C_1 & C_2 & C_3 \end{vmatrix}$$

$$= (A_1\mathbf{i} + A_2\mathbf{j} + A_3\mathbf{k}) \cdot [(B_2C_3 - B_3C_2)\mathbf{i} + (B_3C_1 - B_1C_3)\mathbf{j} + (B_1C_2 - B_2C_1)\mathbf{k}]$$

$$= A_1(B_2C_3 - B_3C_2) + A_2(B_3C_1 - B_1C_3) + A_3(B_1C_2 - B_2C_1)$$

$$= \begin{vmatrix} A_1 & A_2 & A_3 \\ B_1 & B_2 & B_3 \\ C_1 & C_2 & C_3 \end{vmatrix}$$

2.38. Evaluate $(\mathbf{i} + 2\mathbf{j} + 3\mathbf{k}) \cdot (\mathbf{i} + 3\mathbf{j} + 5\mathbf{k}) \times (\mathbf{i} + \mathbf{j} + 6\mathbf{k})$.

Solution

By Problem 2.37, the result is $\begin{vmatrix} 1 & 2 & 3 \\ 1 & 3 & 5 \\ 1 & 1 & 6 \end{vmatrix} = 5$.

2.39. Prove that $\mathbf{A} \cdot (\mathbf{B} \times \mathbf{C}) = \mathbf{B} \cdot (\mathbf{C} \times \mathbf{A}) = \mathbf{C} \cdot (\mathbf{A} \times \mathbf{B})$.

Solution

By Problem 2.37, $\mathbf{A} \cdot (\mathbf{B} \times \mathbf{C}) = \begin{vmatrix} A_1 & A_2 & A_3 \\ B_1 & B_2 & B_3 \\ C_1 & C_2 & C_3 \end{vmatrix}$.

By a theorem of determinants which states that interchange of two rows of a determinant changes its sign, we have

$$\begin{vmatrix} A_1 & A_2 & A_3 \\ B_1 & B_2 & B_3 \\ C_1 & C_2 & C_3 \end{vmatrix} = - \begin{vmatrix} B_1 & B_2 & B_3 \\ A_1 & A_2 & A_3 \\ C_1 & C_2 & C_3 \end{vmatrix} = \begin{vmatrix} B_1 & B_2 & B_3 \\ C_1 & C_2 & C_3 \\ A_1 & A_2 & A_3 \end{vmatrix} = \mathbf{B} \cdot (\mathbf{C} \times \mathbf{A})$$

$$\begin{vmatrix} A_1 & A_2 & A_3 \\ B_1 & B_2 & B_3 \\ C_1 & C_2 & C_3 \end{vmatrix} = - \begin{vmatrix} C_1 & C_2 & C_3 \\ B_1 & B_2 & B_3 \\ A_1 & A_2 & A_3 \end{vmatrix} = \begin{vmatrix} C_1 & C_2 & C_3 \\ A_1 & A_2 & A_3 \\ B_1 & B_2 & B_3 \end{vmatrix} = \mathbf{C} \cdot (\mathbf{A} \times \mathbf{B})$$

2.40. Show that $\mathbf{A} \cdot (\mathbf{B} \times \mathbf{C}) = (\mathbf{A} \times \mathbf{B}) \cdot \mathbf{C}$.

Solution

From Problem 2.39, $\mathbf{A} \cdot (\mathbf{B} \times \mathbf{C}) = \mathbf{C} \cdot (\mathbf{A} \times \mathbf{B}) = (\mathbf{A} \times \mathbf{B}) \cdot \mathbf{C}$.

Occasionally, $\mathbf{A} \cdot (\mathbf{B} \times \mathbf{C})$ is written without parentheses as $\mathbf{A} \cdot \mathbf{B} \times \mathbf{C}$. In such a case, there cannot be any ambiguity since the only possible interpretations are $\mathbf{A} \cdot (\mathbf{B} \times \mathbf{C})$ and $(\mathbf{A} \cdot \mathbf{B}) \times \mathbf{C}$. The latter, however, has no meaning since the cross product of a scalar with a vector is undefined.

The result $\mathbf{A} \cdot \mathbf{B} \times \mathbf{C} = \mathbf{A} \times \mathbf{B} \cdot \mathbf{C}$ is sometimes summarized in the statement that the dot and cross can be interchanged without affecting the result.

2.41. Show that $\mathbf{A} \cdot (\mathbf{A} \times \mathbf{C}) = \mathbf{0}$.

Solution

From Problem 2.40 and that $\mathbf{A} \times \mathbf{A} = \mathbf{0}$, we have $\mathbf{A} \cdot (\mathbf{A} \times \mathbf{C}) = (\mathbf{A} \times \mathbf{A}) \cdot \mathbf{C} = 0$.

2.42. Prove that a necessary and sufficient condition for the vectors $\mathbf{A}$, $\mathbf{B}$, and $\mathbf{C}$ to be coplanar is that $\mathbf{A} \cdot \mathbf{B} \times \mathbf{C} = 0$.

Solution

Note that $\mathbf{A} \cdot \mathbf{B} \times \mathbf{C}$ can have no meaning other than $\mathbf{A} \cdot (\mathbf{B} \times \mathbf{C})$.

If $\mathbf{A}$, $\mathbf{B}$, and $\mathbf{C}$ are coplanar, the volume of the parallelepiped formed by them is zero. Then, by Problem 2.43, $\mathbf{A} \cdot \mathbf{B} \times \mathbf{C} = 0$.

Conversely, if $\mathbf{A} \cdot \mathbf{B} \times \mathbf{C} = 0$, the volume of the parallelepiped formed by vectors $\mathbf{A}$, $\mathbf{B}$, and $\mathbf{C}$ is zero, and so the vectors must lie in a plane.

2.43. Show that the absolute value of the triple product $\mathbf{A} \cdot (\mathbf{B} \times \mathbf{C})$ is the volume of a parallelepiped with sides $\mathbf{A}$, $\mathbf{B}$, and $\mathbf{C}$.

Solution

Let $\mathbf{n}$ be a unit normal to a parallelogram I, having the direction of $\mathbf{B} \times \mathbf{C}$, and let h be the height of the terminal point of $\mathbf{A}$ above the parallelogram I. [See Fig. 2-16.]

$$\text{Volume of parallelepiped} = (\text{height } h)(\text{area of parallelogram } I)$$
$$= (\mathbf{A} \cdot \mathbf{n})(|\mathbf{B} \times \mathbf{C}|)$$
$$= \mathbf{A} \cdot \{|\mathbf{B} \times \mathbf{C}|\mathbf{n}\} = \mathbf{A} \cdot (\mathbf{B} \times \mathbf{C})$$

If $\mathbf{A}$, $\mathbf{B}$, and $\mathbf{C}$ do not form a right-handed system, $\mathbf{A} \cdot \mathbf{n} < 0$ and the volume $= |\mathbf{A} \cdot (\mathbf{B} \times \mathbf{C})|$.

Fig. 2-16

Fig. 2-17

2.44. Let $\mathbf{r}_1 = x_1\mathbf{i} + y_1\mathbf{j} + z_1\mathbf{k}$, $\mathbf{r}_2 = x_2\mathbf{i} + y_2\mathbf{j} + z_2\mathbf{k}$, and $\mathbf{r}_3 = x_3\mathbf{i} + y_3\mathbf{j} + z_3\mathbf{k}$ be the position vectors of points $P_1(x_1, y_1, z_1)$, $P_2(x_2, y_2, z_2)$, and $P_3(x_3, y_3, z_3)$. Find an equation for the plane passing through P_1, P_2, and P_3.

Solution

We assume that P_1, P_2, and P_3 do not lie in the same straight line; hence they determine a plane.

Let $\mathbf{r} = x\mathbf{i} + y\mathbf{j} + z\mathbf{k}$ denote the position vector of any point $P(x, y, z)$ in the plane. Consider vectors $\mathbf{P_1P_2} = \mathbf{r}_2 - \mathbf{r}_1$, $\mathbf{P_1P_3} = \mathbf{r}_3 - \mathbf{r}_1$, and $\mathbf{P_1P} = \mathbf{r} - \mathbf{r}_1$, which all lie in the plane. [See Fig. 2-17.]

By Problem 2.42, $\mathbf{P_1P} \cdot \mathbf{P_1P_2} \times \mathbf{P_1P_3} = 0$ or $(\mathbf{r} - \mathbf{r}_1) \cdot (\mathbf{r}_2 - \mathbf{r}_1) \times (\mathbf{r}_3 - \mathbf{r}_1) = 0$.

In terms of rectangular coordinates, this becomes

$$[(x - x_1)\mathbf{i} + (y - y_1)\mathbf{j} + (z - z_1)\mathbf{k}] \cdot [(x_2 - x_1)\mathbf{i} + (y_2 - y_1)\mathbf{j} + (z_2 - z_1)\mathbf{k}] \times [(x_3 - x_1)\mathbf{i} + (y_3 - y_1)\mathbf{j}$$
$$+ (z_3 - z_1)\mathbf{k}] = 0$$

or, using Problem 2.37,

$$\begin{vmatrix} x - x_1 & y - y_1 & z - z_1 \\ x_2 - x_1 & y_2 - y_1 & z_2 - z_1 \\ x_3 - x_1 & y_3 - y_1 & z_3 - z_1 \end{vmatrix} = 0.$$

2.45. Find an equation for the plane determined by the points $P_1(2, -1, 1)$, $P_2(3, 2, -1)$, and $P_3(-1, 3, 2)$.

Solution

The position vectors of P_1, P_2, P_3 and any point $P(x, y, z)$ are, respectively, $\mathbf{r}_1 = 2\mathbf{i} - \mathbf{j} + \mathbf{k}$, $\mathbf{r}_2 = 3\mathbf{i} + 2\mathbf{j} - \mathbf{k}$, $\mathbf{r}_3 = -\mathbf{i} + 3\mathbf{j} + 2\mathbf{k}$, and $\mathbf{r} = x\mathbf{i} + y\mathbf{j} + z\mathbf{k}$.

Then $\mathbf{PP}_1 = \mathbf{r} - \mathbf{r}_1$, $\mathbf{P_2P_1} = \mathbf{r}_2 - \mathbf{r}_1$, $\mathbf{P_3P_1} = \mathbf{r}_3 - \mathbf{r}_1$ all lie in the required plane, so that

$$(\mathbf{r} - \mathbf{r}_1) \cdot (\mathbf{r}_2 - \mathbf{r}_1) \times (\mathbf{r}_3 - \mathbf{r}_1) = 0$$

that is,

$$[(x - 2)\mathbf{i} + (y + 1)\mathbf{j} + (z - 1)\mathbf{k}] \cdot [\mathbf{i} + 3\mathbf{j} - 2\mathbf{k}] \times [-3\mathbf{i} + 4\mathbf{j} + \mathbf{k}] = 0$$
$$[(x - 2)\mathbf{i} + (y + 1)\mathbf{j} + (z - 1)\mathbf{k}] \cdot [11\mathbf{i} + 5\mathbf{j} + 13\mathbf{k}] = 0$$
$$11(x - 2) + 5(y + 1) + 13(z - 1) = 0 \quad \text{or} \quad 11x + 5y + 13z = 30.$$

2.46. Suppose the points P, Q, and R, not all lying on the same straight line, have position vectors $\mathbf{a}$, $\mathbf{b}$, and $\mathbf{c}$ relative to a given origin. Show that $\mathbf{a} \times \mathbf{b} + \mathbf{b} \times \mathbf{c} + \mathbf{c} \times \mathbf{a}$ is a vector perpendicular to the plane of P, Q, and R.

Solution

Let $\mathbf{r}$ be the position vector of any point in the plane of P, Q, and R. Then the vectors $\mathbf{r} - \mathbf{a}$, $\mathbf{b} - \mathbf{a}$, and $\mathbf{c} - \mathbf{a}$ are coplanar, so that by Problem 2.42

$$(\mathbf{r} - \mathbf{a}) \cdot (\mathbf{b} - \mathbf{a}) \times (\mathbf{c} - \mathbf{a}) = 0 \quad \text{or} \quad (\mathbf{r} - \mathbf{a}) \cdot (\mathbf{a} \times \mathbf{b} + \mathbf{b} \times \mathbf{c} + \mathbf{c} \times \mathbf{a}) = 0.$$

Thus $\mathbf{a} \times \mathbf{b} + \mathbf{b} \times \mathbf{c} + \mathbf{c} \times \mathbf{a}$ is perpendicular to $\mathbf{r} - \mathbf{a}$ and is therefore perpendicular to the plane of P, Q, and R.

2.47. Prove: (a) $\mathbf{A} \times (\mathbf{B} \times \mathbf{C}) = \mathbf{B}(\mathbf{A} \cdot \mathbf{C}) - \mathbf{C}(\mathbf{A} \cdot \mathbf{B})$,   (b) $(\mathbf{A} \times \mathbf{B}) \times \mathbf{C} = \mathbf{B}(\mathbf{A} \cdot \mathbf{C}) - \mathbf{A}(\mathbf{B} \cdot \mathbf{C})$.

Solution

(a)   Let $\mathbf{A} = A_1\mathbf{i} + A_2\mathbf{j} + A_3\mathbf{k}$, $\mathbf{B} = B_1\mathbf{i} + B_2\mathbf{j} + B_3\mathbf{k}$, $\mathbf{C} = C_1\mathbf{i} + C_2\mathbf{j} + C_3\mathbf{k}$. Then

$$\mathbf{A} \times (\mathbf{B} \times \mathbf{C}) = (A_1\mathbf{i} + A_2\mathbf{j} + A_3\mathbf{k}) \times \begin{vmatrix} \mathbf{i} & \mathbf{j} & \mathbf{k} \\ B_1 & B_2 & B_3 \\ C_1 & C_2 & C_3 \end{vmatrix}$$

$$= (A_1\mathbf{i} + A_2\mathbf{j} + A_3\mathbf{k}) \times ([B_2C_3 - B_3C_2]\mathbf{i} + [B_3C_1 - B_1C_3]\mathbf{j} + [B_1C_2 - B_2C_1]\mathbf{k})$$

$$= \begin{vmatrix} \mathbf{i} & \mathbf{j} & \mathbf{k} \\ A_1 & A_2 & A_3 \\ B_2C_3 - B_3C_2 & B_3C_1 - B_1C_3 & B_1C_2 - B_2C_1 \end{vmatrix}$$

$$= (A_2B_1C_2 - A_2B_2C_1 - A_3B_3C_1 + A_3B_1C_3)\mathbf{i} + (A_3B_2C_3 - A_3B_3C_2 - A_1B_1C_2 + A_1B_2C_1)\mathbf{j}$$
$$+ (A_1B_3C_1 - A_1B_1C_3 - A_2B_2C_3 + A_2B_3C_2)\mathbf{k}$$

Also

$$\mathbf{B}(\mathbf{A} \cdot \mathbf{C}) - \mathbf{C}(\mathbf{A} \cdot \mathbf{B}) = (B_1\mathbf{i} + B_2\mathbf{j} + B_3\mathbf{k})(A_1C_1 + A_2C_2 + A_3C_3) - (C_1\mathbf{i} + C_2\mathbf{j} + C_3\mathbf{k})(A_1B_1 + A_2B_2 + A_3B_3)$$

$$= (A_2B_1C_2 + A_3B_1C_3 - A_2C_1B_2 - A_3C_1B_3)\mathbf{i} + (B_2A_1C_1 + B_2A_3C_3 - C_2A_1B_1 - C_2A_3B_3)\mathbf{j}$$
$$+ (B_3A_1C_1 + B_3A_2C_2 - C_3A_1B_1 - C_3A_2B_2)\mathbf{k}$$

and the result follows.

(b) $(\mathbf{A} \times \mathbf{B}) \times \mathbf{C} = -\mathbf{C} \times (\mathbf{A} \times \mathbf{B}) = -\{\mathbf{A}(\mathbf{C} \cdot \mathbf{B}) - \mathbf{B}(\mathbf{C} \cdot \mathbf{A})\} = \mathbf{B}(\mathbf{A} \cdot \mathbf{C}) - \mathbf{A}(\mathbf{B} \cdot \mathbf{C})$ upon replacing $\mathbf{A}$, $\mathbf{B}$, and $\mathbf{C}$ in (a) by $\mathbf{C}$, $\mathbf{A}$, and $\mathbf{B}$, respectively.

    Note that $\mathbf{A} \times (\mathbf{B} \times \mathbf{C}) \neq (\mathbf{A} \times \mathbf{B}) \times \mathbf{C}$, that is, the associative law for vector cross products is not valid for all vectors $\mathbf{A}$, $\mathbf{B}$, and $\mathbf{C}$.

2.48. Prove: $(\mathbf{A} \times \mathbf{B}) \cdot (\mathbf{C} \times \mathbf{D}) = (\mathbf{A} \cdot \mathbf{C})(\mathbf{B} \cdot \mathbf{D}) - (\mathbf{A} \cdot \mathbf{D})(\mathbf{B} \cdot \mathbf{C})$.

Solution

From Problem 2.41, $\mathbf{X} \cdot (\mathbf{C} \times \mathbf{D}) = (\mathbf{X} \times \mathbf{C}) \cdot \mathbf{D}$. Let $\mathbf{X} = \mathbf{A} \times \mathbf{B}$; then

$$(\mathbf{A} \times \mathbf{B}) \cdot (\mathbf{C} \times \mathbf{D}) = \{(\mathbf{A} \times \mathbf{B}) \times \mathbf{C}\} \cdot \mathbf{D}$$
$$= \{\mathbf{B}(\mathbf{A} \cdot \mathbf{C}) - \mathbf{A}(\mathbf{B} \cdot \mathbf{C})\} \cdot \mathbf{D}$$
$$= (\mathbf{A} \cdot \mathbf{C})(\mathbf{B} \cdot \mathbf{D}) - (\mathbf{A} \cdot \mathbf{D})(\mathbf{B} \cdot \mathbf{C}), \quad \text{using Problem 2.47(b).}$$

2.49. Prove: $\mathbf{A} \times (\mathbf{B} \times \mathbf{C}) + \mathbf{B} \times (\mathbf{C} \times \mathbf{A}) + \mathbf{C} \times (\mathbf{A} \times \mathbf{B}) = 0$.

Solution

By Problem 2.47(a), $\mathbf{A} \times (\mathbf{B} \times \mathbf{C}) = \mathbf{B}(\mathbf{A} \cdot \mathbf{C}) - \mathbf{C}(\mathbf{A} \cdot \mathbf{B})$
$$\mathbf{B} \times (\mathbf{C} \times \mathbf{A}) = \mathbf{C}(\mathbf{B} \cdot \mathbf{A}) - \mathbf{A}(\mathbf{B} \cdot \mathbf{C})$$
$$\mathbf{C} \times (\mathbf{A} \times \mathbf{B}) = \mathbf{A}(\mathbf{C} \cdot \mathbf{B}) - \mathbf{B}(\mathbf{C} \cdot \mathbf{A})$$

Adding, the result follows.

2.50. Prove: $(\mathbf{A} \times \mathbf{B}) \times (\mathbf{C} \times \mathbf{D}) = \mathbf{B}(\mathbf{A} \cdot \mathbf{C} \times \mathbf{D}) - \mathbf{A}(\mathbf{B} \cdot \mathbf{C} \times \mathbf{D}) = \mathbf{C}(\mathbf{A} \cdot \mathbf{B} \times \mathbf{D}) - \mathbf{D}(\mathbf{A} \cdot \mathbf{B} \times \mathbf{C})$.

Solution

By Problem 2.47(a), $\mathbf{X} \times (\mathbf{C} \times \mathbf{D}) = \mathbf{C}(\mathbf{X} \cdot \mathbf{D}) - \mathbf{D}(\mathbf{X} \cdot \mathbf{C})$. Let $\mathbf{X} = \mathbf{A} \times \mathbf{B}$; then

$$(\mathbf{A} \times \mathbf{B}) \times (\mathbf{C} \times \mathbf{D}) = \mathbf{C}(\mathbf{A} \times \mathbf{B} \cdot \mathbf{D}) - \mathbf{D}(\mathbf{A} \times \mathbf{B} \cdot \mathbf{C})$$
$$= \mathbf{C}(\mathbf{A} \cdot \mathbf{B} \times \mathbf{D}) - \mathbf{D}(\mathbf{A} \cdot \mathbf{B} \times \mathbf{C})$$

By Problem 2.47(b), $(\mathbf{A} \times \mathbf{B}) \times \mathbf{Y} = \mathbf{B}(\mathbf{A} \cdot \mathbf{Y}) - \mathbf{A}(\mathbf{B} \cdot \mathbf{Y})$. Let $\mathbf{Y} = \mathbf{C} \times \mathbf{D}$; then

$$(\mathbf{A} \times \mathbf{B}) \times (\mathbf{C} \times \mathbf{D}) = \mathbf{B}(\mathbf{A} \cdot \mathbf{C} \times \mathbf{D}) - \mathbf{A}(\mathbf{B} \cdot \mathbf{C} \times \mathbf{D}).$$

2.51. Let PQR be a spherical triangle whose sides p, q, r are arcs of great circles. Prove that

$$\frac{\sin P}{\sin p} = \frac{\sin Q}{\sin q} = \frac{\sin R}{\sin r}$$

Solution

Suppose that the sphere, pictured in Fig. 2-18, has unit radius. Let unit vectors $\mathbf{A}$, $\mathbf{B}$, $\mathbf{C}$ be drawn from the center O of the sphere to P, Q, R, respectively. From Problem 2.50,

$$(\mathbf{A} \times \mathbf{B}) \times (\mathbf{A} \times \mathbf{C}) = (\mathbf{A} \cdot \mathbf{B} \times \mathbf{C})\mathbf{A} \tag{1}$$

A unit vector perpendicular to $\mathbf{A} \times \mathbf{B}$ and $\mathbf{A} \times \mathbf{C}$ is $\mathbf{A}$, so that (1) becomes

$$(\sin r \sin q \sin P) \, \mathbf{A} = (\mathbf{A} \cdot \mathbf{B} \times \mathbf{C})\mathbf{A} \qquad \text{or} \tag{2}$$
$$\sin r \sin q \sin P = \mathbf{A} \cdot \mathbf{B} \times \mathbf{C} \tag{3}$$

By cyclic permutation of p, q, r, P, Q, R and $\mathbf{A}$, $\mathbf{B}$, and $\mathbf{C}$, we obtain

$$\sin p \sin r \sin Q = \mathbf{B} \cdot \mathbf{C} \times \mathbf{A} \tag{4}$$
$$\sin q \sin p \sin R = \mathbf{C} \cdot \mathbf{A} \times \mathbf{B} \tag{5}$$

Then, since the right-hand sides of (3), (4), and (5) are equal (Problem 2.39)

$$\sin r \sin q \sin P = \sin p \sin r \sin Q = \sin q \sin p \sin R$$

from which we find

$$\frac{\sin P}{\sin p} = \frac{\sin Q}{\sin q} = \frac{\sin R}{\sin r}$$

This is called the *law of sines* for spherical triangles.

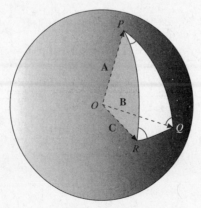

Fig. 2-18

2.52. Prove: $(\mathbf{A} \times \mathbf{B}) \cdot (\mathbf{B} \times \mathbf{C}) \times (\mathbf{C} \times \mathbf{A}) = (\mathbf{A} \cdot \mathbf{B} \times \mathbf{C})^2$.

Solution

By Problem 2.47(a), $\mathbf{X} \times (\mathbf{C} \times \mathbf{A}) = \mathbf{C}(\mathbf{X} \cdot \mathbf{A}) - \mathbf{A}(\mathbf{X} \cdot \mathbf{C})$. Let $\mathbf{X} = \mathbf{B} \times \mathbf{C}$; then

$$(\mathbf{B} \times \mathbf{C}) \times (\mathbf{C} \times \mathbf{A}) = \mathbf{C}(\mathbf{B} \times \mathbf{C} \cdot \mathbf{A}) - \mathbf{A}(\mathbf{B} \times \mathbf{C} \cdot \mathbf{C})$$
$$= \mathbf{C}(\mathbf{A} \cdot \mathbf{B} \times \mathbf{C}) - \mathbf{A}(\mathbf{B} \cdot \mathbf{C} \times \mathbf{C})$$
$$= \mathbf{C}(\mathbf{A} \cdot \mathbf{B} \times \mathbf{C})$$

Thus

$$(\mathbf{A} \times \mathbf{B}) \cdot (\mathbf{B} \times \mathbf{C}) \times (\mathbf{C} \times \mathbf{A}) = (\mathbf{A} \times \mathbf{B}) \cdot \mathbf{C}(\mathbf{A} \cdot \mathbf{B} \times \mathbf{C})$$
$$= (\mathbf{A} \times \mathbf{B} \cdot \mathbf{C})(\mathbf{A} \cdot \mathbf{B} \times \mathbf{C})$$
$$= (\mathbf{A} \cdot \mathbf{B} \times \mathbf{C})^2$$

2.53. Given the vectors $\mathbf{a}' = \dfrac{\mathbf{b} \times \mathbf{c}}{\mathbf{a} \cdot \mathbf{b} \times \mathbf{c}}, \mathbf{b}' = \dfrac{\mathbf{c} \times \mathbf{a}}{\mathbf{a} \cdot \mathbf{b} \times \mathbf{c}}$ and $\mathbf{c}' = \dfrac{\mathbf{a} \times \mathbf{b}}{\mathbf{a} \cdot \mathbf{b} \times \mathbf{c}}$, suppose $\mathbf{a} \cdot \mathbf{b} \times \mathbf{c} \neq 0$. Show that

(a) $\mathbf{a}' \cdot \mathbf{a} = \mathbf{b}' \cdot \mathbf{b} = \mathbf{c}' \cdot \mathbf{c} = 1$,

(b) $\mathbf{a}' \cdot \mathbf{b} = \mathbf{a}' \cdot \mathbf{c} = 0, \mathbf{b}' \cdot \mathbf{a} = \mathbf{b}' \cdot \mathbf{c} = 0, \mathbf{c}' \cdot \mathbf{a} = \mathbf{c}' \cdot \mathbf{b} = 0$,

(c) if $\mathbf{a} \cdot \mathbf{b} \times \mathbf{c} = V$, then $\mathbf{a}' \cdot \mathbf{b}' \times \mathbf{c}' = 1/V$,

(d) $\mathbf{a}', \mathbf{b}'$, and $\mathbf{c}'$ are non-coplanar if $\mathbf{a}, \mathbf{b}$, and $\mathbf{c}$ are non-coplanar.

Solution

(a) $\quad \mathbf{a}' \cdot \mathbf{a} = \mathbf{a} \cdot \mathbf{a}' = \mathbf{a} \cdot \dfrac{\mathbf{b} \times \mathbf{c}}{\mathbf{a} \cdot \mathbf{b} \times \mathbf{c}} = \dfrac{\mathbf{a} \cdot \mathbf{b} \times \mathbf{c}}{\mathbf{a} \cdot \mathbf{b} \times \mathbf{c}} = 1$

$\quad \mathbf{b}' \cdot \mathbf{b} = \mathbf{b} \cdot \mathbf{b}' = \mathbf{b} \cdot \dfrac{\mathbf{c} \times \mathbf{a}}{\mathbf{a} \cdot \mathbf{b} \times \mathbf{c}} = \dfrac{\mathbf{b} \cdot \mathbf{c} \times \mathbf{a}}{\mathbf{a} \cdot \mathbf{b} \times \mathbf{c}} = \dfrac{\mathbf{a} \cdot \mathbf{b} \times \mathbf{c}}{\mathbf{a} \cdot \mathbf{b} \times \mathbf{c}} = 1$

$\quad \mathbf{c}' \cdot \mathbf{c} = \mathbf{c} \cdot \mathbf{c}' = \mathbf{c} \cdot \dfrac{\mathbf{a} \times \mathbf{b}}{\mathbf{a} \cdot \mathbf{b} \times \mathbf{c}} = \dfrac{\mathbf{c} \cdot \mathbf{a} \times \mathbf{b}}{\mathbf{a} \cdot \mathbf{b} \times \mathbf{c}} = \dfrac{\mathbf{a} \cdot \mathbf{b} \times \mathbf{c}}{\mathbf{a} \cdot \mathbf{b} \times \mathbf{c}} = 1$

(b) $\quad \mathbf{a}' \cdot \mathbf{b} = \mathbf{b} \cdot \mathbf{a}' = \mathbf{b} \cdot \dfrac{\mathbf{b} \times \mathbf{c}}{\mathbf{a} \cdot \mathbf{b} \times \mathbf{c}} = \dfrac{\mathbf{b} \cdot \mathbf{b} \times \mathbf{c}}{\mathbf{a} \cdot \mathbf{b} \times \mathbf{c}} = \dfrac{\mathbf{b} \times \mathbf{b} \cdot \mathbf{c}}{\mathbf{a} \cdot \mathbf{b} \times \mathbf{c}} = 0$

Similarly, the other results follow. The results can also be seen by noting, for example, that $\mathbf{a}'$ has the direction of $\mathbf{b} \times \mathbf{c}$ and so must be perpendicular to both $\mathbf{b}$ and $\mathbf{c}$, from which $\mathbf{a}' \cdot \mathbf{b} = 0$ and $\mathbf{a}' \cdot \mathbf{c} = 0$.

From (a) and (b), we see that the sets of vectors $\mathbf{a}, \mathbf{b}, \mathbf{c}$ and $\mathbf{a}', \mathbf{b}', \mathbf{c}'$ are reciprocal vectors. See also Supplementary Problems 2.104 and 2.106.

(c) $\quad \mathbf{a}' = \dfrac{\mathbf{b} \times \mathbf{c}}{V}, \quad \mathbf{b}' = \dfrac{\mathbf{c} \times \mathbf{a}}{V}, \quad \mathbf{c}' = \dfrac{\mathbf{a} \times \mathbf{b}}{V}$

Then $\mathbf{a}' \cdot \mathbf{b}' \times \mathbf{c}' = \dfrac{(\mathbf{b} \times \mathbf{c}) \cdot (\mathbf{c} \times \mathbf{a}) \times (\mathbf{a} \times \mathbf{b})}{V^3} = \dfrac{(\mathbf{a} \times \mathbf{b}) \cdot (\mathbf{b} \times \mathbf{c}) \times (\mathbf{c} \times \mathbf{a})}{V^3}$

$$= \dfrac{(\mathbf{a} \cdot \mathbf{b} \times \mathbf{c})^2}{V^3} = \dfrac{V^2}{V^3} = \dfrac{1}{V} \quad \text{using Problem 2.52.}$$

(d) By Problem 2.42, if $\mathbf{a}, \mathbf{b}$, and $\mathbf{c}$ are non-coplanar $\mathbf{a} \cdot \mathbf{b} \times \mathbf{c} \neq 0$. Then, from part (c), it follows that $\mathbf{a}' \cdot \mathbf{b}' \times \mathbf{c}' \neq 0$, so that $\mathbf{a}', \mathbf{b}'$, and $\mathbf{c}'$ are also non-coplanar.

2.54. Show that any vector **r** can be expressed in terms of the reciprocal vectors of Problem 2.53 as

$$\mathbf{r} = (\mathbf{r} \cdot \mathbf{a'})\mathbf{a} + (\mathbf{r} \cdot \mathbf{b'})\mathbf{b} + (\mathbf{r} \cdot \mathbf{c'})\mathbf{c}.$$

Solution

From Problem 2.50, $\mathbf{B}(\mathbf{A} \cdot \mathbf{C} \times \mathbf{D}) - \mathbf{A}(\mathbf{B} \cdot \mathbf{C} \times \mathbf{D}) = \mathbf{C}(\mathbf{A} \cdot \mathbf{B} \times \mathbf{D}) - \mathbf{D}(\mathbf{A} \cdot \mathbf{B} \times \mathbf{C})$. Then

$$\mathbf{D} = \frac{\mathbf{A}(\mathbf{B} \cdot \mathbf{C} \times \mathbf{D})}{\mathbf{A} \cdot \mathbf{B} \times \mathbf{C}} - \frac{\mathbf{B}(\mathbf{A} \cdot \mathbf{C} \times \mathbf{D})}{\mathbf{A} \cdot \mathbf{B} \times \mathbf{C}} + \frac{\mathbf{C}(\mathbf{A} \cdot \mathbf{B} \times \mathbf{D})}{\mathbf{A} \cdot \mathbf{B} \times \mathbf{C}}$$

Let $\mathbf{A} = \mathbf{a}$, $\mathbf{B} = \mathbf{b}$, $\mathbf{C} = \mathbf{c}$, and $\mathbf{D} = \mathbf{r}$. Then

$$\mathbf{r} = \frac{\mathbf{r} \cdot \mathbf{b} \times \mathbf{c}}{\mathbf{a} \cdot \mathbf{b} \times \mathbf{c}}\mathbf{a} + \frac{\mathbf{r} \cdot \mathbf{c} \times \mathbf{a}}{\mathbf{a} \cdot \mathbf{b} \times \mathbf{c}}\mathbf{b} + \frac{\mathbf{r} \cdot \mathbf{a} \times \mathbf{b}}{\mathbf{a} \cdot \mathbf{b} \times \mathbf{c}}\mathbf{c}$$

$$= \mathbf{r} \cdot \left(\frac{\mathbf{b} \times \mathbf{c}}{\mathbf{a} \cdot \mathbf{b} \times \mathbf{c}}\right)\mathbf{a} + \mathbf{r} \cdot \left(\frac{\mathbf{c} \times \mathbf{a}}{\mathbf{a} \cdot \mathbf{b} \times \mathbf{c}}\right)\mathbf{b} + \mathbf{r} \cdot \left(\frac{\mathbf{a} \times \mathbf{b}}{\mathbf{a} \cdot \mathbf{b} \times \mathbf{c}}\right)\mathbf{c}$$

$$= (\mathbf{r} \cdot \mathbf{a'})\mathbf{a} + (\mathbf{r} \cdot \mathbf{b'})\mathbf{b} + (\mathbf{r} \cdot \mathbf{c'})\mathbf{c}.$$

SUPPLEMENTARY PROBLEMS

2.55. Evaluate: (a) $\mathbf{k} \cdot (\mathbf{i} + \mathbf{j})$,   (b) $(\mathbf{i} - 2\mathbf{k}) \cdot (\mathbf{j} + 3\mathbf{k})$,   (c) $(2\mathbf{i} - \mathbf{j} + 3\mathbf{k}) \cdot (3\mathbf{i} + 2\mathbf{j} - \mathbf{k})$.

2.56. Suppose $\mathbf{A} = \mathbf{i} + 3\mathbf{j} - 2\mathbf{k}$ and $\mathbf{B} = 4\mathbf{i} - 2\mathbf{j} + 4\mathbf{k}$. Find: (a) $\mathbf{A} \cdot \mathbf{B}$,   (b) A,   (c) B,   (d) $|3\mathbf{A} + 2\mathbf{B}|$,   (e) $(2\mathbf{A} + \mathbf{B}) \cdot (\mathbf{A} - 2\mathbf{B})$.

2.57. Find the angle between   (a) $\mathbf{A} = 3\mathbf{i} + 2\mathbf{j} - 6\mathbf{k}$ and $\mathbf{B} = 4\mathbf{i} - 3\mathbf{j} + \mathbf{k}$;   (b) $\mathbf{C} = 4\mathbf{i} - 2\mathbf{j} + 4\mathbf{k}$ and $\mathbf{D} = 3\mathbf{i} - 6\mathbf{j} - 2\mathbf{k}$.

2.58. Find the values of a for which vectors **A** and **B** are perpendicular where:

(a) $\mathbf{A} = a\mathbf{i} - 2\mathbf{j} + \mathbf{k}$ and $\mathbf{B} = 2a\mathbf{i} + a\mathbf{j} - 4\mathbf{k}$,   (b) $\mathbf{A} = 2\mathbf{i} + \mathbf{j} + a\mathbf{k}$ and $\mathbf{B} = 2\mathbf{i} + a\mathbf{j} + \mathbf{k}$.

2.59. Find the acute angles that the line joining the points $(1, -3, 2)$ and $(3, -5, 1)$ makes with the coordinate axes.

2.60. Find the direction cosines of the line joining the points:

(a) $(3, 2, -4)$ and $(1, -1, 2)$,   (b) $(-5, 3, 3)$ and $(-2, 7, 15)$.

2.61. Determine the angles of a triangle where two sides of a triangle are formed by the vectors:

(a) $\mathbf{A} = 3\mathbf{i} - 4\mathbf{j} - \mathbf{k}$ and $\mathbf{B} = 4\mathbf{i} - \mathbf{j} + 3\mathbf{k}$,   (b) $\mathbf{A} = -2\mathbf{i} + 5\mathbf{j} + 6\mathbf{k}$ and $\mathbf{B} = 3\mathbf{i} + \mathbf{j} + 2\mathbf{k}$.

2.62. The diagonals of a parallelogram are given by $\mathbf{A} = 3\mathbf{i} - 4\mathbf{j} - \mathbf{k}$ and $\mathbf{B} = 2\mathbf{i} + 3\mathbf{j} - 6\mathbf{k}$. Show that the parallelogram is a rhombus and determine the length of its sides and angles.

2.63. Find the projection of the vector **A** on the vector **B** where:

(a) $\mathbf{A} = 2\mathbf{i} - 3\mathbf{j} + 6\mathbf{k}$ and $\mathbf{B} = \mathbf{i} + 2\mathbf{j} + 2\mathbf{k}$,   (b) $\mathbf{A} = 2\mathbf{i} + \mathbf{j} - \mathbf{k}$ and $\mathbf{B} = -6\mathbf{i} + 2\mathbf{j} - 3\mathbf{k}$.

2.64. Find the projection of the vector $\mathbf{A} = 4\mathbf{i} - 3\mathbf{j} + \mathbf{k}$ on the line passing through the points $(2, 3, -1)$ and $(-2, -4, 3)$.

2.65. Find a unit vector perpendicular to both vector **A** and vector **B** where:

(a) $\mathbf{A} = 4\mathbf{i} - \mathbf{j} + 3\mathbf{k}$ and $\mathbf{B} = -2\mathbf{i} + \mathbf{j} - 2\mathbf{k}$,   (b) $\mathbf{A} = 6\mathbf{i} + 22\mathbf{j} - 5\mathbf{k}$ and $\mathbf{B} = \mathbf{i} + 6\mathbf{j} - 2\mathbf{k}$.

2.66. Find the acute angle formed by two diagonals of a cube.

2.67. Find a unit vector parallel to the xy-plane and perpendicular to the vector $4\mathbf{i} - 3\mathbf{j} + \mathbf{k}$.

2.68. Show that $\mathbf{A}$, $\mathbf{B}$, and $\mathbf{C}$ are mutually orthogonal unit vectors where:

(a) $\mathbf{A} = (2\mathbf{i} - 2\mathbf{j} + \mathbf{k})/3$, $\mathbf{B} = (\mathbf{i} + 2\mathbf{j} + 2\mathbf{k})/3$,   and $\mathbf{C} = (2\mathbf{i} + \mathbf{j} - 2\mathbf{k})/3$

(b) $\mathbf{A} = (12\mathbf{i} - 4\mathbf{j} - 3\mathbf{k})/13$, $\mathbf{B} = (4\mathbf{i} + 3\mathbf{j} + 12\mathbf{k})/13$,   and $\mathbf{C} = (3\mathbf{i} + 12\mathbf{j} - 4\mathbf{k})/13$.

2.69. Find the work done in moving an object along a straight line:

(a) from $(3, 2, -1)$ to $(2, -1, 4)$ in a force field given by $\mathbf{F} = 4\mathbf{i} - 3\mathbf{j} + 2\mathbf{k}$.

(b) from $(3, 4, 5)$ to $(-1, 9, 9)$ in a force field given by $\mathbf{F} = -3\mathbf{i} + 5\mathbf{j} - 6\mathbf{k}$.

2.70. Let $\mathbf{F}$ be a constant vector field force. Show that the work done in moving an object around any closed polygon in this force field is zero.

2.71. Prove that an angle inscribed in a semicircle is a right angle.

2.72. Let $ABCD$ be a parallelogram. Prove that $\overline{AB}^2 + \overline{BC}^2 + \overline{CD}^2 + \overline{DA}^2 = \overline{AC}^2 + \overline{BD}^2$

2.73. Let $ABCD$ be any quadrilateral where P and Q are the midpoints of its diagonal. Prove that

$$\overline{AB}^2 + \overline{BC}^2 + \overline{CD}^2 + \overline{DA}^2 = \overline{AC}^2 + \overline{BD}^2 + 4\overline{PQ}^2$$

This is a generalization of the preceding problem.

2.74. Consider a plane P perpendicular to a given vector $\mathbf{A}$ and distance p from the origin. (a) Find an equation of the plane P. (b) Express the equation in (a) in rectangular coordinates.

2.75. Let $\mathbf{r}_1$ and $\mathbf{r}_2$ be unit vectors in the xy-plane making angles α and β with the positive x-axis.

(a) Prove that $\mathbf{r}_1 = \cos \alpha \mathbf{i} + \sin \alpha \mathbf{j}$ and $\mathbf{r}_2 = \cos \beta \mathbf{i} + \sin \beta \mathbf{j}$.

(b) By considering $\mathbf{r}_1 \cdot \mathbf{r}_2$, prove the trigonometric formulas

$$\cos(\alpha - \beta) = \cos \alpha \cos \beta + \sin \alpha \sin \beta \quad \text{and} \quad \cos(\alpha + \beta) = \cos \alpha \cos \beta + \sin \alpha \sin \beta$$

2.76. Let $\mathbf{a}$ be the position vector of a given point (x_1, y_1, z_1), and let $\mathbf{r}$ be the position vector of any point (x, y, z). Describe the locus of $\mathbf{r}$ if: (a) $|\mathbf{r} - \mathbf{a}| = 3$,   (b) $(\mathbf{r} - \mathbf{a}) \cdot \mathbf{a} = 0$,   (c) $(\mathbf{r} - \mathbf{a}) \cdot \mathbf{r} = 0$.

2.77. Suppose $\mathbf{A} = 3\mathbf{i} + \mathbf{j} + 2\mathbf{k}$ and $\mathbf{B} = \mathbf{i} - 2\mathbf{j} - 4\mathbf{k}$ are the position vectors of points P and Q, respectively.

(a) Find an equation for the plane passing through Q and perpendicular to the line PQ.

(b) Find the distance from the point $(-1, 1, 1)$ to the plane.

2.78. Evaluate each of the following: (a) $2\mathbf{j} \times (3\mathbf{i} - 4\mathbf{k})$,   (b) $(\mathbf{i} + 2\mathbf{j}) \times \mathbf{k}$,   (c) $(2\mathbf{i} - 4\mathbf{k}) \times (\mathbf{i} + 2\mathbf{j})$, (d) $(4\mathbf{i} + \mathbf{j} - 2\mathbf{k}) \times (3\mathbf{i} + \mathbf{k})$,   (e) $(2\mathbf{i} + \mathbf{j} - \mathbf{k}) \times (3\mathbf{i} - 2\mathbf{j} + 4\mathbf{k})$.

2.79. Suppose $\mathbf{A} = 3\mathbf{i} - \mathbf{j} - 2\mathbf{k}$ and $\mathbf{B} = 2\mathbf{i} + 3\mathbf{j} + \mathbf{k}$.   Find: (a) $|\mathbf{A} \times \mathbf{B}|$,   (b) $(\mathbf{A} + 2\mathbf{B}) \times (2\mathbf{A} - \mathbf{B})$, (c) $|(\mathbf{A} + \mathbf{B}) \times (\mathbf{A} - \mathbf{B})|$.

2.80. Suppose $\mathbf{A} = \mathbf{i} - 2\mathbf{j} - 3\mathbf{k}$, $\mathbf{B} = 2\mathbf{i} + \mathbf{j} - \mathbf{k}$, $\mathbf{C} = \mathbf{i} + 3\mathbf{j} - 2\mathbf{k}$. Find:

(a) $|(\mathbf{A} \times \mathbf{B}) \times \mathbf{C}|$ (c) $\mathbf{A} \cdot (\mathbf{B} \times \mathbf{C})$,   (e) $(\mathbf{A} \times \mathbf{B}) \times (\mathbf{B} \times \mathbf{C})$

(b) $|\mathbf{A} \times (\mathbf{B} \times \mathbf{C})|$ (d) $(\mathbf{A} \times \mathbf{B}) \cdot \mathbf{C}$,   (f) $(\mathbf{A} \times \mathbf{B})(\mathbf{B} \cdot \mathbf{C})$

2.81. Suppose $\mathbf{A} \neq \mathbf{0}$ and both of the following conditions hold simultaneously: (a) $\mathbf{A} \cdot \mathbf{B} = \mathbf{A} \cdot \mathbf{C}$, and (b) $\mathbf{A} \times \mathbf{B} = \mathbf{A} \times \mathbf{C}$. Show that $\mathbf{B} = \mathbf{C}$ but, if only one of the conditions holds, then $\mathbf{B} \neq \mathbf{C}$ necessarily.

2.82. Find the area of a parallelogram having diagonals: (a) $\mathbf{A} = 3\mathbf{i} + \mathbf{j} - 2\mathbf{k}$ and $\mathbf{B} = \mathbf{i} - 3\mathbf{j} - 4\mathbf{k}$, (b) $\mathbf{A} = 2\mathbf{i} + 4\mathbf{j}$ and $\mathbf{B} = -4\mathbf{i} + 4\mathbf{k}$.

2.83. Find the area of a triangle with vertices at: (a) $(3, -1, 2)$, $(1, -1, -3)$, and $(4, -3, 1)$, (b) $(2, -3, -2)$, $(-2, 3, 2)$, and $(4, 3, -1)$.

2.84. Suppose $\mathbf{A} = 2\mathbf{i} + \mathbf{j} - 3\mathbf{k}$ and $\mathbf{B} = \mathbf{i} - 2\mathbf{j} + \mathbf{k}$. Find a vector of magnitude 5 perpendicular to both $\mathbf{A}$ and $\mathbf{B}$.

2.85. Use Problem 2.75 to derive the formulas:

$$\sin(\alpha - \beta) = \sin\alpha\cos\beta - \cos\alpha\sin\beta \quad \text{and} \quad \sin(\alpha + \beta) = \sin\alpha\cos\beta + \cos\alpha\sin\beta$$

2.86. Suppose a force $\mathbf{F} = 3\mathbf{i} + 2\mathbf{j} - 4\mathbf{k}$ is applied at the point $(1, -1, 2)$. Find the moment of $\mathbf{F}$ about the point: (a) $(2, -1, 3)$, (b) $(4, -6, 3)$.

2.87. The angular velocity of a rotating rigid body about an axis of rotation is given by $\omega = 4\mathbf{i} + \mathbf{j} - 2\mathbf{k}$. Find the linear velocity of a point P on the body whose position vector relative to a point on the axis of rotation is $2\mathbf{i} - 3\mathbf{j} + \mathbf{k}$.

2.88. Simplify: (a) $(\mathbf{A} + \mathbf{B}) \cdot (\mathbf{B} + \mathbf{C}) \times (\mathbf{C} + \mathbf{A})$, (b) $\mathbf{A} \cdot (2\mathbf{A} + \mathbf{B}) \times \mathbf{C}$.

2.89. Prove that $(\mathbf{A} \cdot \mathbf{B} \times \mathbf{C})(\mathbf{a} \cdot \mathbf{b} \times \mathbf{c}) = \begin{vmatrix} A \cdot a & A \cdot b & A \cdot c \\ B \cdot a & B \cdot b & B \cdot c \\ C \cdot a & C \cdot b & C \cdot c \end{vmatrix}$

2.90. Find the volume of the parallelepiped whose edges are represented by:

(a) $\mathbf{A} = 2\mathbf{i} - 3\mathbf{j} + 4\mathbf{k}$, $\mathbf{B} = \mathbf{i} + 2\mathbf{j} - \mathbf{k}$, and $\mathbf{C} = 3\mathbf{i} - \mathbf{j} + 2\mathbf{k}$.

(b) $\mathbf{A} = \mathbf{i} - \mathbf{j} + 2\mathbf{k}$, $\mathbf{B} = \mathbf{i} + \mathbf{j} - \mathbf{k}$, and $\mathbf{C} = \mathbf{i} - \mathbf{j} - 4\mathbf{k}$.

2.91. Suppose $\mathbf{A} \cdot \mathbf{B} \times \mathbf{C} = 0$. Show that either (a) $\mathbf{A}$, $\mathbf{B}$, and $\mathbf{C}$ are coplanar but no two of them are collinear, or (b) two of the vectors $\mathbf{A}$, $\mathbf{B}$, and $\mathbf{C}$ are collinear, or (c) all the vectors $\mathbf{A}$, $\mathbf{B}$, and $\mathbf{C}$ are collinear.

2.92. Find the constant a so that the following vectors are coplanar:

(a) $2\mathbf{i} - \mathbf{j} + \mathbf{k}$, $\mathbf{i} + 2\mathbf{j} - 3\mathbf{k}$, $3\mathbf{i} + a\mathbf{j} + 5\mathbf{k}$, (b) $3\mathbf{i} - 3\mathbf{j} - \mathbf{k}$, $-3\mathbf{i} - 2\mathbf{j} + 2\mathbf{k}$, $6\mathbf{i} + a\mathbf{j} - 3\mathbf{k}$.

2.93. Suppose $\mathbf{A} = x_1\mathbf{a} + y_1\mathbf{b} + z_1\mathbf{c}$, $\mathbf{B} = x_2\mathbf{a} + y_2\mathbf{b} + z_2\mathbf{c}$, and $\mathbf{C} = x_3\mathbf{a} + y_3\mathbf{b} + z_3\mathbf{c}$. Prove that

$$\mathbf{A} \cdot \mathbf{B} \times \mathbf{C} = \begin{vmatrix} x_1 & y_1 & z_1 \\ x_2 & y_2 & z_2 \\ x_3 & y_3 & z_3 \end{vmatrix} (\mathbf{a} \cdot \mathbf{b} \times \mathbf{c})$$

2.94. Prove that $(\mathbf{A} \times \mathbf{C}) \times \mathbf{B} = \mathbf{0}$ is a necessary and sufficient condition that $\mathbf{A} \times (\mathbf{B} \times \mathbf{C}) = (\mathbf{A} \times \mathbf{B}) \times \mathbf{C}$. Discuss the cases where $\mathbf{A} \cdot \mathbf{B} = 0$ or $\mathbf{B} \cdot \mathbf{C} = 0$.

2.95. Let points P, Q, and R have position vectors $\mathbf{r}_1 = 3\mathbf{i} - 2\mathbf{j} - \mathbf{k}$, $\mathbf{r}_2 = \mathbf{i} + 3\mathbf{j} + 4\mathbf{k}$ and $\mathbf{r}_3 = 2\mathbf{i} + \mathbf{j} - 2\mathbf{k}$ relative to an origin O. Find the distance from P to the plane OQR.

2.96. Find the shortest distance: (a) from $(6, -4, 4)$ to the line joining $(2, 1, 2)$ and $(3, -1, 4)$, (b) from $(1, -7, 5)$ to the line joining $(13, -12, 5)$ and $(23, 12, 5)$.

2.97. Consider points $P(2, 1, 3)$, $Q(1, 2, 1)$, $R(-1, -1, -2)$, $S(1, -4, 0)$. Find the shortest distance between lines PQ and RS.

2.98. Prove that the perpendiculars from the vertices of a triangle to the opposite sides (extended if necessary) meet at a point (called the *orthocenter* of the triangle).

2.99. Prove that the perpendicular bisectors of the sides of a triangle meet at a point (called the *circumcenter* of the triangle).

2.100. Prove that $(\mathbf{A} \times \mathbf{B}) \cdot (\mathbf{C} \times \mathbf{D}) + (\mathbf{B} \times \mathbf{C}) \cdot (\mathbf{A} \times \mathbf{D}) + (\mathbf{C} \times \mathbf{A}) \cdot (\mathbf{B} \times \mathbf{D}) = 0$.

2.101. Let *PQR* be a spherical triangle whose sides *p*, *q*, *r* are arcs of great circles. Prove the *law of cosines for spherical triangles*,

$$\cos p = \cos q \cos r + \sin q \sin r$$

with analogous formulas for $\cos q$ and $\cos r$ obtained by cyclic permutation of the letters. Hint: Interpret both sides of the identity

$$(\mathbf{A} \times \mathbf{B}) \cdot (\mathbf{A} \times \mathbf{C}) = (\mathbf{B} \cdot \mathbf{C})(\mathbf{A} \cdot \mathbf{A}) - (\mathbf{A} \cdot \mathbf{C})(\mathbf{B} \cdot \mathbf{A})$$

2.102. Find a set of vectors reciprocal to the set vectors:

(a) $2\mathbf{i} + 3\mathbf{j} - \mathbf{k}$, $\mathbf{i} - \mathbf{j} - 2\mathbf{k}$, $-\mathbf{i} + 2\mathbf{j} + 2\mathbf{k}$,   (b) $\mathbf{i} + 2\mathbf{j} + 3\mathbf{k}$, $5\mathbf{i} - \mathbf{j} - \mathbf{k}$, $\mathbf{i} + \mathbf{j} - \mathbf{k}$.

2.103. Suppose $\mathbf{a}' = \dfrac{\mathbf{b} \times \mathbf{c}}{\mathbf{a} \cdot \mathbf{b} \times \mathbf{c}}$, $\mathbf{b}' = \dfrac{\mathbf{c} \times \mathbf{a}}{\mathbf{a} \cdot \mathbf{b} \times \mathbf{c}}$, $\mathbf{c}' = \dfrac{\mathbf{a} \times \mathbf{b}}{\mathbf{a} \cdot \mathbf{b} \times \mathbf{c}}$. Prove that

$$\mathbf{a} = \frac{b' \times c'}{a' \cdot b' \times c'}, \quad \mathbf{b} = \frac{c' \times a'}{a' \cdot b' \times c'}, \quad \mathbf{c} = \frac{a' \times b'}{a' \cdot b' \times c'}$$

2.104. Suppose $\mathbf{a}$, $\mathbf{b}$, $\mathbf{c}$ and $\mathbf{a}'$, $\mathbf{b}'$, $\mathbf{c}'$ have the following properties:

$$\mathbf{a}' \cdot \mathbf{a} = \mathbf{b}' \cdot \mathbf{b} = \mathbf{c}' \cdot \mathbf{c} = 1$$
$$\mathbf{a}' \cdot \mathbf{b} = \mathbf{a}' \cdot \mathbf{c} = \mathbf{b}' \cdot \mathbf{a} = \mathbf{b}' \cdot \mathbf{c} = \mathbf{c}' \cdot \mathbf{a} = \mathbf{c}' \cdot \mathbf{b} = 0$$

Prove that the hypothesis of Problem 2.103 holds, that is,

$$\mathbf{a}' = \frac{\mathbf{b} \times \mathbf{c}}{\mathbf{a} \cdot \mathbf{b} \times \mathbf{c}}, \quad \mathbf{b}' = \frac{\mathbf{c} \times \mathbf{a}}{\mathbf{a} \cdot \mathbf{b} \times \mathbf{c}}, \quad \mathbf{c}' = \frac{\mathbf{a} \times \mathbf{b}}{\mathbf{a} \cdot \mathbf{b} \times \mathbf{c}}.$$

2.105. Prove that the only right-handed self-reciprocal sets of vectors are $\mathbf{i}$, $\mathbf{j}$, $\mathbf{k}$.

2.106. Prove that there is one and only one set of vectors reciprocal to a given set of non-coplanar vectors $\mathbf{a}$, $\mathbf{b}$, $\mathbf{c}$.

ANSWERS TO SUPPLEMENTARY PROBLEMS

2.55. (a) 0, (b) −6, (c) 1   **2.57.** (a) 90°, arc cos 8/21 = 67°36′

2.56. (a) −10, (b) $\sqrt{14}$, (c) 6, (d) $\sqrt{150}$, (e) −14   **2.58.** (a) $a = 2, -1$, (b) $a = 2$

2.59. arc cos 2/3, arc cos 2/3, arc cos 1/3 or 48°12′, 48°12′, 70°32′

2.60. (a) 2/7, 3/7, −6/7 or −2/7, −3/7, 6/7, (b) 3/13, 4/13, 12/13 or −3/13, −4/13, −12/13

2.61. (a) arc cos $7/\sqrt{75}$, arc cos $\sqrt{26}/\sqrt{75}$, 90° or 36°4′, 53°56′, 90°   (b) 68.6°, 83.9°, 27.5°

2.62. $5\sqrt{3}/2$, arc cos 23/75, 180° − arc cos 23/75; or 4.33, 72°8′, 107°52′

2.63. (a) 8/3, (b) −1   **2.66.** arc cos 1/3 or 70°32′

2.64. 1   **2.67.** $\pm(3\mathbf{i} + 4\mathbf{j})/5$

2.65. (a) $\pm(\mathbf{i} - 2\mathbf{j} - 2\mathbf{k})/3$, (b) $\pm(2\mathbf{i} - \mathbf{j} - 2\mathbf{k})/3$ **2.69.** (a) 15, (b) 13

2.74. (a) $\mathbf{r} \cdot \mathbf{n} = p$ where $\mathbf{n} = \mathbf{A}/|\mathbf{A}| = \mathbf{A}/A$, (b) $A_1 x + A_2 y + A_3 z = Ap$

2.76. (a) Sphere with center at (x_1, y_1, z_1) and radius = 3.

      (b) Plane perpendicular to $\mathbf{a}$ and passing through its terminal point.

      (c) Sphere with center at $(x_1/2, y_1/2, z_1/2)$ and radius $\sqrt{x_1^2 + y_1^2 + z_1^2}/2$; or a sphere with $\mathbf{a}$ as diameter.

2.77. (a) $(\mathbf{r} - \mathbf{B}) \cdot (\mathbf{A} - \mathbf{B}) = 0$ or $2x + 3y + 6z = -28$; (b) 5

2.78. (a) $-8\mathbf{i} - 6\mathbf{k}$, (b) $2\mathbf{i} - \mathbf{j}$, (c) $8\mathbf{i} - 4\mathbf{j} + 4\mathbf{k}$, (d) $\mathbf{i} - 10\mathbf{j} - 3\mathbf{k}$, (e) $2\mathbf{i} - 11\mathbf{j} - 7\mathbf{k}$

2.79. (a) $\sqrt{195}$, (b) $-25\mathbf{i} + 34\mathbf{j} - 55\mathbf{k}$, (c) $2\sqrt{195}$

2.80. (a) $5\sqrt{26}$, (b) $3\sqrt{10}$, (c) -20, (d) -20, (e) $-40\mathbf{i} - 20\mathbf{j} + 20\mathbf{k}$, (f) $35\mathbf{i} - 35\mathbf{j} + 35\mathbf{k}$

2.82. (a) $5\sqrt{3}$, (b) 12

2.83. (a) $\sqrt{165}/2$, (b) 21

2.84. $\pm[5\sqrt{3}/3](\mathbf{i} + \mathbf{j} + \mathbf{k})$

2.86. (a) $2\mathbf{i} - 7\mathbf{j} - 2\mathbf{k}$, (b) $-3(6\mathbf{i} + 5\mathbf{j} + 7\mathbf{k})$

2.87. $-5\mathbf{i} - 8\mathbf{j} - 14\mathbf{k}$

2.88. (a) $2\mathbf{A} \cdot \mathbf{B} \times \mathbf{C}$, (b) $\mathbf{A} \cdot \mathbf{B} \times \mathbf{C}$

2.90. (a) 7, (b) 12

2.92. (a) $a = -4$, (b) $a = -13$

2.95. 3

2.96. (a) 3, (b) 13

2.97. $3\sqrt{2}$

2.102. (a) $\dfrac{2}{3}\mathbf{i} + \dfrac{1}{3}\mathbf{k}$, $-\dfrac{8}{3}\mathbf{i} + \mathbf{j} - \dfrac{7}{3}\mathbf{k}$, $-\dfrac{7}{3}\mathbf{i} + \mathbf{j} - \dfrac{5}{3}\mathbf{k}$

(b) $(2\mathbf{i} + 4\mathbf{j} + 6\mathbf{k})/28$, $(5\mathbf{i} - 4\mathbf{j} + \mathbf{k})/28$, $(\mathbf{i} + 9\mathbf{j} - 11\mathbf{k})/28$

CHAPTER 3

Vector Differentiation

3.1 Introduction

The reader is familiar with the differentiation of real valued functions $f(x)$ of one variable. Specifically, we have:

$$f'(x) = \frac{df}{dx} = \lim_{h \to 0} \frac{f(x+h) - f(x)}{h}$$

Here we extend this definition to vector-valued functions of a single variable.

3.2 Ordinary Derivatives of Vector-Valued Functions

Suppose $\mathbf{R}(u)$ is a vector depending on a single scalar variable u. Then

$$\frac{\Delta \mathbf{R}}{\Delta u} = \frac{\mathbf{R}(u + \Delta u) - \mathbf{R}(u)}{\Delta u}$$

where Δu denotes an increment in u as shown in Fig. 3-1.

The ordinary derivative of the vector $\mathbf{R}(u)$ with respect to the scalar u is given as follows when the limit exists:

$$\frac{d\mathbf{R}}{du} = \lim_{\Delta u \to 0} \frac{\Delta \mathbf{R}}{\Delta u} = \lim_{\Delta u \to 0} \frac{\mathbf{R}(u + \Delta u) - \mathbf{R}(u)}{\Delta u}$$

Since $d\mathbf{R}/du$ is itself a vector depending on u, we can consider its derivative with respect to u. If this derivative exists, we denote it by $d^2\mathbf{R}/du^2$. Similarly, higher-order derivatives are described.

Fig. 3-1

Fig. 3-2

Space Curves

Consider now the position vector $\mathbf{r}(u)$ joining the origin O of a coordinate system and any point (x, y, z). Then

$$\mathbf{r}(u) = x(u)\mathbf{i} + y(u)\mathbf{j} + z(u)\mathbf{k}$$

and the specification of the vector function $\mathbf{r}(u)$ defines x, y, and z as functions of u.

As u changes, the terminal point of $\mathbf{r}$ describes a *space curve* having parametric equations

$$x = x(u), \quad y = y(u), \quad z = z(u)$$

Then the following is a vector in the direction of $\Delta\mathbf{r}$ if $\Delta u > 0$ and in the direction of $-\Delta\mathbf{r}$ if $\Delta u < 0$ [as pictured in Fig. 3-2]:

$$\frac{\Delta\mathbf{r}}{\Delta u} = \frac{\mathbf{r}(u + \Delta u) - \mathbf{r}(u)}{\Delta u}$$

Suppose

$$\lim_{\Delta u \to 0} \frac{\Delta\mathbf{r}}{\Delta u} = \frac{d\mathbf{r}}{du}$$

exists. Then the limit will be a vector in the direction of the tangent to the space curve at (x, y, z) and it is given by

$$\frac{d\mathbf{r}}{du} = \frac{dx}{du}\mathbf{i} + \frac{dy}{du}\mathbf{j} + \frac{dz}{du}\mathbf{k}$$

Motion: Velocity and Acceleration

Suppose a particle P moves along a space curve C whose parametric equations are $x = x(t)$, $y = y(t)$, $z = z(t)$, where t represents time. Then the position vector of the particle P along the curve is

$$\mathbf{r}(t) = x(t)\mathbf{i} + y(t)\mathbf{j} + z(t)\mathbf{k}$$

In such a case, the velocity $\mathbf{v}$ and acceleration $\mathbf{a}$ of the particle P is given by:

$$\mathbf{v} = \mathbf{v}(t) = \frac{d\mathbf{r}}{dt} = \frac{dx}{dt}\mathbf{i} + \frac{dy}{dt}\mathbf{j} + \frac{dz}{dt}\mathbf{k}$$

$$\mathbf{a} = \mathbf{a}(t) = \frac{d^2\mathbf{r}}{dt^2} = \frac{d\mathbf{v}}{dt} = \frac{d^2x}{dt^2}\mathbf{i} + \frac{d^2y}{dt^2}\mathbf{j} + \frac{d^2z}{dt^2}\mathbf{k}$$

EXAMPLE 3.1 Suppose a particle P moves along a curve whose parametric equations, where t is time, follows:

$$x = 40t^2 + 8t, \quad y = 2\cos 3t, \quad z = 2\sin 3t$$

(a) Determine its velocity and acceleration at any time.
(b) Find the magnitudes of the velocity and acceleration at $t = 0$.

(a) The position vector of the particle P is

$$\mathbf{r} = x\mathbf{i} + y\mathbf{j} + z\mathbf{k} = (40t^2 + 8t)\mathbf{i} + (2\cos 3t)\mathbf{j} + (2\sin 3t)\mathbf{k}$$

Then the velocity **v** and acceleration **a** of P follow:

$$\mathbf{v} = \frac{d\mathbf{r}}{dt} = (80t + 8)\mathbf{i} + (-6\sin 3t)\mathbf{j} + (6\cos 3t)\mathbf{k}$$

$$\mathbf{a} = \frac{d\mathbf{v}}{dt} = 80\mathbf{i} + (-18\cos 3t)\mathbf{j} + (-18\sin 3t)\mathbf{k}.$$

(b)　At $t = 0$, $\mathbf{v} = 8\mathbf{i} + 6\mathbf{k}$, and $\mathbf{a} = 80\mathbf{i} - 18\mathbf{j}$. Magnitudes of velocity **v** and acceleration **a** follow:

$$|\mathbf{v}| = \sqrt{(8)^2 + (6)^2} = 10 \quad \text{and} \quad |\mathbf{a}| = \sqrt{(80)^2 + (-18)^2} = 82$$

3.3　Continuity and Differentiability

A scalar function $\phi(u)$ is called *continuous* at u if

$$\lim_{\Delta u \to 0} \phi(u + \Delta u) = \phi(u)$$

Equivalently, $\phi(u)$ is continuous at u if, for each positive number ϵ, we can find a positive number δ such that

$$|\phi(u + \Delta u) - \phi(u)| < \epsilon \quad \text{whenever} \quad |\Delta u| < \delta$$

A vector function $\mathbf{R}(u) = R_1(u)\mathbf{i} + R_2(u)\mathbf{j} + R_3(u)\mathbf{k}$ is called *continuous* at u if the three functions $R_1(u)$, $R_2(u)$, $R_3(u)$ are continuous at u or if $\lim_{\Delta u \to 0} \mathbf{R}(u + \Delta u) = \mathbf{R}(u)$. Equivalently, $\mathbf{R}(u)$ is continuous at u if, for each positive number ϵ, we can find a positive number δ such that

$$|\mathbf{R}(u + \Delta u) - \mathbf{R}(u)| < \epsilon \quad \text{whenever} \quad |\Delta u| < \delta$$

A scalar or vector function of u is called *differentiable of order n* if its nth derivative exists. A function that is differentiable is necessarily continuous but the converse is not true. Unless otherwise stated, we assume that all functions considered are differentiable to any order needed in a particular discussion.

The following proposition applies.

PROPOSITION 2.1　　Suppose **A**, **B**, and **C** are differentiable vector functions of a scalar u, and ϕ is a differentiable scalar function of u. Then the following laws hold:

(i)　$\dfrac{d}{du}(\mathbf{A} + \mathbf{B}) = \dfrac{d\mathbf{A}}{du} + \dfrac{d\mathbf{B}}{du}$

(ii)　$\dfrac{d}{du}(\mathbf{A} \cdot \mathbf{B}) = \mathbf{A} \cdot \dfrac{d\mathbf{B}}{du} + \dfrac{d\mathbf{A}}{du} \cdot \mathbf{B}$

(iii)　$\dfrac{d}{du}(\mathbf{A} \times \mathbf{B}) = \mathbf{A} \times \dfrac{d\mathbf{B}}{du} + \dfrac{d\mathbf{A}}{du} \times \mathbf{B}$

(iv)　$\dfrac{d}{du}(\phi\mathbf{A}) = \phi\dfrac{d\mathbf{A}}{du} + \dfrac{d\phi}{du}\mathbf{A}$

(v)　$\dfrac{d}{du}(\mathbf{A} \cdot \mathbf{B} \times \mathbf{C}) = \mathbf{A} \cdot \mathbf{B} \times \dfrac{d\mathbf{C}}{du} + \mathbf{A} \cdot \dfrac{d\mathbf{B}}{du} \times \mathbf{C} + \dfrac{d\mathbf{A}}{du} \cdot \mathbf{B} \times \mathbf{C}$

(vi)　$\dfrac{d}{du}\{\mathbf{A} \times (\mathbf{B} \times \mathbf{C})\} = \mathbf{A} \times \left(\mathbf{B} \times \dfrac{d\mathbf{C}}{du}\right) + \mathbf{A} \times \left(\dfrac{d\mathbf{B}}{du} \times \mathbf{C}\right) + \dfrac{d\mathbf{A}}{du} \times (\mathbf{B} \times \mathbf{C})$

The order of the products in Proposition 2.1 may be important.

EXAMPLE 3.2 Suppose $\mathbf{A} = 5u^2\mathbf{i} + u\mathbf{j} - u^3\mathbf{k}$ and $\mathbf{B} = \sin u\mathbf{i} - \cos u\mathbf{j}$. Find $\dfrac{d}{du}(\mathbf{A} \cdot \mathbf{B})$

$$\frac{d}{du}(\mathbf{A} \cdot \mathbf{B}) = \mathbf{A} \cdot \frac{d\mathbf{B}}{du} + \frac{d\mathbf{A}}{du} \cdot \mathbf{B}$$

$$= (5u^2\mathbf{i} + u\mathbf{j} - u^3\mathbf{k}) \cdot (\cos u\mathbf{i} + \sin u\mathbf{j}) + (10u\mathbf{i} + \mathbf{j} - 3u^2\mathbf{k}) \cdot (\sin u\mathbf{i} - \cos u\mathbf{j})$$

$$= [5u^2 \cos u + u \sin u] + [10u \sin u - \cos u]$$

$$= (5u^2 - 1)\cos u + 11u \sin u$$

Another Method

$$\mathbf{A} \cdot \mathbf{B} = 5u^2 \sin u - u \cos u. \text{ Then}$$

$$\frac{d}{du}(\mathbf{A} \cdot \mathbf{B}) = \frac{d}{du}(5u^2 \sin u - ut \cos u) = 5u^2 \cos u + 10u \sin u + u \sin u - \cos u$$

$$= (5u^2 - 1)\cos u + 11u \sin u$$

3.4 Partial Derivative of Vectors

Suppose $\mathbf{A}$ is a vector depending on more than one variable, say x, y, z, for example. Then we write $\mathbf{A} = \mathbf{A}(x, y, z)$. The partial derivative of $\mathbf{A}$ with respect to x is denoted and defined as follows when the limit exists:

$$\frac{\partial \mathbf{A}}{\partial x} = \lim_{\Delta x \to 0} \frac{\mathbf{A}(x + \Delta x, y, z) - \mathbf{A}(x, y, z)}{\Delta x}$$

Similarly, the following are the partial derivatives of $\mathbf{A}$ with respect to y and z, respectively, when the limits exist:

$$\frac{\partial \mathbf{A}}{\partial y} = \lim_{\Delta y \to 0} \frac{\mathbf{A}(x, y + \Delta y, z) - \mathbf{A}(x, y, z)}{\Delta y}$$

$$\frac{\partial \mathbf{A}}{\partial z} = \lim_{\Delta z \to 0} \frac{\mathbf{A}(x, y, z + \Delta z) - \mathbf{A}(x, y, z)}{\Delta z}$$

The remarks on continuity and differentiability of functions of one variable can be extended to functions of two or more variables. For example, $\phi(x, y)$ is called continuous at (x, y) if

$$\lim_{\substack{\Delta x \to 0 \\ \Delta y \to 0}} \phi(x + \Delta x, y + \Delta y) = \phi(x, y)$$

or if for each positive number ϵ we can find a positive number δ such that

$$|\phi(x + \Delta x, y + \Delta y) - \phi(x, y)| < \epsilon \quad \text{whenever} \quad |\Delta x| < \delta \quad \text{and} \quad |\Delta y| < \delta$$

Similar definitions hold for vector functions of more than two variables.

For functions of two or more variables, we use the term *differentiable* to mean the function has continuous first partial derivatives. (The term is used by others in a slightly weaker sense.)

Higher derivatives can be defined as in calculus. Thus, for example:

$$\frac{\partial^2 \mathbf{A}}{\partial x^2} = \frac{\partial}{\partial x}\left(\frac{\partial \mathbf{A}}{\partial x}\right), \quad \frac{\partial^2 \mathbf{A}}{\partial y^2} = \frac{\partial}{\partial y}\left(\frac{\partial \mathbf{A}}{\partial y}\right), \quad \frac{\partial^2 \mathbf{A}}{\partial z^2} = \frac{\partial}{\partial z}\left(\frac{\partial \mathbf{A}}{\partial z}\right)$$

$$\frac{\partial^2 \mathbf{A}}{\partial x \partial y} = \frac{\partial}{\partial x}\left(\frac{\partial \mathbf{A}}{\partial y}\right), \quad \frac{\partial^2 \mathbf{A}}{\partial y \partial x} = \frac{\partial}{\partial y}\left(\frac{\partial \mathbf{A}}{\partial x}\right), \quad \frac{\partial^3 \mathbf{A}}{\partial x \partial z^2} = \frac{\partial}{\partial x}\left(\frac{\partial^2 \mathbf{A}}{\partial z^2}\right)$$

In the case that **A** has continuous partial derivatives of the second order at least, we have

$$\frac{\partial^2 \mathbf{A}}{\partial x \partial y} = \frac{\partial^2 \mathbf{A}}{\partial y \partial x}$$

That is, the order of differentiation does not matter.

EXAMPLE 3.3   Suppose $\phi(x, y, z) = xy^2z$ and $\mathbf{A} = x\mathbf{i} + \mathbf{j} + xy\mathbf{k}$. Find $\dfrac{\partial^3}{\partial x^2 \partial z}(\phi A)$ at the point $P(1, 2, 2)$.

$$\phi A = x^2 y^2 z \mathbf{i} + x y^2 z \mathbf{j} + x^2 y^3 z \mathbf{k}$$

$$\frac{\partial}{\partial z}(\phi A) = x^2 y^2 \mathbf{i} + x y^2 \mathbf{j} + x^2 y^3 \mathbf{k}$$

$$\frac{\partial^2}{\partial x \partial z}(\phi A) = 2x y^2 \mathbf{i} + y^2 \mathbf{j} + 2x y^3 \mathbf{k}$$

$$\frac{\partial^3}{\partial x^2 \partial z}(\phi A) = 2y^2 \mathbf{i} + 2y^3 \mathbf{k}$$

When $x = 1$, $y = 2$, and $z = 2$, $\dfrac{\partial^3}{\partial x^2 \partial z}(\phi A) = 8\mathbf{i} + 16\mathbf{k}$.

Rules for partial differentiation of vectors are similar to those in elementary calculus for scalar functions. In particular, the following proposition applies.

PROPOSITION 3.2   Suppose **A** and **B** are vector functions of x, y, z. Then the following laws hold:

(i) $\dfrac{\partial}{\partial x}(\mathbf{A} \cdot \mathbf{B}) = \mathbf{A} \cdot \dfrac{\partial \mathbf{B}}{\partial x} + \dfrac{\partial \mathbf{A}}{\partial x} \cdot \mathbf{B}$

(ii) $\dfrac{\partial}{\partial x}(\mathbf{A} \times \mathbf{B}) = \mathbf{A} \times \dfrac{\partial \mathbf{B}}{\partial x} + \dfrac{\partial \mathbf{A}}{\partial x} \times \mathbf{B}$

(iii) $\dfrac{\partial^2}{\partial y \partial x}(\mathbf{A} \cdot \mathbf{B}) = \dfrac{\partial}{\partial y}\left\{\dfrac{\partial}{\partial x}(\mathbf{A} \cdot \mathbf{B})\right\} = \dfrac{\partial}{\partial y}\left\{\mathbf{A} \cdot \dfrac{\partial \mathbf{B}}{\partial x} + \dfrac{\partial \mathbf{A}}{\partial x} \cdot \mathbf{B}\right\}$

$$= \mathbf{A} \cdot \frac{\partial^2 \mathbf{B}}{\partial y \partial x} + \frac{\partial \mathbf{A}}{\partial y} \cdot \frac{\partial \mathbf{B}}{\partial x} + \frac{\partial \mathbf{A}}{\partial x} \cdot \frac{\partial \mathbf{B}}{\partial y} + \frac{\partial^2 \mathbf{A}}{\partial y \partial x} \cdot \mathbf{B}, \text{ and so on.}$$

The rules for the differentials of vectors are essentially the same as those of elementary calculus as seen in the following proposition.

PROPOSITION 3.3   Suppose **A** and **B** are functions of x, y, z. Then the following laws hold.

(i)   If $\mathbf{A} = A_1\mathbf{i} + A_2\mathbf{j} + A_3\mathbf{k}$, then $d\mathbf{A} = dA_1\mathbf{i} + dA_2\mathbf{j} + dA_3\mathbf{k}$

(ii) $d(\mathbf{A} \cdot \mathbf{B}) = \mathbf{A} \cdot d\mathbf{B} + d\mathbf{A} \cdot \mathbf{B}$

(iii) $d(\mathbf{A} \times \mathbf{B}) = \mathbf{A} \times d\mathbf{B} + d\mathbf{A} \times \mathbf{B}$

(iv)   If $\mathbf{A} = \mathbf{A}(x, y, z)$, then $d\mathbf{A} = \dfrac{\partial \mathbf{A}}{\partial x}dx + \dfrac{\partial \mathbf{A}}{\partial y}dy + \dfrac{\partial \mathbf{A}}{\partial z}dz$, and so on.

3.5   Differential Geometry

Differential geometry involves the study of curves and surfaces. Suppose C is a space curve defined by the function $\mathbf{r}(u)$. Then, we have seen that $d\mathbf{r}/du$ is a vector in the direction of the tangent to C. Suppose the scalar u is taken as the arc length s measured from some fixed point on C. Then $d\mathbf{r}/ds$ is a unit tangent vector to C and it is denoted by **T** (see Fig. 3-3). The rate at which **T** changes with respect to s is a measure of the curvature of C and is given by $d\mathbf{T}/ds$. The direction of $d\mathbf{T}/ds$ at any given point on C is normal to the curve at that point (see Problem 3.9). If **N** is a unit vector in this normal direction, it is called the *principal normal* to the curve. Then $d\mathbf{T}/ds = \kappa\mathbf{N}$, where κ is called the *curvature* of C at the specified point. The quantity $\rho = 1/\kappa$ is called the *radius* of *curvature*.

Fig. 3-3

A unit vector $\mathbf{B}$ perpendicular to the plane of $\mathbf{T}$ and $\mathbf{N}$ and such that $\mathbf{B} = \mathbf{T} \times \mathbf{N}$, is called the *binormal* to the curve. It follows that directions $\mathbf{T}$, $\mathbf{N}$, $\mathbf{B}$ form a localized right-handed rectangular coordinate system at any specified point of C. This coordinate system is called the *trihedral* or *triad* at the point. As s changes, the coordinate system moves and is known as the *moving trihedral*.

Frenet–Serret Formulas

A set of relations involving derivatives of the fundamental vectors $\mathbf{T}$, $\mathbf{N}$, and $\mathbf{B}$ is known collectively as the *Frenet–Serret formulas* given by

$$\frac{d\mathbf{T}}{ds} = \kappa\mathbf{N}, \qquad \frac{d\mathbf{N}}{ds} = \tau\mathbf{B} - \kappa\mathbf{T}, \qquad \frac{d\mathbf{B}}{ds} = -\tau\mathbf{N}$$

where τ is a scalar called the *torsion*. The quantity $\sigma = 1/\tau$ is called the *radius of torsion*.

The *osculating plane* to a curve at a point P is the plane containing the tangent and principal normal at P. The *normal plane* is the plane through P perpendicular to the tangent. The *rectifying plane* is the plane through P, which is perpendicular to the principal normal.

Mechanics

Mechanics often includes the study of the motion of particles along curves. (This study being known as *kinematics*.) In this area, some of the results of differential geometry can be of value.

A study of forces on moving objects is considered in *dynamics*. Fundamental to this study is Newton's famous law which states that if $\mathbf{F}$ is the net force acting on an object of mass m moving with velocity $\mathbf{v}$, then

$$\mathbf{F} = \frac{d}{dt}(m\mathbf{v})$$

where $m\mathbf{v}$ is the momentum of the object. If m is constant, this becomes $\mathbf{F} = m(d\mathbf{v}/dt) = m\mathbf{a}$, where $\mathbf{a}$ is the acceleration of the object.

SOLVED PROBLEMS

3.1. Suppose $\mathbf{R}(u) = x(u)\mathbf{i} + y(u)\mathbf{j} + z(u)\mathbf{k}$, where x, y, and z are differentiable functions of a scalar u. Prove that

$$\frac{d\mathbf{R}}{du} = \frac{dx}{du}\mathbf{i} + \frac{dy}{du}\mathbf{j} + \frac{dz}{du}\mathbf{k}.$$

Solution

$$\frac{d\mathbf{R}}{du} = \lim_{\Delta u \to 0} \frac{\mathbf{R}(u + \Delta u) - \mathbf{R}(u)}{\Delta u}$$

$$= \lim_{\Delta u \to 0} \frac{[x(u + \Delta u)\mathbf{i} + y(u + \Delta u)\mathbf{j} + z(u + \Delta u)\mathbf{k}] - [x(u)\mathbf{i} + y(u)\mathbf{j} + z(u)\mathbf{k}]}{\Delta u}$$

$$= \lim_{\Delta u \to 0} \frac{x(u + \Delta u) - x(u)}{\Delta u}\mathbf{i} + \frac{y(u + \Delta u) - y(u)}{\Delta u}\mathbf{j} + \frac{z(u + \Delta u) - z(u)}{\Delta u}\mathbf{k}$$

$$= \frac{dx}{du}\mathbf{i} + \frac{dy}{du}\mathbf{j} + \frac{dz}{du}\mathbf{k}$$

3.2. Given $\mathbf{R} = (3\cos t)\mathbf{i} + (3\sin t)\mathbf{j} + (4t)\mathbf{k}$. Find: (a) $\dfrac{d\mathbf{R}}{dt}$, (b) $\dfrac{d^2\mathbf{R}}{dt}$, (c) $\left|\dfrac{d\mathbf{R}}{dt}\right|$, (d) $\left|\dfrac{d^2\mathbf{R}}{dt}\right|$.

Solution

(a) $\dfrac{d\mathbf{R}}{dt} = \dfrac{d}{dt}(3\cos t)\mathbf{i} + \dfrac{d}{dt}(3\sin t)\mathbf{j} + \dfrac{d}{dt}(4t)\mathbf{k} = (-3\sin t)\mathbf{i} + (3\cos t)\mathbf{j} + 4\mathbf{k}$.

(b) $\dfrac{d^2\mathbf{R}}{dt} = \dfrac{d}{dt}\left(\dfrac{d\mathbf{R}}{dt}\right) = \dfrac{d}{dt}(-3\sin t)\mathbf{i} + \dfrac{d}{dt}(3\cos t)\mathbf{j} + \dfrac{d}{dt}(4)\mathbf{k} = (-3\cos t)\mathbf{i} + (-3\sin t)\mathbf{j}$.

(c) $\left|\dfrac{d\mathbf{R}}{dt}\right| = [(-3\sin t)^2 + (3\cos t)^2 + (4)^2]^{1/2} = 5$.

(d) $\left|\dfrac{d^2\mathbf{R}}{dt}\right| = [(-3\cos t)^2 + (-3\sin t)^2 + (0)^2]^{1/2} = 3$.

3.3. A particle moves along a curve whose parametric equations are $x = e^{-t}$, $y = 2\cos 3t$, $z = 2\sin 3t$, where t is the time.

(a) Determine its velocity and acceleration at any time.

(b) Find the magnitudes of the velocity and acceleration at $t = 0$.

Solution

(a) The position vector $\mathbf{r}$ of the particle is $\mathbf{r} = x\mathbf{i} + y\mathbf{j} + z\mathbf{k} = e^{-t}\mathbf{i} + 2\cos 3t\mathbf{j} + 2\sin 3t\mathbf{k}$. Then the velocity is $\mathbf{v} = d\mathbf{r}/dt = -e^{-t}\mathbf{i} - 6\sin 3t\mathbf{j} + 6\cos 3t\mathbf{k}$ and the acceleration is $\mathbf{a} = d^2\mathbf{r}/dt^2 = e^{-t}\mathbf{i} - 18\cos 3t\mathbf{j} - 18\sin 3t\mathbf{k}$.

(b) At $t = 0$, $d\mathbf{r}/dt = -\mathbf{i} + 6\mathbf{k}$ and $d^2\mathbf{r}/dt^2 = \mathbf{i} - 18\mathbf{j}$. Then

$$\text{magnitude of velocity at } t = 0 \quad \text{is} \quad \sqrt{(-1)^2 + (6)^2} = \sqrt{37}$$

$$\text{magnitude of acceleration at } t = 0 \quad \text{is} \quad \sqrt{(1)^2 + (-18)^2} = \sqrt{325}.$$

3.4. A particle moves along the curve $x = 2t^2$, $y = t^2 - 4t$, $z = -t - 5$ where t is the time. Find the components of its velocity and acceleration at time $t = 1$ in the direction $\mathbf{i} - 2\mathbf{j} + 2\mathbf{k}$.

Solution

$$\text{Velocity} = \frac{d\mathbf{r}}{dt} = \frac{d}{dt}[(2t^2)\mathbf{i} + (t^2 - 4t)\mathbf{j} + (-t - 5)\mathbf{k}]$$

$$= (4t)\mathbf{i} + (2t - 4)\mathbf{j} - \mathbf{k} = 4\mathbf{i} - 2\mathbf{j} - \mathbf{k} \quad \text{at } t = 1.$$

$$\text{Unit vector in direction } \mathbf{i} - 2\mathbf{j} + 2\mathbf{k} \text{ is } \frac{\mathbf{i} - 2\mathbf{j} + 2\mathbf{k}}{\sqrt{(1)^2 + (-2)^2 + (2)^2}} = \tfrac{1}{3}\mathbf{i} - \tfrac{2}{3}\mathbf{j} + \tfrac{2}{3}\mathbf{k}.$$

Then the component in the given direction is $(4\mathbf{i} - 2\mathbf{j} - \mathbf{k}) \cdot \left(\frac{1}{3}\mathbf{i} - \frac{2}{3}\mathbf{j} + \frac{2}{3}\mathbf{k}\right) = 2$

$$\text{Acceleration} = \frac{d^2\mathbf{r}}{dt^2} = \frac{d}{dt}\left(\frac{d\mathbf{r}}{dt}\right) = \frac{d}{dt}[(4t)\mathbf{i} + (2t - 4)\mathbf{j} - \mathbf{k}] = 4\mathbf{i} + 2\mathbf{j}.$$

Then the component of the acceleration in the given direction is $(4\mathbf{i} + 2\mathbf{j}) \cdot \left(\frac{1}{3}\mathbf{i} - \frac{2}{3}\mathbf{j} + \frac{2}{3}\mathbf{k}\right) = 0$.

3.5. A curve C is defined by parametric equations $x = x(s)$, $y = y(s)$, $z = z(s)$, where s is the arc length of C measured from a fixed point on C. If $\mathbf{r}$ is the position vector of any point on C, show that $d\mathbf{r}/ds$ is a unit vector tangent to C.

Solution

The vector

$$\frac{d\mathbf{r}}{ds} = \frac{d}{ds}(x\mathbf{i} + y\mathbf{j} + z\mathbf{k}) = \frac{dx}{ds}\mathbf{i} + \frac{dy}{ds}\mathbf{j} + \frac{dz}{ds}\mathbf{k}$$

is tangent to the curve $x = x(s)$, $y = y(s)$, $z = z(s)$. To show that it has unit magnitude, we note that

$$\left|\frac{d\mathbf{r}}{ds}\right| = \sqrt{\left(\frac{dx}{ds}\right)^2 + \left(\frac{dy}{ds}\right)^2 + \left(\frac{dz}{ds}\right)^2} = \sqrt{\frac{(dx)^2 + (dy)^2 + (dz)^2}{(ds)^2}} = 1$$

since $(ds)^2 = (dx)^2 + (dy)^2 + (dz)^2$ from the calculus.

3.6. (a) Find the unit tangent vector to any point on the curve $x = t^2 - t$, $y = 4t - 3$, $z = 2t^2 - 8t$.

(b) Determine the unit tangent at the point where $t = 2$.

Solution

(a) A tangent to the curve at any point is

$$\frac{d\mathbf{r}}{dt} = \frac{d}{dt}[(t^2 - t)\mathbf{i} + (4t - 3)\mathbf{j} + (2t^2 - 8t)\mathbf{k}] = (2t - 1)\mathbf{i} + 4\mathbf{j} + (4t - 8)\mathbf{k}.$$

The magnitude of the vector is $|d\mathbf{r}/dt| = [(2t - 1)^2 + (4)^2 + (4t - 8)^2]^{1/2}$. Then the required unit tangent vector is

$$\mathbf{T} = [(2t - 1)\mathbf{i} + 4\mathbf{j} + (4t - 8)\mathbf{k}]/[(2t - 1)^2 + (4)^2 + (4t - 8)^2]^{1/2}.$$

Note that since $|d\mathbf{r}/dt| = ds/dt$, we have

$$\mathbf{T} = \frac{d\mathbf{r}/dt}{ds/dt} = \frac{d\mathbf{r}}{ds}.$$

(b) At $t = 2$, the unit tangent vector is $\mathbf{T} = \dfrac{3\mathbf{i} + 4\mathbf{j}}{\sqrt{(3)^2 + (4)^2 + 0^2}} = \frac{3}{5}\mathbf{i} + \frac{4}{5}\mathbf{j}$.

3.7. Suppose $\mathbf{A}$ and $\mathbf{B}$ are differentiable functions of a scalar u. Prove:

(a) $\dfrac{d}{du}(\mathbf{A} \cdot \mathbf{B}) = \mathbf{A} \cdot \dfrac{d\mathbf{B}}{du} + \dfrac{d\mathbf{A}}{du} \cdot \mathbf{B}$,   (b) $\dfrac{d}{du}(\mathbf{A} \times \mathbf{B}) = \mathbf{A} \times \dfrac{d\mathbf{B}}{du} + \dfrac{d\mathbf{A}}{du} \times \mathbf{B}$

Solution

(a) $\dfrac{d}{du}(\mathbf{A}\cdot\mathbf{B}) = \lim\limits_{\Delta u\to 0} \dfrac{(\mathbf{A}+\Delta\mathbf{A})\cdot(\mathbf{B}+\Delta\mathbf{B}) - \mathbf{A}\cdot\mathbf{B}}{\Delta u}$

$= \lim\limits_{\Delta u\to 0}\dfrac{\mathbf{A}\cdot\Delta\mathbf{B}+\Delta\mathbf{A}\cdot\mathbf{B}+\Delta\mathbf{A}\cdot\Delta\mathbf{B}}{\Delta u}$

$= \lim\limits_{\Delta u\to 0}\mathbf{A}\cdot\dfrac{\Delta\mathbf{B}}{\Delta u}+\dfrac{\Delta\mathbf{A}}{\Delta u}\cdot\mathbf{B}+\dfrac{\Delta\mathbf{A}}{\Delta u}\cdot\Delta\mathbf{B} = \mathbf{A}\cdot\dfrac{d\mathbf{B}}{du}+\dfrac{d\mathbf{A}}{du}\cdot\mathbf{B}$

Another Method

Let $\mathbf{A} = A_1\mathbf{i}+A_2\mathbf{j}+A_3\mathbf{k}$, $\mathbf{B} = B_1\mathbf{i}+B_2\mathbf{j}+B_3\mathbf{k}$. Then

$\dfrac{d}{du}(\mathbf{A}\cdot\mathbf{B}) = \dfrac{d}{du}(A_1B_1+A_2B_2+A_3B_3)$

$= \left(A_1\dfrac{dB_1}{du}+A_2\dfrac{dB_2}{du}+A_3\dfrac{dB_3}{du}\right)+\left(\dfrac{dA_1}{du}B_1+\dfrac{dA_2}{du}B_2+\dfrac{dA_3}{du}B_3\right)$

$= \mathbf{A}\cdot\dfrac{d\mathbf{B}}{du}+\dfrac{d\mathbf{A}}{du}\cdot\mathbf{B}$

(b) $\dfrac{d}{du}(\mathbf{A}\times\mathbf{B}) = \lim\limits_{\Delta u\to 0}\dfrac{(\mathbf{A}+\Delta\mathbf{A})\times(\mathbf{B}+\Delta\mathbf{B})-\mathbf{A}\times\mathbf{B}}{\Delta u}$

$= \lim\limits_{\Delta u\to 0}\dfrac{\mathbf{A}\times\Delta\mathbf{B}+\Delta\mathbf{A}\times\mathbf{B}+\Delta\mathbf{A}\times\Delta\mathbf{B}}{\Delta u}$

$= \lim\limits_{\Delta u\to 0}\mathbf{A}\times\dfrac{\Delta\mathbf{B}}{\Delta u}+\dfrac{\Delta\mathbf{A}}{\Delta u}\times\mathbf{B}+\dfrac{\Delta\mathbf{A}}{\Delta u}\times\Delta\mathbf{B}$

$= \mathbf{A}\times\dfrac{d\mathbf{B}}{du}+\dfrac{d\mathbf{A}}{du}\times\mathbf{B}$

Another Method

$$\dfrac{d}{du}(\mathbf{A}\times\mathbf{B}) = \dfrac{d}{du}\begin{vmatrix} \mathbf{i} & \mathbf{j} & \mathbf{k} \\ A_1 & A_2 & A_3 \\ B_1 & B_2 & B_3 \end{vmatrix}$$

Using a theorem on differentiation of a determinant, this becomes

$$\begin{vmatrix} \mathbf{i} & \mathbf{j} & \mathbf{k} \\ A_1 & A_2 & A_3 \\ \dfrac{dB_1}{du} & \dfrac{dB_2}{du} & \dfrac{dB_3}{du} \end{vmatrix} + \begin{vmatrix} \mathbf{i} & \mathbf{j} & \mathbf{k} \\ \dfrac{dA_1}{du} & \dfrac{dA_2}{du} & \dfrac{dA_3}{du} \\ B_1 & B_2 & B_3 \end{vmatrix} = \mathbf{A}\times\dfrac{d\mathbf{B}}{du}+\dfrac{d\mathbf{A}}{du}\times\mathbf{B}$$

3.8. Suppose $\mathbf{A} = 5t^2\mathbf{i}+t\mathbf{j}-t^3\mathbf{k}$ and $\mathbf{B} = \sin t\,\mathbf{i}-\cos t\,\mathbf{j}$. Find: (a) $\dfrac{d}{dt}(\mathbf{A}\times\mathbf{B})$, (b) $\dfrac{d}{dt}(\mathbf{A}\cdot\mathbf{A})$.

Solution

(a) $\dfrac{d}{dt}(\mathbf{A}\cdot\mathbf{B}) = \mathbf{A}\cdot\dfrac{d\mathbf{B}}{dt}+\dfrac{d\mathbf{A}}{dt}\cdot\mathbf{B}$

$= (5t^2\mathbf{i}+t\mathbf{j}-t^3\mathbf{k})\cdot(\cos t\,\mathbf{i}+\sin t\,\mathbf{j})+(10t\mathbf{i}+\mathbf{j}-3t^2\mathbf{k})\cdot(\sin t\,\mathbf{i}-\cos t\,\mathbf{j})$

$= 5t^2\cos t+t\sin t+10t\sin t-\cos t = (5t^2-1)\cos t+11t\sin t$

Another Method

$\mathbf{A} \cdot \mathbf{B} = 5t^2 \sin t - t \cos t$. Then

$$\frac{d}{dt}(\mathbf{A} \cdot \mathbf{B}) = \frac{d}{dt}(5t^2 \sin t - t \cos t) = 5t^2 \cos t + 10t \sin t + t \sin t - \cos t$$

$$= (5t^2 - 1) \cos t + 11t \sin t$$

(b) $\dfrac{d}{dt}(\mathbf{A} \times \mathbf{B}) = \mathbf{A} \times \dfrac{d\mathbf{B}}{dt} + \dfrac{d\mathbf{A}}{dt} \times \mathbf{B} = \begin{vmatrix} \mathbf{i} & \mathbf{j} & \mathbf{k} \\ 5t^2 & t & -t^3 \\ \cos t & \sin t & 0 \end{vmatrix} + \begin{vmatrix} \mathbf{i} & \mathbf{j} & \mathbf{k} \\ 10t & 1 & -3t^2 \\ \sin t & -\cos t & 0 \end{vmatrix}$

$$= [t^3 \sin t\mathbf{i} - t^3 \cos t\mathbf{j} + (5t^2 \sin t - t \cos t)\mathbf{k}]$$
$$+ [-3t^2 \cos t\mathbf{i} - 3t^2 \sin t\mathbf{j} + (-10t \cos t - \sin t)\mathbf{k}]$$
$$= (t^3 \sin t - 3t^2 \cos t)\mathbf{i} - (t^3 \cos t + 3t^2 \sin t)\mathbf{j} + (5t^2 \sin t - \sin t - 11t \cos t)\mathbf{k}$$

Another Method

$$\mathbf{A} \times \mathbf{B} = \begin{vmatrix} \mathbf{i} & \mathbf{j} & \mathbf{k} \\ 5t^2 & t & -t^3 \\ \sin t & -\cos t & 0 \end{vmatrix} = -t^3 \cos t\mathbf{i} - t^3 \sin t\mathbf{j} + (-5t^2 \cos t - t \sin t)\mathbf{k}$$

Then $\dfrac{d}{dt}(\mathbf{A} \times \mathbf{B}) = (t^3 \sin t - 3t^2 \cos t)\mathbf{i} - (t^3 \cos t + 3t^2 \sin t)\mathbf{j} + (5t^2 \sin t - 11t \cos t - \sin t)\mathbf{k}$

(c) $\dfrac{d}{dt}(\mathbf{A} \cdot \mathbf{A}) = \mathbf{A} \cdot \dfrac{d\mathbf{A}}{dt} + \dfrac{d\mathbf{A}}{dt} \cdot \mathbf{A} = 2\mathbf{A} \cdot \dfrac{d\mathbf{A}}{dt}$

$$= 2(5t^2\mathbf{i} + t\mathbf{j} - t^3\mathbf{k}) \cdot (10t\mathbf{i} + \mathbf{j} - 3t^2\mathbf{k}) = 100t^3 + 2t + 6t^5$$

Another Method

$\mathbf{A} \cdot \mathbf{A} = (5t^2)^2 + (t)^2 + (-t^3)^2 = 25t^4 + t^2 + t^6$

Then $\dfrac{d}{dt}(25t^4 + t^2 + t^6) = 100t^3 + 2t + 6t^5$.

3.9. Suppose $\mathbf{A}$ has constant magnitude. Show that $\mathbf{A} \cdot d\mathbf{A}/dt = 0$ and that $\mathbf{A}$ and $d\mathbf{A}/dt$ are perpendicular provided $|d\mathbf{A}/dt| \neq 0$.

Solution

Since $\mathbf{A}$ has constant magnitude, $\mathbf{A} \cdot \mathbf{A} = $ constant.

Then $\dfrac{d}{dt}(\mathbf{A} \cdot \mathbf{A}) = \mathbf{A} \cdot \dfrac{d\mathbf{A}}{dt} + \dfrac{d\mathbf{A}}{dt} \cdot \mathbf{A} = 2\mathbf{A} \cdot \dfrac{d\mathbf{A}}{dt} = 0.$

Thus $\mathbf{A} \cdot \dfrac{d\mathbf{A}}{dt} = 0$ and $\mathbf{A}$ is perpendicular to $\dfrac{d\mathbf{A}}{dt}$ provided $\left|\dfrac{d\mathbf{A}}{dt}\right| \neq 0$.

3.10. Suppose $\mathbf{A}$, $\mathbf{B}$, and $\mathbf{C}$ are differentiable functions of a scalar u. Prove

$$\frac{d}{du}(\mathbf{A} \cdot \mathbf{B} \times \mathbf{C}) = \mathbf{A} \cdot \mathbf{B} \times \frac{d\mathbf{C}}{du} + \mathbf{A} \cdot \frac{d\mathbf{B}}{du} \times \mathbf{C} + \frac{d\mathbf{A}}{du} \cdot \mathbf{B} \times \mathbf{C}.$$

Solution

By Problems 3.7(a) and 3.7(b), $\dfrac{d}{du}\mathbf{A} \cdot (\mathbf{B} \times \mathbf{C}) = \mathbf{A} \cdot \dfrac{d}{du}(\mathbf{B} \times \mathbf{C}) + \dfrac{d\mathbf{A}}{du} \cdot \mathbf{B} \times \mathbf{C}$

$$= \mathbf{A} \cdot \left[\mathbf{B} \times \frac{d\mathbf{C}}{du} + \frac{d\mathbf{B}}{du} \times \mathbf{C}\right] + \frac{d\mathbf{A}}{du} \cdot \mathbf{B} \times \mathbf{C}$$

$$= \mathbf{A} \cdot \mathbf{B} \times \frac{d\mathbf{C}}{du} + \mathbf{A} \cdot \frac{d\mathbf{B}}{du} \times \mathbf{C} + \frac{d\mathbf{A}}{du} \cdot \mathbf{B} \times \mathbf{C}$$

3.11. A particle moves so that its position vector is given by $\mathbf{r} = \cos\omega t\,\mathbf{i} + \sin\omega t\,\mathbf{j}$ where ω is a constant. Show that (a) the velocity $\mathbf{v}$ of the particle is perpendicular to $\mathbf{r}$, (b) the acceleration $\mathbf{a}$ is directed toward the origin and has magnitude proportional to the distance from the origin, (c) $\mathbf{r} \times \mathbf{v} = \mathbf{a}$ constant vector.

Solution

(a) $\mathbf{v} = \dfrac{d\mathbf{r}}{dt} = -\omega\sin\omega t\,\mathbf{i} + \omega\cos\omega t\,\mathbf{j}$. Then

$$\mathbf{r} \cdot \mathbf{v} = [\cos\omega t\,\mathbf{i} + \sin\omega t\,\mathbf{j}] \cdot [-\omega\sin\omega t\,\mathbf{i} + \omega\cos\omega t\,\mathbf{j}]$$
$$= (\cos\omega t)(-\omega\sin\omega t) + (\sin\omega t)(\omega\cos\omega t) = 0$$

and $\mathbf{r}$ and $\mathbf{v}$ are perpendicular.

(b) $\dfrac{d^2\mathbf{r}}{dt^2} = \dfrac{d\mathbf{v}}{dt} = -\omega^2\cos\omega t\,\mathbf{i} - \omega^2\sin\omega t\,\mathbf{j}$

$$= -\omega^2[\cos\omega t\,\mathbf{i} + \sin\omega t\,\mathbf{j}] = -\omega^2\mathbf{r}$$

Then the acceleration is opposite to the direction of $\mathbf{r}$, that is, it is directed toward the origin. Its magnitude is proportional to $|\mathbf{r}|$, which is the distance from the origin.

(c)
$$\mathbf{r} \times \mathbf{v} = [\cos\omega t\,\mathbf{i} + \sin\omega t\,\mathbf{j}] \times [-\omega\sin\omega t\,\mathbf{i} + \omega\cos\omega t\,\mathbf{j}]$$

$$= \begin{vmatrix} \mathbf{i} & \mathbf{j} & \mathbf{k} \\ \cos\omega t & \sin\omega t & 0 \\ -\omega\sin\omega t & \omega\cos\omega t & 0 \end{vmatrix}$$

$$= \omega(\cos^2\omega t + \sin^2\omega t)\mathbf{k}$$
$$= \omega\mathbf{k}, \quad \text{a constant vector.}$$

Physically, the motion is that of a particle moving on the circumference of a circle with constant angular speed ω. The acceleration, directed toward the center of the circle, is the *centripetal acceleration*.

3.12. Show that $\mathbf{A} \cdot \dfrac{d\mathbf{A}}{dt} = A\dfrac{dA}{dt}$.

Solution

Let $\mathbf{A} = A_1\mathbf{i} + A_2\mathbf{j} + A_3\mathbf{k}$. Then $A = \sqrt{A_1^2 + A_2^2 + A_3^2}$.

$$\frac{dA}{dt} = \frac{1}{2}(A_1^2 + A_2^2 + A_3^2)^{-1/2}\left(2A_1\frac{dA_1}{dt} + 2A_2\frac{dA_2}{dt} + 2A_3\frac{dA_3}{dt}\right)$$

$$= \frac{A_1\dfrac{dA_1}{dt} + A_2\dfrac{dA_2}{dt} + A_3\dfrac{dA_3}{dt}}{(A_1^2 + A_2^2 + A_3^2)^{1/2}}$$

$$= \frac{\mathbf{A} \cdot \dfrac{d\mathbf{A}}{dt}}{A}, \qquad \text{i.e.,} \quad A\frac{dA}{dt} = \mathbf{A} \cdot \frac{d\mathbf{A}}{dt}$$

Another Method

Since $\mathbf{A} \cdot \mathbf{A} = A^2$, $\dfrac{d}{dt}(\mathbf{A} \cdot \mathbf{A}) = \dfrac{d}{dt}(A^2)$.

$$\frac{d}{dt}(\mathbf{A} \cdot \mathbf{A}) = \mathbf{A} \cdot \frac{d\mathbf{A}}{dt} + \frac{d\mathbf{A}}{dt} \cdot \mathbf{A} = 2\mathbf{A} \cdot \frac{d\mathbf{A}}{dt} \quad \text{and} \quad \frac{d}{dt}(A^2) = 2A\frac{dA}{dt}$$

Then $\quad 2\mathbf{A} \cdot \dfrac{d\mathbf{A}}{dt} = 2A\dfrac{dA}{dt} \quad$ or $\quad \mathbf{A} \cdot \dfrac{d\mathbf{A}}{dt} = A\dfrac{dA}{dt}$.

Note that if $\mathbf{A}$ is a constant vector $\mathbf{A} \cdot \dfrac{d\mathbf{A}}{dt} = 0$ as in Problem 3.9.

3.13. Let $\mathbf{A} = (2x^2y - x^4)\mathbf{i} + (e^{xy} - y\sin x)\mathbf{j} + (x^2\cos y)\mathbf{k}$. Find: (a) $\dfrac{\partial\mathbf{A}}{\partial x}$, (b) $\dfrac{\partial\mathbf{A}}{\partial y}$.

Solution

(a) $\dfrac{\partial\mathbf{A}}{\partial x} = \dfrac{\partial}{\partial x}(2x^2y - x^4)\mathbf{i} + \dfrac{\partial}{\partial x}(e^{xy} - y\sin x)\mathbf{j} + \dfrac{\partial}{\partial x}(x^2\cos y)\mathbf{k}$

$= (4xy - 4x^3)\mathbf{i} + (ye^{xy} - y\cos x)\mathbf{j} + 2x\cos y\,\mathbf{k}$

(b) $\dfrac{\partial\mathbf{A}}{\partial y} = \dfrac{\partial}{\partial y}(2x^2y - x^4)\mathbf{i} + \dfrac{\partial}{\partial y}(e^{xy} - y\sin x)\mathbf{j} + \dfrac{\partial}{\partial y}(x^2\cos y)\mathbf{k}$

$= 2x^2\mathbf{i} + (xe^{xy} - \sin x)\mathbf{j} - x^2\sin y\,\mathbf{k}$

3.14. Let $\mathbf{A}$ be the vector in Problem 3.13. Find: (a) $\dfrac{\partial^2\mathbf{A}}{\partial x^2}$, (b) $\dfrac{\partial^2\mathbf{A}}{\partial y^2}$.

Solution

(a) $\dfrac{\partial^2\mathbf{A}}{\partial x^2} = \dfrac{\partial}{\partial x}(4xy - 4x^3)\mathbf{i} + \dfrac{\partial}{\partial x}(ye^{xy} - y\cos x)\mathbf{j} + \dfrac{\partial}{\partial x}(2x\cos y)\mathbf{k}$

$= (4y - 12x^2)\mathbf{i} + (y^2e^{xy} + y\sin x)\mathbf{j} + 2\cos y\,\mathbf{k}$

(b) $\dfrac{\partial^2\mathbf{A}}{\partial y^2} = \dfrac{\partial}{\partial y}(2x^2)\mathbf{i} + \dfrac{\partial}{\partial y}(xe^{xy} - \sin x)\mathbf{j} - \dfrac{\partial}{\partial y}(x^2\sin y)\mathbf{k}$

$= 0 + x^2e^{xy}\mathbf{j} - x^2\cos y\,\mathbf{k} = x^2e^{xy}\mathbf{j} - x^2\cos y\,\mathbf{k}$

3.15. Let $\mathbf{A}$ be the vector in Problem 3.13. Find: (a) $\dfrac{\partial^2\mathbf{A}}{\partial x\partial y}$, (b) $\dfrac{\partial^2\mathbf{A}}{\partial y\partial x}$.

Solution

(a) $\dfrac{\partial^2\mathbf{A}}{\partial x\partial y} = \dfrac{\partial}{\partial x}\left(\dfrac{\partial\mathbf{A}}{\partial y}\right) = \dfrac{\partial}{\partial x}(2x^2)\mathbf{i} + \dfrac{\partial}{\partial x}(xe^{xy} - \sin x)\mathbf{j} - \dfrac{\partial}{\partial x}(x^2\sin y)\mathbf{k}$

$= 4x\mathbf{i} + (xye^{xy} + e^{xy} - \cos x)\mathbf{j} - 2x\sin y\,\mathbf{k}$

(b) $\dfrac{\partial^2\mathbf{A}}{\partial y\partial x} = \dfrac{\partial}{\partial y}\left(\dfrac{\partial\mathbf{A}}{\partial x}\right) = \dfrac{\partial}{\partial y}(4xy - 4x^3)\mathbf{i} + \dfrac{\partial}{\partial y}(ye^{xy} - y\cos x)\mathbf{j} + \dfrac{\partial}{\partial y}(2x\cos y)\mathbf{k}$

$= 4x\mathbf{i} + (xye^{xy} + e^{xy} - \cos x)\mathbf{j} - 2x\sin y\,\mathbf{k}$

Note that $\partial^2\mathbf{A}/\partial y\partial x = \partial^2\mathbf{A}/\partial x\partial y$, that is, the order of differentiation is immaterial. This is true in general if $\mathbf{A}$ has continuous partial derivatives of the second order at least.

3.16. Suppose $\phi(x, y, z) = xy^2z$ and $\mathbf{A} = xz\mathbf{i} - xy^2\mathbf{j} + yz^2\mathbf{k}$. Find $\dfrac{\partial^3}{\partial x^2\partial z}(\phi\mathbf{A})$ at the point $(2, -1, 1)$.

Solution

$$\phi\mathbf{A} = (xy^2z)(xz\mathbf{i} - xy^2\mathbf{j} + yz^2\mathbf{k}) = x^2y^2z^2\mathbf{i} - x^2y^4z\mathbf{j} + xy^3z^3\mathbf{k}$$

$$\frac{\partial}{\partial z}(\phi\mathbf{A}) = \frac{\partial}{\partial z}(x^2y^2z^2\mathbf{i} - x^2y^4z\mathbf{j} + xy^3z^3\mathbf{k}) = 2x^2y^2z\mathbf{i} - x^2y^4\mathbf{j} + 3xy^3z^2\mathbf{k}$$

$$\frac{\partial^2}{\partial x\partial z}(\phi\mathbf{A}) = \frac{\partial}{\partial x}(2x^2y^2z\mathbf{i} - x^2y^4\mathbf{j} + 3xy^3z^2\mathbf{k}) = 4xy^2z\mathbf{i} - 2xy^4\mathbf{j} + 3y^3z^2\mathbf{k}$$

$$\frac{\partial^3}{\partial x^2\partial z}(\phi\mathbf{A}) = \frac{\partial}{\partial x}(4xy^2z\mathbf{i} - 2xy^4\mathbf{j} + 3y^3z^2\mathbf{k}) = 4y^2z\mathbf{i} - 2y^4\mathbf{j}$$

If $x = 2$, $y = -1$, and $z = 1$, this becomes $4(-1)^2(1)\mathbf{i} - 2(-1)^4\mathbf{j} = 4\mathbf{i} - 2\mathbf{j}$.

3.17. Let **F** depend on x, y, z, t where x, y, and z depend on t. Prove that

$$\frac{d\mathbf{F}}{dt} = \frac{\partial \mathbf{F}}{\partial t} + \frac{\partial \mathbf{F}}{\partial x}\frac{dx}{dt} + \frac{\partial \mathbf{F}}{\partial y}\frac{dy}{dt} + \frac{\partial \mathbf{F}}{\partial z}\frac{dz}{dt}$$

under suitable assumptions of differentiability.

Solution

Suppose that $\mathbf{F} = F_1(x, y, z, t)\mathbf{i} + F_2(x, y, z, t)\mathbf{j} + F_3(x, y, z, t)\mathbf{k}$. Then

$$dF = dF_1\mathbf{i} + dF_2\mathbf{j} + dF_3\mathbf{k}$$

$$= \left[\frac{\partial F_1}{\partial t}dt + \frac{\partial F_1}{\partial x}dx + \frac{\partial F_1}{\partial y}dy + \frac{\partial F_1}{\partial z}dz\right]\mathbf{i} + \left[\frac{\partial F_2}{\partial t}dt + \frac{\partial F_2}{\partial x}dx + \frac{\partial F_2}{\partial y}dy + \frac{\partial F_2}{\partial z}dz\right]\mathbf{j}$$

$$+ \left[\frac{\partial F_3}{\partial t}dt + \frac{\partial F_3}{\partial x}dx + \frac{\partial F_3}{\partial y}dy + \frac{\partial F_3}{\partial z}dz\right]\mathbf{k}$$

$$= \left(\frac{\partial F_1}{\partial t}\mathbf{i} + \frac{\partial F_2}{\partial t}\mathbf{j} + \frac{\partial F_3}{\partial t}\mathbf{k}\right)dt + \left(\frac{\partial F_1}{\partial x}\mathbf{i} + \frac{\partial F_2}{\partial x}\mathbf{j} + \frac{\partial F_3}{\partial x}\mathbf{k}\right)dx$$

$$+ \left(\frac{\partial F_1}{\partial y}\mathbf{i} + \frac{\partial F_2}{\partial y}\mathbf{j} + \frac{\partial F_3}{\partial y}\mathbf{k}\right)dy + \left(\frac{\partial F_1}{\partial z}\mathbf{i} + \frac{\partial F_2}{\partial z}\mathbf{j} + \frac{\partial F_3}{\partial z}\mathbf{k}\right)dz$$

$$= \frac{\partial \mathbf{F}}{\partial t}dt + \frac{\partial \mathbf{F}}{\partial x}dx + \frac{\partial \mathbf{F}}{\partial y}dy + \frac{\partial \mathbf{F}}{\partial z}dz$$

and so $\dfrac{d\mathbf{F}}{dt} = \dfrac{\partial \mathbf{F}}{\partial t} + \dfrac{\partial \mathbf{F}}{\partial x}\dfrac{dx}{dt} + \dfrac{\partial \mathbf{F}}{\partial y}\dfrac{dy}{dt} + \dfrac{\partial \mathbf{F}}{\partial z}\dfrac{dz}{dt}$.

Differential Geometry

3.18. Prove the Frenet–Serret formulas: (a) $\dfrac{d\mathbf{T}}{ds} = \kappa\mathbf{N}$, (b) $\dfrac{d\mathbf{B}}{ds} = -\tau\mathbf{N}$, (c) $\dfrac{d\mathbf{N}}{ds} = \tau\mathbf{B} - \kappa\mathbf{T}$.

Solution

(a) Since $\mathbf{T} \cdot \mathbf{T} = 1$, it follows from Problem 3.9 that $\mathbf{T} \cdot \dfrac{d\mathbf{T}}{ds} = 0$, that is, $\dfrac{d\mathbf{T}}{ds}$ is perpendicular to $\mathbf{T}$.

If $\mathbf{N}$ is a unit vector in the direction $\dfrac{d\mathbf{T}}{ds}$, then $\dfrac{d\mathbf{T}}{ds} = \kappa\mathbf{N}$. We call $\mathbf{N}$ the *principal normal*, κ the *curvature* and $\rho = 1/\kappa$ the *radius of curvature*.

(b) Let $\mathbf{B} = \mathbf{T} \times \mathbf{N}$, so that $\dfrac{d\mathbf{B}}{ds} = \mathbf{T} \times \dfrac{d\mathbf{N}}{ds} + \dfrac{d\mathbf{T}}{ds} \times \mathbf{N} = \mathbf{T} \times \dfrac{d\mathbf{N}}{ds} + \kappa\mathbf{N} \times \mathbf{N} = \mathbf{T} \times \dfrac{d\mathbf{N}}{ds}$.

Then $\mathbf{T} \cdot \dfrac{d\mathbf{B}}{ds} = \mathbf{T} \cdot \mathbf{T} \times \dfrac{d\mathbf{N}}{ds} = 0$, so that $\mathbf{T}$ is perpendicular to $\dfrac{d\mathbf{B}}{ds}$.

But from $\mathbf{B} \cdot \mathbf{B} = 1$, it follows that $\mathbf{B} \cdot \dfrac{d\mathbf{B}}{ds} = 0$ (Problem 3.9), so that $\dfrac{d\mathbf{B}}{ds}$ is perpendicular to $\mathbf{B}$ and is thus in the plane of $\mathbf{T}$ and $\mathbf{N}$.

Since $\dfrac{d\mathbf{B}}{ds}$ is in the plane of $\mathbf{T}$ and $\mathbf{N}$ and is perpendicular to $\mathbf{T}$, it must be parallel to $\mathbf{N}$; then $\dfrac{d\mathbf{B}}{ds} = -\tau\mathbf{N}$. We call $\mathbf{B}$ the *binormal*, τ the *torsion*, and $\sigma = 1/\tau$ the *radius of torsion*.

(c) Since $\mathbf{T}, \mathbf{N}, \mathbf{B}$ form a right-handed system, so do $\mathbf{N}, \mathbf{B},$ and $\mathbf{T}$, that is, $\mathbf{N} = \mathbf{B} \times \mathbf{T}$.

Then $\dfrac{d\mathbf{N}}{ds} = \mathbf{B} \times \dfrac{d\mathbf{T}}{ds} + \dfrac{d\mathbf{B}}{ds} \times \mathbf{T} = \mathbf{B} \times \kappa\mathbf{N} - \tau\mathbf{N} \times \mathbf{T} = -\kappa\mathbf{T} + \tau\mathbf{B} = \tau\mathbf{B} - \kappa\mathbf{T}$.

3.19. Prove that the radius of curvature of the curve with parametric equations $x = x(s), y = y(s)$, and $z = z(s)$ is given by

$$\rho = \left[\left(\frac{d^2x}{ds^2}\right)^2 + \left(\frac{d^2y}{ds^2}\right)^2 + \left(\frac{d^2z}{ds^2}\right)^2\right]^{-1/2}.$$

Solution

The position vector of any point on the curve is $\mathbf{r} = x(s)\mathbf{i} + y(s)\mathbf{j} + z(s)\mathbf{k}$. Then

$$\mathbf{T} = \frac{d\mathbf{r}}{ds} = \frac{dx}{ds}\mathbf{i} + \frac{dy}{ds}\mathbf{j} + \frac{dz}{ds}\mathbf{k} \quad \text{and} \quad \frac{d\mathbf{T}}{ds} = \frac{d^2x}{ds^2}\mathbf{i} + \frac{d^2y}{ds^2}\mathbf{j} + \frac{d^2z}{ds^2}\mathbf{k}.$$

But $d\mathbf{T}/ds = \kappa\mathbf{N}$ so that

$$\kappa = \left|\frac{d\mathbf{T}}{ds}\right| = \sqrt{\left(\frac{d^2x}{ds^2}\right)^2 + \left(\frac{d^2y}{ds^2}\right)^2 + \left(\frac{d^2z}{ds^2}\right)^2}$$

and the result follows since $\rho = 1/\kappa$.

3.20. Show that $\dfrac{d\mathbf{r}}{ds} \cdot \dfrac{d^2\mathbf{r}}{ds^2} \times \dfrac{d^3\mathbf{r}}{ds^3} = \dfrac{\tau}{\rho^2}$.

Solution

$$\frac{d\mathbf{r}}{ds} = \mathbf{T}, \quad \frac{d^2\mathbf{r}}{ds^2} = \frac{d\mathbf{T}}{ds} = \kappa\mathbf{N}, \quad \frac{d^3\mathbf{r}}{ds^3} = \kappa\frac{d\mathbf{N}}{ds} + \frac{d\kappa}{ds}\mathbf{N} = \kappa(\tau\mathbf{B} - \kappa\mathbf{T}) + \frac{d\kappa}{ds}\mathbf{N} = \kappa\tau\mathbf{B} - \kappa^2\mathbf{T} + \frac{d\kappa}{ds}\mathbf{N}$$

$$\frac{d\mathbf{r}}{ds} \cdot \frac{d^2\mathbf{r}}{ds^2} \times \frac{d^3\mathbf{r}}{ds^3} = \mathbf{T} \cdot \kappa\mathbf{N} \times \left(\kappa\tau\mathbf{B} - \kappa^2\mathbf{T} + \frac{d\kappa}{ds}\mathbf{N}\right)$$

$$= \mathbf{T} \cdot \left(\kappa^2\tau\mathbf{N} \times \mathbf{B} - \kappa^3\mathbf{N} \times \mathbf{T} + \kappa\frac{d\kappa}{ds}\mathbf{N} \times \mathbf{N}\right)$$

$$= \mathbf{T} \cdot (\kappa^2\tau\mathbf{T} + \kappa^3\mathbf{B})$$

$$= \kappa^2\tau = \frac{\tau}{\rho^2}$$

The result can be written

$$\tau = [(x'')^2 + (y'')^2 + (z'')^2]^{-1}\begin{vmatrix} x' & y' & z' \\ x'' & y'' & z'' \\ x''' & y''' & z''' \end{vmatrix}$$

where primes denote derivatives with respect to s, by using the result of Problem 3.19.

3.21. Given the space curve $x = t, y = t^2, z = \frac{2}{3}t^3$. Find: (a) the curvature κ, (b) the torsion τ.

Solution

(a) The position vector is $\mathbf{r} = t\mathbf{i} + t^2\mathbf{j} + \frac{2}{3}t^3\mathbf{k}$. Then

$$\frac{d\mathbf{r}}{dt} = \mathbf{i} + 2t\mathbf{j} + 2t^2\mathbf{k} \quad \text{and} \quad \frac{ds}{dt} = \left|\frac{d\mathbf{r}}{dt}\right| = \sqrt{\frac{d\mathbf{r}}{dt} \cdot \frac{d\mathbf{r}}{dt}} = \sqrt{(1)^2 + (2t)^2 + (2t^2)^2} = 1 + 2t^2$$

and

$$\mathbf{T} = \frac{d\mathbf{r}}{ds} = \frac{d\mathbf{r}/dt}{ds/dt} = \frac{\mathbf{i} + 2t\mathbf{j} + 2t^2\mathbf{k}}{1 + 2t^2}$$

$$\frac{d\mathbf{T}}{dt} = \frac{(1 + 2t^2)(2\mathbf{j} + 4t\mathbf{k}) - (\mathbf{i} + 2t\mathbf{j} + 2t^2\mathbf{k})(4t)}{(1 + 2t^2)^2} = \frac{-4t\mathbf{i} + (2 - 4t^2)\mathbf{j} + 4t\mathbf{k}}{(1 + 2t^2)^2}$$

Then

$$\frac{d\mathbf{T}}{ds} = \frac{d\mathbf{T}/dt}{ds/dt} = \frac{-4t\mathbf{i} + (2 - 4t^2)\mathbf{j} + 4t\mathbf{k}}{(1 + 2t^2)^3}.$$

Since $d\mathbf{T}/ds = \kappa\mathbf{N}$,

$$\kappa = \left|\frac{d\mathbf{T}}{ds}\right| = \frac{\sqrt{(-4t)^2 + (2 - 4t^2)^2 + (4t)^2}}{(1 + 2t^2)^3} = \frac{2}{(1 + 2t^2)^2}$$

(b) From (a), $\mathbf{N} = \dfrac{1}{\kappa}\dfrac{d\mathbf{T}}{ds} = \dfrac{-2t\mathbf{i} + (1 - 2t^2)\mathbf{j} + 2t\mathbf{k}}{1 + 2t^2}$

Then

$$\mathbf{B} = \mathbf{T} \times \mathbf{N} = \begin{vmatrix} \mathbf{i} & \mathbf{j} & \mathbf{k} \\ \dfrac{1}{1 + 2t^2} & \dfrac{2t}{1 + 2t^2} & \dfrac{2t^2}{1 + 2t^2} \\ \dfrac{-2t}{1 + 2t^2} & \dfrac{1 - 2t^2}{1 + 2t^2} & \dfrac{2t}{1 + 2t^2} \end{vmatrix} = \frac{2t^2\mathbf{i} - 2t\mathbf{j} + \mathbf{k}}{1 + 2t^2}$$

Now $\dfrac{d\mathbf{B}}{dt} = \dfrac{4t\mathbf{i} + (4t^2 - 2)\mathbf{j} - 4t\mathbf{k}}{(1 + 2t^2)^2}$ and $\dfrac{d\mathbf{B}}{ds} = \dfrac{d\mathbf{B}/dt}{ds/dt} = \dfrac{4t\mathbf{i} + (4t^2 - 2)\mathbf{j} - 4t\mathbf{k}}{(1 + 2t^2)^3}$

Also, $-\tau\mathbf{N} = -\tau\left[\dfrac{-2t\mathbf{i} + (1 - 2t^2)\mathbf{j} + 2t\mathbf{k}}{1 + 2t^2}\right]$. Since $\dfrac{d\mathbf{B}}{ds} = -\tau\mathbf{N}$, we find $\tau = \dfrac{2}{(1 + 2t^2)^2}$.

Note that $\kappa = \tau$ for this curve.

3.22. Find equations in vector and rectangular form for the (a) tangent, (b) principal normal, and (c) binormal to the curve of Problem 3.21 at the point where $t = 1$.

Solution

Let $\mathbf{T}_O, \mathbf{N}_O,$ and $\mathbf{B}_O$ denote the tangent, principal normal, and binormal vectors at the required point. Then, from Problem 3.21,

$$\mathbf{T}_O = \frac{\mathbf{i} + 2\mathbf{j} + 2\mathbf{k}}{3}, \quad \mathbf{N}_O = \frac{-2\mathbf{i} - \mathbf{j} + 2\mathbf{k}}{3}, \quad \mathbf{B}_O = \frac{2\mathbf{i} - 2\mathbf{j} + \mathbf{k}}{3}$$

If $\mathbf{A}$ denotes a given vector while $\mathbf{r}_O$ and $\mathbf{r}$ denote, respectively, the position vectors of the initial point and an arbitrary point of $\mathbf{A}$, then $\mathbf{r} - \mathbf{r}_O$ is parallel to $\mathbf{A}$ and so the equation of $\mathbf{A}$ is $(\mathbf{r} - \mathbf{r}_O) \times \mathbf{A} = \mathbf{0}$. Then

Equation of tangent is	$(\mathbf{r} - \mathbf{r}_O) \times \mathbf{T}_O = \mathbf{0}$
Equation of principal normal is	$(\mathbf{r} - \mathbf{r}_O) \times \mathbf{N}_O = \mathbf{0}$
Equation of binormal is	$(\mathbf{r} - \mathbf{r}_O) \times \mathbf{B}_O = \mathbf{0}$

In rectangular form, with $\mathbf{r} = x\mathbf{i} + y\mathbf{j} + z\mathbf{k}, \mathbf{r}_O = \mathbf{i} + \mathbf{j} + \frac{2}{3}\mathbf{k}$, these become, respectively,

$$\frac{x-1}{1} = \frac{y-1}{2} = \frac{z-2/3}{2}, \quad \frac{x-1}{-2} = \frac{y-1}{-1} = \frac{z-2/3}{2}, \quad \frac{x-1}{2} = \frac{y-1}{-2} = \frac{z-2/3}{1}.$$

These equations can also be written in parametric form (see Problem 1.28, Chapter 1).

3.23. Sketch the space curve $x = 3\cos t, y = 3\sin t, z = 4t$ and find (a) the unit tangent $\mathbf{T}$, (b) the principal normal $\mathbf{N}$, curvature κ, and radius of curvature ρ, (c) the binormal $\mathbf{B}$, torsion τ, and radius of torsion σ.

Solution

The space curve is a *circular helix* (see Fig. 3-4). Since $t = z/4$, the curve has equations $x = 3\cos(z/4)$, $y = 3\sin(z/4)$ and therefore lies on the cylinder $x^2 + y^2 = 9$.

(a) The position vector for any point on the curve is

$$\mathbf{r} = 3\cos t\mathbf{i} + 3\sin t\mathbf{j} + 4t\mathbf{k}$$

Then $\dfrac{d\mathbf{r}}{dt} = -3\sin t\mathbf{i} + 3\cos t\mathbf{j} + 4\mathbf{k}$

$$\frac{ds}{dt} = \left|\frac{d\mathbf{r}}{dt}\right| = \sqrt{\frac{d\mathbf{r}}{dt} \cdot \frac{d\mathbf{r}}{dt}} = \sqrt{(-3\sin t)^2 + (3\cos t)^2 + 4^2} = 5$$

Thus $\mathbf{T} = \dfrac{d\mathbf{r}}{ds} = \dfrac{d\mathbf{r}/dt}{ds/dt} = -\dfrac{3}{5}\sin t\mathbf{i} + \dfrac{3}{5}\cos t\mathbf{j} + \dfrac{4}{5}\mathbf{k}.$

(b) $\dfrac{d\mathbf{T}}{dt} = \dfrac{d}{dt}\left(-\dfrac{3}{5}\sin t\mathbf{i} + \dfrac{3}{5}\cos t\mathbf{j} + \dfrac{4}{5}\mathbf{k}\right) = -\dfrac{3}{5}\cos t\mathbf{i} - \dfrac{3}{5}\sin t\mathbf{j}$

$\dfrac{d\mathbf{T}}{ds} = \dfrac{d\mathbf{T}/dt}{ds/dt} = -\dfrac{3}{25}\cos t\mathbf{i} - \dfrac{3}{25}\sin t\mathbf{j}$

Since $\dfrac{d\mathbf{T}}{ds} = \kappa\mathbf{N}$, $\left|\dfrac{d\mathbf{T}}{ds}\right| = |\kappa||\mathbf{N}| = \kappa$ as $\kappa \geqq 0$.

Then $\kappa = \left|\dfrac{d\mathbf{T}}{ds}\right| = \sqrt{\left(-\dfrac{3}{25}\cos t\right)^2 + \left(-\dfrac{3}{25}\sin t\right)^2} = \dfrac{3}{25}$ and $\rho = \dfrac{1}{\kappa} = \dfrac{25}{3}.$

From $\dfrac{d\mathbf{T}}{ds} = \kappa\mathbf{N}$, we obtain $\mathbf{N} = \dfrac{1}{\kappa}\dfrac{d\mathbf{T}}{ds} = -\cos t\mathbf{i} - \sin t\mathbf{j}.$

(c) $\mathbf{B} = \mathbf{T} \times \mathbf{N} = \begin{vmatrix} \mathbf{i} & \mathbf{j} & \mathbf{k} \\ -\frac{3}{5}\sin t & \frac{3}{5}\cos t & \frac{4}{5} \\ -\cos t & -\sin t & 0 \end{vmatrix} = \dfrac{4}{5}\sin t\mathbf{i} - \dfrac{4}{5}\cos t\mathbf{j} + \dfrac{3}{5}\mathbf{k}$

$\dfrac{d\mathbf{B}}{dt} = \dfrac{4}{5}\cos t\mathbf{i} + \dfrac{4}{5}\sin t\mathbf{j}, \quad \dfrac{d\mathbf{B}}{ds} = \dfrac{d\mathbf{B}/dt}{ds/dt} = \dfrac{4}{25}\cos t\mathbf{i} + \dfrac{4}{25}\sin t\mathbf{j}$

$-\tau\mathbf{N} = -\tau(-\cos t\mathbf{i} - \sin t\mathbf{j}) = \dfrac{4}{25}\cos t\mathbf{i} + \dfrac{4}{25}\sin t\mathbf{j}$ or $\tau = \dfrac{4}{25}$ and $\sigma = \dfrac{1}{\tau} = \dfrac{25}{4}.$

Fig. 3-4

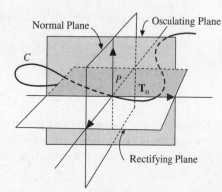

Fig. 3-5

3.24. Find equations in vector and rectangular form for the (a) osculating plane, (b) normal plane, and (c) rectifying plane to the curve of Problems 3.21 and 3.22 at the point where $t = 1$.

Solution

(a) The osculating plane is the plane which contains the tangent and principal normal. If $\mathbf{r}$ is the position vector of any point in this plane and $\mathbf{r}_O$ is the position vector of the point $t = 1$, then $\mathbf{r} - \mathbf{r}_O$ is perpendicular to $\mathbf{B}_O$, the binormal at the point $t = 1$, i.e. $(\mathbf{r} - \mathbf{r}_O) \cdot \mathbf{B}_O = 0$.

(b) The normal plane is the plane which is perpendicular to the tangent vector at the given point. Then the required equation is $(\mathbf{r} - \mathbf{r}_O) \cdot \mathbf{T}_O = 0$.

(c) The rectifying plane is the plane which is perpendicular to the principal normal at the given point. The required equation is $(\mathbf{r} - \mathbf{r}_O) \cdot \mathbf{N}_O = 0$.

In rectangular form the equation of (a), (b) and (c) become respectively,

$$2(x - 1) - 2(y - 1) + 1(z - 2/3) = 0,$$
$$1(x - 1) + 2(y - 1) + 2(z - 2/3) = 0,$$
$$-2(x - 1) - 1(y - 1) + 2(z - 2/3) = 0.$$

Fig. 3-5 shows the osculating, normal and rectifying planes to a curve C at the point P.

3.25. (a) Show that the equation $\mathbf{r} = \mathbf{r}(u, v)$ represents a surface.

(b) Show that $\dfrac{\partial \mathbf{r}}{\partial u} \times \dfrac{\partial \mathbf{r}}{\partial v}$ represents a vector normal to the surface.

Solution

(a) If we consider u to have a fixed value, say u_O, then $\mathbf{r} = \mathbf{r}(u_O, v)$ represents a curve which can be denoted by $u = u_O$. Similarly $u = u_1$ defines another curve $\mathbf{r} = \mathbf{r}(u_1, v)$. As u varies, therefore, $\mathbf{r} = \mathbf{r}(u, v)$ represents a curve which moves in space and generates a surface S. Then $\mathbf{r} = \mathbf{r}(u, v)$ represents the surface S thus generated, as shown in Fig. 3-6(a).

The curves $u = u_O, u = u_1, \ldots$ represent definite curves on the surface. Similarly $v = v_O, v = v_1, \ldots$ represent curves on the surface.

By assigning definite values to u and v, we obtain a point on the surface. Thus curves $u = u_O$ and $v = v_O$, for example, intersect and define the point (u_O, v_O) on the surface. We speak of the pair of numbers (u, v) as defining the *curvilinear coordinates* on the surface. If all the curves $u = $ constant and $v = $ constant are perpendicular at each point of intersection, we call the curvilinear coordinate system *orthogonal*. For further discussion of curvilinear coordinates see Chapter 7.

Fig. 3-6a   Fig. 3-6b

(b) Consider point P having coordinates (u_O, v_O) on a surface S, as shown in Fig. 3-6(b). The vector $\partial \mathbf{r}/\partial u$ at P is obtained by differentiating $\mathbf{r}$ with respect to u, keeping $v = $ constant $= v_O$. From the theory of space curves, it follows that $\partial \mathbf{r}/\partial u$ at P represents a vector tangent to the curve $v = v_O$ at P, as shown in the adjoining figure. Similarly, $\partial\mathbf{r}/\partial v$ at P represents a vector tangent to the curve $u = \text{constant} = u_O$. Since $\partial\mathbf{r}/\partial u$ and $\partial\mathbf{r}/\partial v$ represent vectors at P tangent to curves which lie on the surface S at P, it follows that these vectors are tangent to the surface at P. Hence it follows that $\dfrac{\partial\mathbf{r}}{\partial u} \times \dfrac{\partial\mathbf{r}}{\partial v}$ is a vector normal to S at P.

3.26. Determine a unit normal to the following surface, where $a > 9$:

$$\mathbf{r} = a\cos u \sin v\mathbf{i} + a\sin u \sin v\mathbf{j} + a\cos v\mathbf{k}$$

Solution

$$\frac{\partial\mathbf{r}}{\partial u} = -a\sin u \sin v\mathbf{i} + a\cos u \sin v\mathbf{j}$$

$$\frac{\partial\mathbf{r}}{\partial v} = a\cos u \cos v\mathbf{i} + a\sin u \cos v\mathbf{j} - a\sin v\mathbf{k}$$

Then

$$\frac{\partial\mathbf{r}}{\partial u} \times \frac{\partial\mathbf{r}}{\partial v} = \begin{vmatrix} \mathbf{i} & \mathbf{j} & \mathbf{k} \\ -a\sin u \sin v & a\cos u \sin v & 0 \\ a\cos u \cos v & a\sin u \cos v & -a\sin v \end{vmatrix}$$

$$= -a^2\cos u \sin^2 v\mathbf{i} - a^2\sin u \sin^2 v\mathbf{j} - a^2\sin v \cos v\mathbf{k}$$

represents a vector normal to the surface at any point (u, v).

A unit normal is obtained by dividing $\dfrac{\partial\mathbf{r}}{\partial u} \times \dfrac{\partial\mathbf{r}}{\partial v}$ by its magnitude, $\left|\dfrac{\partial\mathbf{r}}{\partial \mathbf{u}} \times \dfrac{\partial\mathbf{r}}{\partial \mathbf{v}}\right|$, given by

$$\sqrt{a^4\cos^2 u \sin^4 v + a^4\sin^2 u \sin^4 v + a^4\sin^2 v \cos^2 v} = \sqrt{a^4(\cos^2 u + \sin^2 u)\sin^4 v + a^4\sin^2 v \cos^2 v}$$

$$= \sqrt{a^4\sin^2 v(\sin^2 v + \cos^2 v)}$$

$$= \begin{cases} a^2\sin v & \text{if } \sin v > 0 \\ -a^2\sin v & \text{if } \sin v < 0 \end{cases}$$

Then there are two unit normals given by

$$\pm(\cos u \sin v\mathbf{i} + \sin u \sin v\mathbf{j} + \cos v\mathbf{k}) = \pm\mathbf{n}$$

It should be noted that the given surface is defined by $x = a\cos u \sin v$, $y = a\sin u \sin v$, $z = a\cos v$ from which it is seen that $x^2 + y^2 + z^2 = a^2$, which is a sphere of radius a. Since $\mathbf{r} = a\mathbf{n}$, it follows that

$$\mathbf{n} = \cos u \sin v\mathbf{i} + \sin u \sin v\mathbf{j} + \cos v\mathbf{k}$$

is the *outward drawn unit normal* to the sphere at the point (u, v).

3.27. Find an equation of the tangent plane to the surface $x^2 + 2xy^2 - 3z^3 = 6$ at the point $P(1, 2, 1)$.

Solution

The normal direction $\mathbf{N}$ to a surface $F(x, y, z) = k$, where k is a constant, follows:

$$\mathbf{N} = [F_x, F_y, F_z]$$

We have $F_x = 2x + 2y^2$, $F_y = 2x$, $F_z = 3z^2$. Thus, at the point P, the normal to the surface (and the tangent plane) is $\mathbf{N}(P) = [10, 2, 3]$.

The tangent plane E at P has the form $10x + 2y + 3z = b$. Substituting P in the equation gives $b = 10 + 4 + 3 = 17$. Thus $10x + 2y + 3z = 17$ is an equation for the tangent plane at P.

Mechanics

3.28. Show that the acceleration **a** of a particle which travels along a space curve with velocity **v** is given by

$$\mathbf{a} = \frac{dv}{dt}\mathbf{T} + \frac{v^2}{\rho}\mathbf{N}$$

where **T** is the unit tangent vector to the space curve, **N** is its unit principal normal, and ρ is the radius of curvature.

Solution

Velocity **v** = magnitude of **v** multiplied by unit tangent vector **T**

or $\qquad \mathbf{v} = v\mathbf{T}$

Differentiating,

$$\mathbf{a} = \frac{d\mathbf{v}}{dt} = \frac{d}{dt}(v\mathbf{T}) = \frac{dv}{dt}\mathbf{T} + v\frac{d\mathbf{T}}{dt}$$

But by Problem 3.18(a),

$$\frac{d\mathbf{T}}{dt} = \frac{d\mathbf{T}}{ds}\frac{ds}{dt} = \kappa\mathbf{N}\frac{ds}{dt} = \kappa v\mathbf{N} = \frac{v\mathbf{N}}{\rho}$$

Then

$$\mathbf{a} = \frac{dv}{dt}\mathbf{T} + v\left(\frac{v\mathbf{N}}{\rho}\right) = \frac{dv}{dt}\mathbf{T} + \frac{v^2}{\rho}\mathbf{N}$$

This shows that the component of the acceleration is dv/dt in a direction tangent to the path and v^2/ρ in a direction of the principal normal to the path. The latter acceleration is often called the *centripetal acceleration*. For a special case of this problem see Problem 3.11.

3.29. If **r** is the position vector of a particle of mass m relative to point O and **F** is the external force on the particle, then $\mathbf{r} \times \mathbf{F} = \mathbf{M}$ is the torque or moment of **F** about O. Show that $\mathbf{M} = d\mathbf{H}/dt$, where $\mathbf{H} = \mathbf{r} \times m\mathbf{v}$ and **v** is the velocity of the particle.

Solution

$$\mathbf{M} = \mathbf{r} \times \mathbf{F} = \mathbf{r} \times \frac{d}{dt}(m\mathbf{v}) \quad \text{by Newton's law.}$$

$$\text{But} \quad \frac{d}{dt}(\mathbf{r} \times m\mathbf{v}) = \mathbf{r} \times \frac{d}{dt}(m\mathbf{v}) + \frac{d\mathbf{r}}{dt} \times m\mathbf{v}$$

$$= \mathbf{r} \times \frac{d}{dt}(m\mathbf{v}) + \mathbf{v} \times m\mathbf{v} = \mathbf{r} \times \frac{d}{dt}(m\mathbf{v}) + 0$$

i.e., $\qquad \mathbf{M} = \frac{d}{dt}(\mathbf{r} \times m\mathbf{v}) = \frac{d\mathbf{H}}{dt}$

Note that the result holds whether m is constant or not. **H** is called the *angular momentum*. The result states that the torque is equal to the time rate of change of angular momentum.

This result is easily extended to a system of n particles having respective masses $m_1, m_2, \ldots, m_n$ and position vectors $\mathbf{r}_1, \mathbf{r}_2, \ldots, \mathbf{r}_n$ with external forces $\mathbf{F}_1, \mathbf{F}_2, \ldots, \mathbf{F}_n$. For this case, $\mathbf{H} = \sum_{k=1}^{n} m_k \mathbf{r}_k \times \mathbf{v}_k$ is the total angular momentum, $\mathbf{M} = \sum_{k=1}^{n} \mathbf{r}_k \times \mathbf{F}_k$ is the total torque, and the result is $\mathbf{M} = \frac{d\mathbf{H}}{dt}$ as before.

3.30. An observer stationed at a point which is fixed relative to an xyz coordinate system with origin O, as shown in Fig 3-7, observes a vector $\mathbf{A} = A_1\mathbf{i} + A_2\mathbf{j} + A_3\mathbf{k}$ and calculates its time derivative to be $\frac{dA_1}{dt}\mathbf{i} + \frac{dA_2}{dt}\mathbf{j} + \frac{dA_3}{dt}\mathbf{k}$. Later, he finds out that he and his coordinate system are actually rotating with respect to an XYZ coordinate system taken as fixed in space and having origin also at O. He asks, 'What would be the time derivative of **A** for an observer who is fixed relative to the *XYZ* coordinate system?'

Fig. 3-7

(a) Let $\dfrac{d\mathbf{A}}{dt}\Big|_f$ and $\dfrac{d\mathbf{A}}{dt}\Big|_m$ denote respectively the time derivatives of **A** with respect to the fixed and moving systems. Show that there exists a vector quantity $\boldsymbol{\omega}$ such that

$$\frac{d\mathbf{A}}{dt}\Big|_f = \frac{d\mathbf{A}}{dt}\Big|_m + \boldsymbol{\omega} \times \mathbf{A}$$

(b) Let D_f and D_m be symbolic time derivative operators in the fixed and moving systems respectively. Demonstrate the operator equivalence

$$D_f \equiv D_m + \boldsymbol{\omega}\times$$

Solution

(a) To the fixed observer the unit vectors **i**, **j**, **k** actually change with time. Hence such an observer would compute the time derivative of **A** as

$$\frac{d\mathbf{A}}{dt} = \frac{dA_1}{dt}\mathbf{i} + \frac{dA_2}{dt}\mathbf{j} + \frac{dA_3}{dt}\mathbf{k} + A_1\frac{d\mathbf{i}}{dt} + A_2\frac{d\mathbf{j}}{dt} + A_3\frac{d\mathbf{k}}{dt} \tag{1}$$

that is, $$\frac{d\mathbf{A}}{dt}\Big|_f = \frac{d\mathbf{A}}{dt}\Big|_m + A_1\frac{d\mathbf{i}}{dt} + A_2\frac{d\mathbf{j}}{dt} + A_3\frac{d\mathbf{k}}{dt} \tag{2}$$

Since **i** is a unit vector, $d\mathbf{i}/dt$ is perpendicular to **i** (see Problem 3.9) and must therefore lie in the plane of **j** and **k**. Then

$$\frac{d\mathbf{i}}{dt} = \alpha_1\mathbf{j} + \alpha_2\mathbf{k} \tag{3}$$

Similarly,

$$\frac{d\mathbf{j}}{dt} = \alpha_3\mathbf{k} + \alpha_4\mathbf{i} \tag{4}$$

$$\frac{d\mathbf{k}}{dt} = \alpha_5\mathbf{i} + \alpha_6\mathbf{j} \tag{5}$$

From $\mathbf{i}\cdot\mathbf{j} = 0$, differentiation yields $\mathbf{i}\cdot\dfrac{d\mathbf{j}}{dt} + \dfrac{d\mathbf{i}}{dt}\cdot\mathbf{j} = 0$. But $\mathbf{i}\cdot\dfrac{d\mathbf{j}}{dt} = \alpha_4$ from (4), and $\dfrac{d\mathbf{i}}{dt}\cdot\mathbf{j} = \alpha_1$ from (3); then $\alpha_4 = -\alpha_1$.

Similarly from $\mathbf{i}\cdot\mathbf{k} = 0$, $\mathbf{i}\cdot\dfrac{d\mathbf{k}}{dt} + \dfrac{d\mathbf{i}}{dt}\cdot\mathbf{k} = 0$ and $\alpha_5 = -\alpha_2$;

from $\mathbf{j}\cdot\mathbf{k} = 0$, $\mathbf{j}\cdot\dfrac{d\mathbf{k}}{dt} + \dfrac{d\mathbf{j}}{dt}\cdot\mathbf{k} = 0$ and $\alpha_6 = -\alpha_3$.

Then $\quad \dfrac{d\mathbf{i}}{dt} = \alpha_1\mathbf{j} + \alpha_2\mathbf{k}, \quad \dfrac{d\mathbf{j}}{dt} = \alpha_3\mathbf{k} - \alpha_1\mathbf{i}, \quad \dfrac{d\mathbf{k}}{dt} = -\alpha_2\mathbf{i} - \alpha_3\mathbf{j} \quad$ and

$$A_1\frac{d\mathbf{i}}{dt} + A_2\frac{d\mathbf{j}}{dt} + A_3\frac{d\mathbf{k}}{dt} = (-\alpha_1 A_2 - \alpha_2 A_3)\mathbf{i} + (\alpha_1 A_1 - \alpha_3 A_3)\mathbf{j} + (\alpha_2 A_1 + \alpha_3 A_2)\mathbf{k} \text{ which can be written as}$$

$$\begin{vmatrix} \mathbf{i} & \mathbf{j} & \mathbf{k} \\ \alpha_3 & -\alpha_2 & \alpha_1 \\ A_1 & A_2 & A_3 \end{vmatrix}$$

Then if we choose $\alpha_3 = \omega_1, -\alpha_2 = \omega_2, \alpha_1 = \omega_3$ the determinant becomes

$$\begin{vmatrix} \mathbf{i} & \mathbf{j} & \mathbf{k} \\ \omega_1 & \omega_2 & \omega_3 \\ A_1 & A_2 & A_3 \end{vmatrix} = \boldsymbol{\omega} \times \mathbf{A}$$

where $\boldsymbol{\omega} = \omega_1\mathbf{i} + \omega_2\mathbf{j} + \omega_3\mathbf{k}$. The quantity $\boldsymbol{\omega}$ is the angular velocity vector of the moving system with respect to the fixed system.

(b) By definition

$$D_f\mathbf{A} = \left.\frac{d\mathbf{A}}{dt}\right|_f = \text{derivative in fixed system}$$

$$D_m\mathbf{A} = \left.\frac{d\mathbf{A}}{dt}\right|_m = \text{derivative in moving system.}$$

From (a),

$$D_f\mathbf{A} = D_m\mathbf{A} + \boldsymbol{\omega} \times \mathbf{A} = (D_m + \boldsymbol{\omega} \times)\mathbf{A}$$

and shows the equivalence of the operators $D_f \equiv D_m + \boldsymbol{\omega} \times$.

SUPPLEMENTARY PROBLEMS

3.31. Suppose $\mathbf{R} = e^{-t}\mathbf{i} + \ln(t^2 + 1)\mathbf{j} - \tan t\mathbf{k}$. Find: (a) $d\mathbf{R}/dt$, (b) $d^2\mathbf{R}/dt^2$, (c) $|d\mathbf{R}/dt|$, (d) $|d^2\mathbf{R}/dt^2|$ at $t = 0$.

3.32. Suppose a particle moves along the curve $x = 2\sin 3t, y = 2\cos 3t, z = 8t$ at any time $t > 0$.
(a) Find the velocity and acceleration of the particle.
(b) Find the magnitude of the velocity and acceleration.

3.33. Find a unit tangent vector to any point on the curve $x = a\cos \omega t, \quad y = a\sin \omega t, \quad z = bt$ where $a, b,$ and ω are constants.

3.34. Suppose $\mathbf{A} = t^2\mathbf{i} - t\mathbf{j} + (2t + 1)\mathbf{k}$ and $\mathbf{B} = (2t - 3)\mathbf{i} + \mathbf{j} - t\mathbf{k}$. Find
(a) $\dfrac{d}{dt}(\mathbf{A} \cdot \mathbf{B})$, (b) $\dfrac{d}{dt}(\mathbf{A} \times \mathbf{B})$, (c) $\dfrac{d}{dt}|\mathbf{A} + \mathbf{B}|$, (d) $\dfrac{d}{dt}\left(\mathbf{A} \times \dfrac{d\mathbf{B}}{du}\right)$ at $t = 1$.

3.35. Suppose $\mathbf{A} = \sin u\mathbf{i} + \cos u\mathbf{j} + u\mathbf{k}, \quad \mathbf{B} = \cos u\mathbf{i} - \sin u\mathbf{j} - 3\mathbf{k},$ and $\quad \mathbf{C} = 2\mathbf{i} + 3\mathbf{j} - \mathbf{k}$.
Find $\dfrac{d}{du}(\mathbf{A} \times (\mathbf{B} \times \mathbf{C}))$ at $u = 0$.

3.36. Show: (a) $\dfrac{d}{ds}\left(\mathbf{A} \cdot \dfrac{d\mathbf{B}}{ds} - \dfrac{d\mathbf{A}}{ds} \cdot \mathbf{B}\right) = \mathbf{A}\dfrac{d^2\mathbf{B}}{ds^2} - \dfrac{d^2\mathbf{A}}{ds^2} \cdot \mathbf{B}$ where $\mathbf{A}$ and $\mathbf{B}$ are differential functions of s.

(b) $\dfrac{d}{ds}\left(\mathbf{A} \times \dfrac{d\mathbf{B}}{ds} - \dfrac{d\mathbf{A}}{ds} \times \mathbf{B}\right) = \mathbf{A} \times \dfrac{d^2\mathbf{B}}{ds^2} - \dfrac{d^2\mathbf{A}}{ds^2} \times \mathbf{B}$

(c) $\dfrac{d}{dt}\left(\mathbf{V} \cdot \dfrac{d\mathbf{V}}{dt} \times \dfrac{d^2\mathbf{V}}{dt^2}\right) = \mathbf{V} \cdot \dfrac{d\mathbf{V}}{dt} \times \dfrac{d^3\mathbf{V}}{dt^3}$

3.37. Suppose $\mathbf{A}(t) = 3t^2\mathbf{i} - (t + 4)\mathbf{j} + (t^2 - 2t)\mathbf{k}$ and $\mathbf{B}(t) = \sin t\mathbf{i} + 3e^{-t}\mathbf{j} - 3\cos t\mathbf{k}$. Find $\dfrac{d^2}{dt^2}(\mathbf{A} \times \mathbf{B})$ at $t = 0$.

3.38. Let $\dfrac{d^2\mathbf{A}}{dt^2} = 6t\mathbf{i} - 24t^2\mathbf{j} + 4\sin t\mathbf{k}$. Find $\mathbf{A}$ given that $\mathbf{A} = 2\mathbf{i} + \mathbf{j}$ and $\dfrac{d\mathbf{A}}{dt} = -\mathbf{i} - 3\mathbf{k}$ at $t = 0$.

3.39. Show that $\mathbf{r} = e^{-t}(\mathbf{C}_1\cos 2t + \mathbf{C}_2\sin 2t)$, where $\mathbf{C}_1$ and $\mathbf{C}_2$ are constant vectors, is a solution of the differential equation $\dfrac{d^2\mathbf{r}}{dt^2} + 2\dfrac{d\mathbf{r}}{dt} + 5\mathbf{r} = 0$.

3.40. Show that the general solution of the differential equation $\dfrac{d^2\mathbf{r}}{dt^2} + 2\alpha\dfrac{d\mathbf{r}}{dt} + \omega^2\mathbf{r} = 0$, where α and ω are constants, is

(a) $\mathbf{r} = e^{-at}\left(\mathbf{C}_1 e^{\sqrt{\alpha^2-\omega^2}\,t} + \mathbf{C}_2 e^{-\sqrt{\alpha^2-\omega^2}\,t}\right)$ if $\alpha^2 - \omega^2 > 0$

(b) $\mathbf{r} = e^{-at}(\mathbf{C}_1 \sin\sqrt{\omega^2-\alpha^2}\,t + \mathbf{C}_2 \cos\sqrt{\omega^2-\alpha^2}\,t)$ if $\alpha^2 - \omega^2 < 0$.

(c) $\mathbf{r} = e^{-at}(\mathbf{C}_1 + \mathbf{C}_2 t)$ if $\alpha^2 - \omega^2 = 0$.

where $\mathbf{C}_1$ and $\mathbf{C}_2$ are arbitrary constant vectors.

3.41. Solve: (a) $\dfrac{d^2\mathbf{r}}{dt^2} - 4\dfrac{d\mathbf{r}}{dt} - 5\mathbf{r} = 0$, (b) $\dfrac{d^2\mathbf{r}}{dt^2} + 2\dfrac{d\mathbf{r}}{dt} + \mathbf{r} = 0$, (c) $\dfrac{d^2\mathbf{r}}{dt^2} + 4\mathbf{r} = 0$.

3.42. Solve $\dfrac{d\mathbf{Y}}{dt} = \mathbf{X}$, $\dfrac{d\mathbf{X}}{dt} = -\mathbf{Y}$.

3.43. Suppose $\mathbf{A} = \cos xy\,\mathbf{i} + (3xy - 2x^2)\mathbf{j} - (3x + 2y)\mathbf{k}$. Find $\dfrac{\partial\mathbf{A}}{\partial x}, \dfrac{\partial\mathbf{A}}{\partial y}, \dfrac{\partial^2\mathbf{A}}{\partial x^2}, \dfrac{\partial^2\mathbf{A}}{\partial y^2}, \dfrac{\partial^2\mathbf{A}}{\partial x\partial y}, \dfrac{\partial^2\mathbf{A}}{\partial y\partial x}$.

3.44. Suppose $\mathbf{A} = x^2yz\,\mathbf{i} - 2xz^3\mathbf{j} + xz^2\mathbf{k}$ and $\mathbf{B} = 2z\mathbf{i} + y\mathbf{j} - x^2\mathbf{k}$. Find $\dfrac{\partial^2}{\partial x\partial y}(\mathbf{A}\times\mathbf{B})$ at $(1, 0, -2)$.

3.45. Suppose $\mathbf{C}_1$ and $\mathbf{C}_2$ are constant vectors and λ is a constant scalar. Show that

$\mathbf{H} = e^{-\lambda x}(\mathbf{C}_1 \sin\lambda y + \mathbf{C}_2 \cos\lambda y)$ satisfies the partial differential equation $\dfrac{\partial^2\mathbf{H}}{\partial x^2} + \dfrac{\partial^2\mathbf{H}}{\partial y^2} = 0$.

3.46. Suppose $\mathbf{p}_0$ is a constant vector, ω and c are constant scalars and $i = \sqrt{-1}$. Prove that $\mathbf{A} = [\mathbf{p}_0 e^{i\omega(t-r/c)}]/r$ satisfies the equation $\dfrac{\partial^2\mathbf{A}}{\partial r^2} + \dfrac{2}{r}\dfrac{\partial\mathbf{A}}{\partial r} = \dfrac{1}{c^2}\dfrac{\partial^2\mathbf{A}}{\partial t^2}$. [This result is of importance in electromagnetic theory.]

Differential Geometry

3.47. Consider the space curve $x = t - t^3/3, y = t^2, z = t + t^3/3$. Find: (a) the unit tangent $\mathbf{T}$, (b) the curvature κ, (c) the principal normal $\mathbf{N}$, (d) the binormal $\mathbf{B}$, (e) the torsion τ.

3.48. Suppose a space curve is defined in terms of the arc length parameter s by the equations

$$x = \arctan s, \quad y = \tfrac{1}{2}\sqrt{2}(s^2 + 1), \quad z = s - \arctan s$$

Find (a) $\mathbf{T}$, (b) $\mathbf{N}$, (c) $\mathbf{B}$, (d) κ, (e) τ, (f) ρ, (g) σ.

3.49. Consider the space curve $x = t, y = t^2, z = t^3$ (called the twisted cubic). Find κ and τ.

3.50. Show that for a plane curve the torsion $\tau = 0$.

3.51. Consider the radius of curvature $\rho = 1/\kappa$ of a plane curve with equations $y = f(x), z = 0$, that is, a curve in the xy plane. Show that $\rho = \{[1 + (y')^2]^{3/2}\}/|y''|$.

3.52. Consider the curve with position vector $\mathbf{r} = a\cos u\,\mathbf{i} + b\sin u\,\mathbf{j}$, where a and b are positive constants. Find its curvature κ and radius of curvature $\rho = 1/\kappa$. Interpret the case where $a = b$.

3.53. Show that the Frenet–Serret formulas can be written in the form $\dfrac{d\mathbf{T}}{ds} = \boldsymbol{\omega}\times\mathbf{T}$, $\dfrac{d\mathbf{N}}{ds} = \boldsymbol{\omega}\times\mathbf{N}$, $\dfrac{d\mathbf{B}}{ds} = \boldsymbol{\omega}\times\mathbf{B}$. Also, determine $\boldsymbol{\omega}$.

3.54. Prove that the curvature of the space curve $\mathbf{r} = \mathbf{r}(t)$ is given numerically by $\kappa = \dfrac{|\dot{\mathbf{r}}\times\ddot{\mathbf{r}}|}{|\dot{\mathbf{r}}|^3}$, where dots denote differentiation with respect to t.

3.55. (a) Consider the space curve $\mathbf{r} = \mathbf{r}(t)$. Prove that $\tau = \dfrac{\dot{\mathbf{r}}\cdot\ddot{\mathbf{r}}\times\dddot{\mathbf{r}}}{|\dot{\mathbf{r}}\times\ddot{\mathbf{r}}|^2}$ for the space curve $\mathbf{r} = \mathbf{r}(t)$.

(b) Suppose the parameter t is the arc length s. Show that

$$\tau = \frac{\dfrac{d\mathbf{r}}{ds}\cdot\dfrac{d^2\mathbf{r}}{ds^2}\times\dfrac{d^3\mathbf{r}}{ds^3}}{(d^2\mathbf{r}/ds^2)^2}.$$

3.56. Let $\mathbf{Q} = \dot{\mathbf{r}} \times \ddot{\mathbf{r}}$. Show that $\kappa = \dfrac{Q}{|\dot{\mathbf{r}}|^3}, \tau = \dfrac{\mathbf{Q} \cdot \dddot{\mathbf{r}}}{Q^2}$.

3.57. Find κ and τ for the space curve $x = \theta - \sin \theta, \quad y = 1 - \cos \theta, \quad z = 4 \sin(\theta/2)$.

3.58. Find the torsion of the curve $x = \dfrac{2t + 1}{t - 1}, \quad y = \dfrac{t^2}{t - 1}, \quad z = t + 2$. Explain your answer.

3.59. Consider the equations of the tangent line, principal normal, and binormal to the space curve $\mathbf{r} = \mathbf{r}(t)$ at the point $t = t_0$. Show they can be written, respectively, $r = r_0 + t\mathbf{T}_0$, $r = r_0 + t\mathbf{N}_0$, $r = r_0 + t\mathbf{B}_0$ where t is a parameter.

3.60. Consider the curve $x = 3 \cos t, y = 3 \sin t, z = 4t$. Find equations for the (a) tangent, (b) principal normal and (c) binormal at the point where $t = \pi$.

3.61. Find equations for the (a) osculating plane, (b) normal plane, and (c) rectifying plane to the curve $x = 3t - t^3, y = 3t^2, z = 3t + t^3$ at the point where $t = 1$.

3.62. (a) Show that the differential of arc length on the surface $\mathbf{r} = \mathbf{r}(u, v)$ is given by

$$ds^2 = E\, du^2 + 2F\, du\, dv + G\, dv^2$$

where $E = \dfrac{\partial \mathbf{r}}{\partial u} \cdot \dfrac{\partial \mathbf{r}}{\partial u} = \left(\dfrac{\partial \mathbf{r}}{\partial u} \right)^2, \quad F = \dfrac{\partial \mathbf{r}}{\partial u} \cdot \dfrac{\partial \mathbf{r}}{\partial v}, \quad G = \dfrac{\partial \mathbf{r}}{\partial v} \cdot \dfrac{\partial \mathbf{r}}{\partial v} = \left(\dfrac{\partial \mathbf{r}}{\partial v} \right)^2.$

(b) Prove that a necessary and sufficient condition that the u, v curvilinear coordinate system be orthogonal is $F \equiv 0$.

3.63. Find an equation of the tangent plane to the surface $z = xy$ at the point $(2, 3, 6)$.

3.64. Find equations of the tangent plane and normal line to the surface $4z = x^2 - y^2$ at the point $(3, 1, 2)$.

3.65. Assuming E, F, and G are defined as in Problem 3.62, prove that a unit normal to the surface $\mathbf{r} = \mathbf{r}(u, v)$ is

$$\mathbf{n} = \pm \frac{\dfrac{\partial \mathbf{r}}{\partial \mathbf{u}} \times \dfrac{\partial \mathbf{r}}{\partial \mathbf{v}}}{\sqrt{EG - F^2}}$$

Mechanics

3.66. Suppose a particle moves along a curve $\mathbf{r} = (t^3 - 4t)\mathbf{i} + (t^2 + 4t)\mathbf{j} + (8t^2 - 3t^3)\mathbf{k}$. Find the magnitudes of the tangential and normal components of its acceleration when $t = 2$.

3.67. Suppose a particle has velocity $\mathbf{v}$ and acceleration $\mathbf{a}$ along a space curve C. Prove that the radius of curvature ρ of its path is given numerically by $\rho = \dfrac{v^3}{|\mathbf{v} \times \mathbf{a}|}$.

3.68. An object is attracted to a fixed point O with a force $\mathbf{F} = f(r)\mathbf{r}$, called a *central force*, where $\mathbf{r}$ is the position vector of the object relative to O. Show that $\mathbf{r} \times \mathbf{v} = \mathbf{h}$ where $\mathbf{h}$ is a constant vector. Prove that the angular momentum is constant.

3.69. Prove that the acceleration vector of a particle moving along a space curve always lies in the osculating plane.

3.70. (a) Find the acceleration of a particle moving in the xy plane in terms of polar coordinates (ρ, ϕ).

(b) What are the components of the acceleration parallel and perpendicular to ρ?

3.71. Determine the (a) velocity and (b) acceleration of a moving particle as seen by the two observers in Problem 3.30.

ANSWERS TO SUPPLEMENTARY PROBLEMS

3.31. (a) $-\mathbf{i} - \mathbf{k}$, (b) $\mathbf{i} + 2\mathbf{j}$, (c) $\sqrt{2}$, (d) $\sqrt{5}$

3.32. $\mathbf{v} = 6\cos 3t\mathbf{i} - 6\sin 3t\mathbf{j} + 8\mathbf{k}$, $\mathbf{a} = -18\sin 3t\mathbf{i} - 18\cos 3t\mathbf{j}$, $|\mathbf{v}| = 10$, $|\mathbf{a}| = 18$

3.33. $\dfrac{-a\omega\sin\omega t\mathbf{i} + a\omega\cos\omega t\mathbf{j} + b\mathbf{k}}{\sqrt{a^2\omega^2 + b^2}}$ **3.35.** $7\mathbf{i} + 6\mathbf{j} - 6\mathbf{k}$

3.34. (a) -6, (b) $7\mathbf{j} + 3\mathbf{k}$, (c) 1, (d) $\mathbf{i} + 6\mathbf{j} + 2\mathbf{k}$ **3.37.** $-30\mathbf{i} + 14\mathbf{j} + 20\mathbf{k}$

3.38. $\mathbf{A} = (t^3 - t + 2)\mathbf{i} + (1 - 2t^4)\mathbf{j} + (t - 4\sin t)\mathbf{k}$

3.41. (a) $\mathbf{r} = C_1 e^{5t} + C_2 e^{-t}$, (b) $\mathbf{r} = e^{-t}(C_1 + C_2 t)$, (c) $\mathbf{r} = C_1\cos 2t + C_2\sin 2t$

3.42. $X = C_1\cos t + C_2\sin t$, $Y = C_1\sin t - C_2\cos t$

3.43.

$$\frac{\partial \mathbf{A}}{\partial x} = -y\sin xy\mathbf{i} + (3y - 4x)\mathbf{j} - 3\mathbf{k}, \qquad \frac{\partial \mathbf{A}}{\partial y} = -x\sin xy\mathbf{i} + 3x\mathbf{j} - 2\mathbf{k},$$

$$\frac{\partial^2 \mathbf{A}}{\partial x^2} = -y^2\cos xy\mathbf{i} - 4\mathbf{j}, \quad \frac{\partial^2 \mathbf{A}}{\partial y^2} = -x^2\cos xy\mathbf{i}, \quad \frac{\partial^2 \mathbf{A}}{\partial x\partial y} = \frac{\partial^2 \mathbf{A}}{\partial y\partial x} = -(xy\cos xy + \sin xy)\mathbf{i} + 3\mathbf{j}$$

3.44. $-4\mathbf{i} - 8\mathbf{j}$

3.47. (a) $\mathbf{T} = \dfrac{(1 - t^2)\mathbf{i} + 2t\mathbf{j} + (1 + t^2)\mathbf{k}}{\sqrt{2}(1 + t^2)}$ (c) $\mathbf{N} = -\dfrac{2t}{1 + t^2}\mathbf{i} + \dfrac{1 - t^2}{1 + t^2}\mathbf{j}$

      (b) $\kappa = \dfrac{1}{(1 + t^2)^2}$ (d) $\mathbf{B} = \dfrac{(t^2 - 1)\mathbf{i} - 2t\mathbf{j} + (t^2 + 1)\mathbf{k}}{\sqrt{2}(1 + t^2)}$ (e) $\tau = \dfrac{1}{(1 + t^2)^2}$

3.48. (a) $\mathbf{T} = \dfrac{\mathbf{i} + \sqrt{2}s\mathbf{j} + s^2\mathbf{k}}{s^2 + 1}$ (d) $\kappa = \dfrac{\sqrt{2}}{s^2 + 1}$

      (b) $\mathbf{N} = \dfrac{-\sqrt{2}s\mathbf{i} + (1 - s^2)\mathbf{j} + \sqrt{2}s\mathbf{k}}{s^2 + 1}$ (e) $\tau = \dfrac{\sqrt{2}}{s^2 + 1}$ (g) $\sigma = \dfrac{s^2 + 1}{\sqrt{2}}$

      (c) $\mathbf{B} = \dfrac{s^2\mathbf{i} - \sqrt{2}s\mathbf{j} + \mathbf{k}}{s^2 + 1}$ (f) $\rho = \dfrac{s^2 + 1}{\sqrt{2}}$

3.49. $\kappa = \dfrac{2\sqrt{9t^4 + 9t^2 + 1}}{(9t^4 + 4t^2 + 1)^{3/2}}$, $\tau = \dfrac{3}{9t^4 + 9t^2 + 1}$

3.52. $\kappa = \dfrac{ab}{(a^2\sin^2 u + b^2\cos^2 u)^{3/2}} = \dfrac{1}{\rho}$; if $a = b$, the given curve, which is an ellipse, becomes a circle of radius a and its radius of curvature $\rho = a$.

3.53. $\omega = \tau\mathbf{T} + \kappa\mathbf{B}$

3.57. $\kappa = \frac{1}{8}\sqrt{6 - 2\cos\theta}$, $\tau = \dfrac{(3 + \cos\theta)\cos\theta/2 + 2\sin\theta\sin\theta/2}{12\cos\theta - 4}$

3.58. $\tau = 0$. The curve lies on the plane $x - 3y + 3z = 5$.

3.60. (a) Tangent: $\mathbf{r} = -3\mathbf{i} + 4\pi\mathbf{k} + t\left(-\frac{3}{5}\mathbf{j} + \frac{4}{5}\mathbf{k}\right)$ or $x = -3$, $y = -\frac{3}{5}t$, $z = 4\pi + \frac{4}{5}t$.

      (b) Normal: $\mathbf{r} = -3\mathbf{i} + 4\pi\mathbf{j} + t\mathbf{i}$ or $x = -3 + t$, $y = 4\pi$, $z = 0$.

      (c) Binormal: $\mathbf{r} = -3\mathbf{i} + 4\pi\mathbf{j} + t\left(\frac{4}{5}\mathbf{j} + \frac{3}{5}\mathbf{k}\right)$ or $x = -3$, $y = 4\pi + \frac{4}{5}t$, $z = \frac{3}{5}t$.

3.61. (a) $y - z + 1 = 0$, (b) $y + z - 7 = 0$, (c) $x = 2$. $z = 2 - 2t$.

3.63. $3x + 2y - z = 6$.

3.64. $3x - y - 2z = 4$; $x = 3t + 3$, $y = 1 - t$,

3.66. Tangential, 16; normal $2\sqrt{73}$.

3.70. (a) $\ddot{\mathbf{r}} = [(\ddot{\rho} - \rho\dot{\phi}^2)\cos\phi - (\rho\ddot{\phi} + 2\dot{\rho}\dot{\phi})\sin\phi]\mathbf{i} + [(\ddot{\rho} - \rho\dot{\phi}^2)\sin\phi + (\rho\ddot{\phi} + 2\dot{\rho}\dot{\phi})\cos\phi]\mathbf{j}$

(b) $\ddot{\rho} - \rho\dot{\phi}^2$, $\rho\ddot{\phi} + 2\dot{\rho}\dot{\phi}$

3.71. (a) $\mathbf{v}_{p|f} = \mathbf{v}_{p|m} + \boldsymbol{\omega} \times \mathbf{r}$, (b) $\mathbf{a}_{p|f} = \mathbf{a}_{p|m} + \mathbf{a}_{m|f}$. For many cases, $\boldsymbol{\omega}$ is a constant, i.e., the rotation proceeds with constant angular velocity. Then $\mathbf{D}_m\boldsymbol{\omega} = 0$ and

$$\mathbf{a}_{m|f} = 2\boldsymbol{\omega} \times \mathbf{D}_m\mathbf{r} + \boldsymbol{\omega} \times (\boldsymbol{\omega} \times \mathbf{r}) = 2\boldsymbol{\omega} \times \mathbf{v}_m + \boldsymbol{\omega} \times (\boldsymbol{\omega} \times \mathbf{r})$$

The quantity $2\boldsymbol{\omega} \times \mathbf{v}_m$ is called the *Cortiolis acceleration* and $\boldsymbol{\omega} \times (\boldsymbol{\omega} \times \mathbf{r})$ is called the *centripetal acceleration*.

CHAPTER 4

Gradient, Divergence, Curl

4.1   Introduction

The vector differential operator del, written ∇, is defined as follows:

$$\nabla = \frac{\partial}{\partial x}\mathbf{i} + \frac{\partial}{\partial y}\mathbf{j} + \frac{\partial}{\partial z}\mathbf{k} = \mathbf{i}\frac{\partial}{\partial x} + \mathbf{j}\frac{\partial}{\partial y} + \mathbf{k}\frac{\partial}{\partial z}$$

This vector operator possesses properties analogous to those of ordinary vectors. It is useful in defining three quantities that appear in applications and which are known as the *gradient*, the *divergence*, and the *curl*. The operator ∇ is also known as *nabla*.

4.2   Gradient

Let $\phi(x, y, z)$ be a scalar function defined and differentiable at each point (x, y, z) in a certain region of space. [That is, ϕ defines a differentiable scalar field.] Then the gradient of ϕ, written $\nabla\phi$ or grad ϕ is defined as follows:

$$\nabla\phi = \left(\frac{\partial}{\partial x}\mathbf{i} + \frac{\partial}{\partial y}\mathbf{j} + \frac{\partial}{\partial z}\mathbf{k}\right)\phi = \frac{\partial\phi}{\partial x}\mathbf{i} + \frac{\partial\phi}{\partial y}\mathbf{j} + \frac{\partial\phi}{\partial z}\mathbf{k}$$

Note that $\nabla\phi$ defines a vector field.

EXAMPLE 4.1   Suppose $\phi(x, y, z) = 3xy^3 - y^2z^2$. Find $\nabla\phi$ (or grad ϕ) at the point $P(1, 1, 2)$.

$$\nabla\phi = \left(\frac{\partial}{\partial x}\mathbf{i} + \frac{\partial}{\partial y}\mathbf{j} + \frac{\partial}{\partial z}\mathbf{k}\right)(3xy^3 - y^2z^2)$$

$$= 3y^3\mathbf{i} + (9xy^2 - 2yz^2)\mathbf{j} - 2y^2z\mathbf{k}$$

Therefore $\nabla\phi(1, 1, 2) = 3(1)^3\mathbf{i} + [9(1)(1)^2 - 2(1)(2)^2]\mathbf{j} - 2(1)^2(2)\mathbf{k} = 3\mathbf{i} + \mathbf{j} - 4\mathbf{k}$.

Directional Derivatives

Consider a scalar function $\phi = \phi(x, y, z)$. Then the directional derivative of ϕ in the direction of a vector $\mathbf{A}$ is denoted by $D_{\mathbf{A}}(\phi)$. Letting $\mathbf{a} = \mathbf{A}/|\mathbf{A}|$, the unit vector in the direction of $\mathbf{A}$,

$$D_{\mathbf{A}}(\phi) = \nabla\phi \cdot \mathbf{a}$$

We emphasize that $\mathbf{a}$ must be a unit vector.

EXAMPLE 4.2 Consider the scalar function $\phi(x, y, z) = x^2 + y^2 + xz$.
(a) Find grad ϕ. (b) Find grad ϕ at the point $P = P(2, -1, 3)$. (c) Find the direction derivative of ϕ at the point P in the direction of $\mathbf{A} = \mathbf{i} + 2\mathbf{j} + \mathbf{k}$.

(a) grad $\phi = \left(\dfrac{\partial}{\partial x}\mathbf{i} + \dfrac{\partial}{\partial y}\mathbf{j} + \dfrac{\partial}{\partial z}\mathbf{k}\right)(x^2 + y^2 + xz) = (2x + z)\mathbf{i} + 2y\mathbf{j} + x\mathbf{k}$.

(b) At $P(2, -1, 3)$, grad $\phi = 7\mathbf{i} - 2\mathbf{j} + 3\mathbf{k}$.

(c) First we find the unit vector $\mathbf{a} = \mathbf{A}/|\mathbf{A}| = (\mathbf{i} + 2\mathbf{j} + \mathbf{k})/\sqrt{6}$ in the direction of $\mathbf{A}$. Then the directional derivative of ϕ at the point $P(2, -1, 3)$ in the direction of $\mathbf{A}$ follows:

$$\nabla\phi \cdot \mathbf{a} = (7\mathbf{i} - 2\mathbf{j} + 3\mathbf{k}) \cdot \left[(\mathbf{i} + 2\mathbf{j} + \mathbf{k})/\sqrt{6}\right] = 6/\sqrt{6} = \sqrt{6}/6.$$

Lagrange Multiplier

Here we want to find the points (x, y) that give the extrema (maximum or minimum value) of a function $f(x, y)$ subject to the constraint $g(x, y) = d$, where d is a constant. [More generally, we want to find the points $(x_1, x_2, \ldots, x_n)$ that give the extrema (maximum or minimum value) of a function $f(x_1, x_2, \ldots, x_n)$ subject to the constraint $g(x_1, x_2, \ldots, x_n) = d$, where d is a constant.]

This will occur only when the gradients ∇f and ∇g (directional derivatives) are orthogonal to the given curve [surface] $g(x, y) = d$. Thus ∇f and ∇g are parallel; and hence there must is a constant λ such that $\nabla f = \lambda \nabla g$.

The Greek letter λ (lamda) introduced above is called a *Lagrange multiplier*. The condition $\nabla f = \lambda \nabla g$ together with the original constraint yield three $(n + 1)$ equations in the unknowns x, y and λ:

$$f_x(x, y) = \lambda g_x(x, y), \quad f_y(x, y) = \lambda g_y(x, y), \quad g(x, y) = d$$

Solutions of the system for x and y give the candidates for the extrema of $f(x, y)$ subject to the constraint $g(x, y) = d$.

EXAMPLE 4.3 Minimize the function $f(x, y) = x^2 + 2y^2$ subject to the constraint $g(x, y) = 2x + y = 9$.

Using the condition that $\nabla f = \lambda \nabla g$ and the constraint, we obtain the three equations

$$2x = 2\lambda, \quad 4y = \lambda, \quad 2x + y = 9$$

Eliminating λ from the first two equations, we obtain $x = 4y$. This and $2x + y = 9$ gives $9y = 9$. Thus we obtain the solution $y = 1$ and $x = 4$. Thus $f(4, 1) = 16 + 2 = 18$ is the minimum value of f subject to the constraint $2x + y = 9$.

4.3 Divergence

Suppose $\mathbf{V}(x, y, z) = V_1\mathbf{i} + V_2\mathbf{j} + V_3\mathbf{k}$ is defined and differentiable at each point (x, y, z) in a region of space. (That is, $\mathbf{V}$ defines a differentiable vector field.) Then the *divergence* of $\mathbf{V}$, written $\nabla \cdot \mathbf{V}$ or div $\mathbf{V}$ is defined as follows:

$$\nabla \cdot \mathbf{V} = \left(\dfrac{\partial}{\partial x}\mathbf{i} + \dfrac{\partial}{\partial y}\mathbf{j} + \dfrac{\partial}{\partial z}\mathbf{k}\right) \cdot (V_1\mathbf{i} + V_2\mathbf{j} + V_3\mathbf{k})$$

$$= \dfrac{\partial V_1}{\partial x} + \dfrac{\partial V_2}{\partial y} + \dfrac{\partial V_3}{\partial z}$$

Although $\mathbf{V}$ is a vector, $\nabla \cdot \mathbf{V}$ is a scalar.

EXAMPLE 4.4 Suppose $\mathbf{A} = x^2z^2\mathbf{i} - 2y^2z^2\mathbf{j} + xy^2z\mathbf{k}$. Find $\nabla \cdot \mathbf{A}$ (or div $\mathbf{A}$) at the point $P(1, -1, 1)$.

$$\nabla \cdot \mathbf{A} = \left(\frac{\partial}{\partial x}\mathbf{i} + \frac{\partial}{\partial y}\mathbf{j} + \frac{\partial}{\partial z}\mathbf{k}\right) \cdot (x^2z^2\mathbf{i} - 2y^2z^2\mathbf{j} + xy^2z\mathbf{k})$$

$$= \frac{\partial}{\partial x}(x^2z^2) + \frac{\partial}{\partial y}(-2y^2z^2) + \frac{\partial}{\partial z}(xy^2z) = 2xz^2 - 4yz^2 + xy^2$$

At the point $P(1, -1, 1)$,

$$\nabla \cdot \mathbf{A} = 2(1)(1)^2 - 4(-1)(1)^2 + (1)(-1)^2 = 7$$

4.4 Curl

Suppose $\mathbf{V}(x, y, z) = V_1\mathbf{i} + V_2\mathbf{j} + V_3\mathbf{k}$ is a differentiable vector field. Then the *curl* or *rotation* of $\mathbf{V}$, written $\nabla \times \mathbf{V}$, curl $\mathbf{V}$ or rot $\mathbf{V}$, is defined as follows:

$$\nabla \times \mathbf{V} = \left(\frac{\partial}{\partial x}\mathbf{i} + \frac{\partial}{\partial y}\mathbf{j} + \frac{\partial}{\partial z}\mathbf{k}\right) \times (V_1\mathbf{i} + V_2\mathbf{j} + V_3\mathbf{k})$$

$$= \begin{vmatrix} \mathbf{i} & \mathbf{j} & \mathbf{k} \\ \dfrac{\partial}{\partial x} & \dfrac{\partial}{\partial y} & \dfrac{\partial}{\partial z} \\ V_1 & V_2 & V_3 \end{vmatrix}$$

$$= \begin{vmatrix} \dfrac{\partial}{\partial y} & \dfrac{\partial}{\partial z} \\ V_2 & V_3 \end{vmatrix}\mathbf{i} - \begin{vmatrix} \dfrac{\partial}{\partial x} & \dfrac{\partial}{\partial z} \\ V_1 & V_3 \end{vmatrix}\mathbf{j} + \begin{vmatrix} \dfrac{\partial}{\partial x} & \dfrac{\partial}{\partial y} \\ V_1 & V_2 \end{vmatrix}\mathbf{k}$$

$$= \left(\frac{\partial V_3}{\partial y} - \frac{\partial V_2}{\partial z}\right)\mathbf{i} + \left(\frac{\partial V_1}{\partial z} - \frac{\partial V_3}{\partial x}\right)\mathbf{j} + \left(\frac{\partial V_2}{\partial x} - \frac{\partial V_1}{\partial y}\right)\mathbf{k}$$

Note that in the expansion of the determinant the operators $\dfrac{\partial}{\partial x}$, $\dfrac{\partial}{\partial y}$, $\dfrac{\partial}{\partial z}$ must *precede* V_1, V_2, V_3.

EXAMPLE 4.5 Suppose $\mathbf{A} = x^2z^2\mathbf{i} - 2y^2z^2\mathbf{j} + xy^2z\mathbf{k}$. Find $\nabla \times \mathbf{A}$ (or curl $\mathbf{A}$) at the point $P(1, -1, 1)$.

$$\nabla \times \mathbf{A} = \left(\frac{\partial}{\partial x}\mathbf{i} + \frac{\partial}{\partial y}\mathbf{j} + \frac{\partial}{\partial z}\mathbf{k}\right) \times (x^2z^2\mathbf{i} - 2y^2z^2\mathbf{j} + xy^2z\mathbf{k})$$

$$= \begin{vmatrix} \mathbf{i} & \mathbf{j} & \mathbf{k} \\ \dfrac{\partial}{\partial x} & \dfrac{\partial}{\partial y} & \dfrac{\partial}{\partial z} \\ x^2z^2 & -2y^2z^2 & xy^2z \end{vmatrix}$$

$$= \left[\frac{\partial}{\partial y}(xy^2z) - \frac{\partial}{\partial z}(-2y^2z^2)\right]\mathbf{i} - \left[\frac{\partial}{\partial x}(xy^2z) - \frac{\partial}{\partial z}(x^2z^2)\right]\mathbf{j} + \left[\frac{\partial}{\partial x}(-2y^2z^2) - \frac{\partial}{\partial y}(x^2z^2)\right]\mathbf{k}$$

$$= (2xyz + 4y^2z)\mathbf{i} - (y^2z - 2x^2z)\mathbf{j} + 0\mathbf{k}$$

At the point $P(1, -1, 1)$, $\nabla \times \mathbf{A} = 2\mathbf{i} + \mathbf{j}$.

4.5 Formulas Involving ∇

The following propositions give many of the properties of the del operator ∇.

PROPOSITION 4.1: Suppose $\mathbf{A}$ and $\mathbf{B}$ are differentiable vector functions, and ϕ and ψ are differentiable scalar functions of position (x, y, z). Then the following laws hold.

(i) $\nabla(\phi + \psi) = \nabla\phi + \nabla\psi$ or $\text{grad}(\phi + \psi) = \text{grad }\phi + \text{grad }\psi$

(ii) $\nabla \cdot (\mathbf{A} + \mathbf{B}) = \nabla \cdot \mathbf{A} + \nabla \cdot \mathbf{B}$ or $\text{div}(\mathbf{A} + \mathbf{B}) = \text{div }\mathbf{A} + \text{div }\mathbf{B}$

(iii) $\nabla \times (\mathbf{A} + \mathbf{B}) = \nabla \times \mathbf{A} + \nabla \times \mathbf{B}$ or $\text{curl}(\mathbf{A} + \mathbf{B}) = \text{curl}\,\mathbf{A} + \text{curl}\,\mathbf{B}$

(iv) $\nabla \cdot (\phi \mathbf{A}) = (\nabla \phi) \cdot \mathbf{A} + \phi(\nabla \cdot \mathbf{A})$

(v) $\nabla \times (\phi \mathbf{A}) = (\nabla \phi) \times \mathbf{A} + \phi(\nabla \times \mathbf{A})$

(vi) $\nabla \cdot (\mathbf{A} \times \mathbf{B}) = \mathbf{B} \cdot (\nabla \times \mathbf{A}) - \mathbf{A} \cdot (\nabla \times \mathbf{B})$

(vii) $\nabla \times (\mathbf{A} \times \mathbf{B}) = (\mathbf{B} \cdot \nabla)\mathbf{A} - \mathbf{B}(\nabla \cdot \mathbf{A}) - (\mathbf{A} \cdot \nabla)\mathbf{B} + \mathbf{A}(\nabla \cdot \mathbf{B})$

(viii) $\nabla(\mathbf{A} \cdot \mathbf{B}) = (\mathbf{B} \cdot \nabla)\mathbf{A} + (\mathbf{A} \cdot \nabla)\mathbf{B} + \mathbf{B} \times (\nabla \times \mathbf{A}) + \mathbf{A} \times (\nabla \times \mathbf{B})$

PROPOSITION 4.2:   Suppose ϕ and $\mathbf{A}$ are differentiable scalar and vector functions, respectively, and both have continuous second partial derivatives. Then the following laws hold.

(i) $\nabla \cdot (\nabla \phi) = \nabla^2 \phi = \dfrac{\partial^2 \phi}{\partial x^2} + \dfrac{\partial^2 \phi}{\partial y^2} + \dfrac{\partial^2 \phi}{\partial z^2}$

where $\nabla^2 = \dfrac{\partial^2}{\partial x^2} + \dfrac{\partial^2}{\partial y^2} + \dfrac{\partial^2}{\partial z^2}$ is called the *Laplacian operator*.

(ii) $\nabla \times (\nabla \phi) = 0$. The curl of the gradient of ϕ is zero.

(iii) $\nabla \cdot (\nabla \times \mathbf{A}) = 0$. The divergence of the curl of $\mathbf{A}$ is zero.

(iv) $\nabla \times (\nabla \times \mathbf{A}) = \nabla(\nabla \cdot \mathbf{A}) - \nabla^2 \mathbf{A}$.

4.6   Invariance

Consider two rectangular coordinate systems or frames of reference xyz and $x'y'z'$ having the same origin O but with axes rotated with respect to each other. (See Fig. 4-1.)

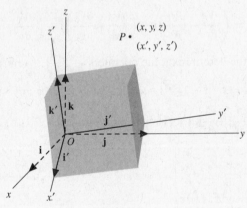

Fig. 4-1

A point P in space has coordinates (x, y, z) or (x', y', z') relative to these coordinate systems. The equations of transformation between coordinates of both systems or the *coordinate transformations* are given as follows:

$$\begin{aligned}
x' &= l_{11}x + l_{12}y + l_{13}z \\
y' &= l_{21}x + l_{22}y + l_{23}z \\
z' &= l_{31}x + l_{32}y + l_{33}z
\end{aligned} \tag{1}$$

Here l_{jk}, j, $k = 1, 2, 3$ represent direction cosines of the x', y', z' axes with respect to the x, y, z axes. (See Problem 4.38.) In case the origins of the two coordinate systems are not coincident the equations of transformation become

$$x' = l_{11}x + l_{12}y + l_{13}z + a_1'$$
$$y' = l_{21}x + l_{22}y + l_{23}z + a_2' \qquad (2)$$
$$z' = l_{31}x + l_{32}y + l_{33}z + a_3'$$

where the origin O of the xyz coordinate system is located at (a_1', a_2', a_3') relative to the $x'y'z'$ coordinate system.

The transformation equations (1) define a *pure rotation*, while equations (2) define a *rotation plus translation*. Any rigid body motion has the effect of a translation followed by a rotation. The transformation (1) is also called an *orthogonal transformation*. A general linear transformation is called an *affine transformation*.

Physically, a scalar point function or scalar field $\phi(x, y, z)$ evaluated at a particular point should be independent of the coordinates of the point. Thus the temperature at a point is not dependent on whether coordinates (x, y, z) or (x', y', z') are used. Then, if $\phi(x, y, z)$ is the temperature at point P with coordinates (x, y, z) while $\phi'(x', y', z')$ is the temperature at the same point P with coordinates (x', y', z'), we must have $\phi(x, y, z) = \phi'(x', y', z')$. If $\phi(x, y, z) = \phi'(x', y', z')$, where x, y, z and x', y', z' are related by the transformation equations (1) or (2), we call $\phi(x, y, z)$ an *invariant* with respect to the transformation. For example, $x^2 + y^2 + z^2$ is invariant under the transformation of rotation (1), since $x^2 + y^2 + z^2 = x'^2 + y'^2 + z'^2$.

Similarly, a vector point function or vector field $\mathbf{A}(x, y, z)$ is called an *invariant* if $\mathbf{A}(x, y, z) = \mathbf{A}'(x', y', z')$. This will be true if

$$A_1(x, y, z)\mathbf{i} + A_2(x, y, z)\mathbf{j} + A_3(x, y, z)\mathbf{k} = A_1'(x', y', z')\mathbf{i}' + A_2'(x', y', z')\mathbf{j}' + A_3'(x', y', z')\mathbf{k}'$$

In Chapters 7 and 8, more general transformations are considered and the above concepts are extended.

It can be shown (see Problem 4.41) that the gradient of an invariant scalar field is an invariant vector field with respect to the transformations (1) or (2). Similarly, the divergence and curl of an invariant vector field are invariant under this transformation.

SOLVED PROBLEMS

4.1. Suppose $\phi(x, y, z) = 3x^2y - y^2z^2$. Find $\nabla\phi$ (or grad ϕ) at the point $(1, -2, -1)$.

Solution

$$\nabla\phi = \left(\frac{\partial}{\partial x}\mathbf{i} + \frac{\partial}{\partial y}\mathbf{j} + \frac{\partial}{\partial z}\mathbf{k}\right)(3x^2y - y^3z^2)$$

$$= \mathbf{i}\frac{\partial}{\partial x}\left(3x^2y - y^3z^2\right) + \mathbf{j}\frac{\partial}{\partial y}\left(3x^2y - y^3z^2\right) + \mathbf{k}\frac{\partial}{\partial z}\left(3x^2y - y^3z^2\right)$$

$$= 6xy\mathbf{i} + \left(3x^2 - 3y^2z^2\right)\mathbf{j} - 2y^3z\mathbf{k}$$

$$= 6(1)(-2)\mathbf{i} + \left\{3(1)^2 - 3(-2)^2(-1)^2\right\}\mathbf{j} - 2(-2)^3(-1)\mathbf{k}$$

$$= -12\mathbf{i} - 9\mathbf{j} - 16\mathbf{k}$$

4.2. Suppose F and G are differentiable scalar functions of x, y and z. Prove (a) $\nabla(F+G) = \nabla F + \nabla G$, (b) $\nabla(FG) = F\nabla G + G\nabla F$.

Solution

(a) $\quad \nabla(F+G) = \left(\dfrac{\partial}{\partial x}\mathbf{i} + \dfrac{\partial}{\partial y}\mathbf{j} + \dfrac{\partial}{\partial z}\mathbf{k} \right)(F+G)$

$$= \mathbf{i}\dfrac{\partial}{\partial x}(F+G) + \mathbf{j}\dfrac{\partial}{\partial y}(F+G) + \mathbf{k}\dfrac{\partial}{\partial z}(F+G)$$

$$= \mathbf{i}\dfrac{\partial F}{\partial x} + \mathbf{i}\dfrac{\partial G}{\partial x} + \mathbf{j}\dfrac{\partial F}{\partial y} + \mathbf{j}\dfrac{\partial G}{\partial y} + \mathbf{k}\dfrac{\partial F}{\partial z} + \mathbf{k}\dfrac{\partial G}{\partial z}$$

$$= \mathbf{i}\dfrac{\partial F}{\partial x} + \mathbf{j}\dfrac{\partial F}{\partial y} + \mathbf{k}\dfrac{\partial F}{\partial z} + \mathbf{i}\dfrac{\partial G}{\partial x} + \mathbf{j}\dfrac{\partial G}{\partial y} + \mathbf{k}\dfrac{\partial G}{\partial z}$$

$$= \left(\mathbf{i}\dfrac{\partial}{\partial x} + \mathbf{j}\dfrac{\partial}{\partial y} + \mathbf{k}\dfrac{\partial}{\partial z} \right)F + \left(\mathbf{i}\dfrac{\partial}{\partial x} + \mathbf{j}\dfrac{\partial}{\partial y} + \mathbf{k}\dfrac{\partial}{\partial z} \right)G = \nabla F + \nabla G$$

(b) $\quad \nabla(FG) = \left(\dfrac{\partial}{\partial x}\mathbf{i} + \dfrac{\partial}{\partial y}\mathbf{j} + \dfrac{\partial}{\partial z}\mathbf{k} \right)(FG) = \dfrac{\partial}{\partial x}(FG)\mathbf{i} + \dfrac{\partial}{\partial y}(FG)\mathbf{j} + \dfrac{\partial}{\partial z}(FG)\mathbf{k}$

$$= \left(F\dfrac{\partial G}{\partial x} + G\dfrac{\partial F}{\partial x} \right)\mathbf{i} + \left(F\dfrac{\partial G}{\partial y} + G\dfrac{\partial F}{\partial y} \right)\mathbf{j} + \left(F\dfrac{\partial G}{\partial z} + G\dfrac{\partial F}{\partial z} \right)\mathbf{k}$$

$$= F\left(\dfrac{\partial G}{\partial x}\mathbf{i} + \dfrac{\partial G}{\partial y}\mathbf{j} + \dfrac{\partial G}{\partial z}\mathbf{k} \right) + G\left(\dfrac{\partial F}{\partial x}\mathbf{i} + \dfrac{\partial F}{\partial y}\mathbf{j} + \dfrac{\partial F}{\partial z}\mathbf{k} \right) = F\nabla G + G\nabla F$$

4.3. Find $\nabla\phi$ if (a) $\phi = \ln |\mathbf{r}|$, (b) $\phi = \dfrac{1}{r}$.

Solution

(a) $\quad \mathbf{r} = x\mathbf{i} + y\mathbf{j} + z\mathbf{k}$. Then $|\mathbf{r}| = \sqrt{x^2 + y^2 + z^2}$ and $\phi = \ln|\mathbf{r}| = \tfrac{1}{2}\ln(x^2 + y^2 + z^2)$.

$$\nabla\phi = \frac{1}{2}\nabla \ln(x^2 + y^2 + z^2)$$

$$= \frac{1}{2}\left\{ \mathbf{i}\frac{\partial}{\partial x}\ln(x^2+y^2+z^2) + \mathbf{j}\frac{\partial}{\partial y}\ln(x^2+y^2+z^2) + \mathbf{k}\frac{\partial}{\partial z}\ln(x^2+y^2+z^2) \right\}$$

$$= \frac{1}{2}\left\{ \mathbf{i}\frac{2x}{x^2+y^2+z^2} + \mathbf{j}\frac{2y}{x^2+y^2+z^2} + \mathbf{k}\frac{2z}{x^2+y^2+z^2} \right\} = \frac{x\mathbf{i}+y\mathbf{j}+z\mathbf{k}}{x^2+y^2+z^2} = \frac{\mathbf{r}}{r^2}$$

(b) $\quad \nabla\phi = \nabla\left(\dfrac{1}{r}\right) = \nabla\left(\dfrac{1}{\sqrt{x^2+y^2+z^2}} \right) = \nabla\left\{ (x^2+y^2+z^2)^{-1/2} \right\}$

$$= \mathbf{i}\frac{\partial}{\partial x}(x^2+y^2+z^2)^{-1/2} + \mathbf{j}\frac{\partial}{\partial y}(x^2+y^2+z^2)^{-1/2} + \mathbf{k}\frac{\partial}{\partial z}(x^2+y^2+z^2)^{-1/2}$$

$$= \mathbf{i}\left\{ -\frac{1}{2}(x^2+y^2+z^2)^{-3/2}2x \right\} + \mathbf{j}\left\{ -\frac{1}{2}(x^2+y^2+z^2)^{-3/2}2y \right\} + \mathbf{k}\left\{ -\frac{1}{2}(x^2+y^2+z^2)^{-3/2}2z \right\}$$

$$= \frac{-x\mathbf{i}-y\mathbf{j}-z\mathbf{k}}{(x^2+y^2+z^2)^{3/2}} = -\frac{\mathbf{r}}{r^3}$$

4.4. Show that $\nabla r^n = nr^{n-2}\,\mathbf{r}$.

Solution

$$\nabla r^n = \nabla\left(\sqrt{x^2 + y^2 + z^2}\right)^n = \nabla(x^2 + y^2 + z^2)^{n/2}$$

$$= \mathbf{i}\frac{\partial}{\partial x}\{(x^2 + y^2 + z^2)^{n/2}\} + \mathbf{j}\frac{\partial}{\partial y}\{(x^2 + y^2 + z^2)^{n/2}\} + \mathbf{k}\frac{\partial}{\partial z}\{(x^2 + y^2 + z^2)^{n/2}\}$$

$$= \mathbf{i}\left\{\frac{n}{2}(x^2 + y^2 + z^2)^{n/2-1}\,2x\right\} + \mathbf{j}\left\{\frac{n}{2}(x^2 + y^2 + z^2)^{n/2-1}\,2y\right\} + \mathbf{k}\left\{\frac{n}{2}(x^2 + y^2 + z^2)^{n/2-1}\,2z\right\}$$

$$= n(x^2 + y^2 + z^2)^{n/2-1}(x\mathbf{i} + y\mathbf{j} + z\mathbf{k}) = n(r^2)^{n/2-1}\mathbf{r} = nr^{n-2}\mathbf{r}$$

Note that if $\mathbf{r} = r\mathbf{r}_1$ where $\mathbf{r}_1$ is a unit vector in the direction $\mathbf{r}$, then $\nabla r^n = nr^{n-1}\,\mathbf{r}_1$.

4.5. Show that $\nabla\phi$ is a vector perpendicular to the surface $\phi(x, y, z) = c$ where c is a constant.

Solution

Let $\mathbf{r} = x\mathbf{i} + y\mathbf{j} + z\mathbf{k}$ be the position vector to any point $P(x, y, z)$ on the surface. Then $d\mathbf{r} = dx\mathbf{i} + dy\mathbf{j} + dz\mathbf{k}$ lies in the tangent plane to the surface at P.

But $\quad d\phi = \dfrac{\partial\phi}{\partial x}dx + \dfrac{\partial\phi}{\partial y}dy + \dfrac{\partial\phi}{\partial z}dz = 0 \quad$ or $\quad \left(\dfrac{\partial\phi}{\partial x}\mathbf{i} + \dfrac{\partial\phi}{\partial y}\mathbf{j} + \dfrac{\partial\phi}{\partial z}\mathbf{k}\right)\cdot(dx\mathbf{i} + dy\mathbf{j} + dz\mathbf{k}) = 0, \quad$ that is,

$\nabla\phi\cdot d\mathbf{r} = 0$, so that $\nabla\phi$ is perpendicular to $d\mathbf{r}$ and therefore to the surface.

4.6. Find a unit normal to the surface $-x^2yz^2 + 2xy^2z = 1$ at the point $P(1, 1, 1)$.

Solution

Let $\phi = -x^2yz^2 + 2xy^2z$. Using Problem 4.5, $\nabla\phi(1, 1, 1)$ is normal to the surface $-x^2yz^2 + 2xy^2z = 1$ at the point $P(1, 1, 1)$; hence, $\dfrac{\nabla\phi(1, 1, 1)}{|\nabla\phi(1, 1, 1)|}$ will suffice.

$$\nabla\phi = (-2xyz^2)\mathbf{i} + (-x^2z^2 + 4xyz)\mathbf{j} + (-2x^2yz + 2xy^2)\mathbf{k}.$$

Then $\nabla\phi(1, 1, 1) = 3\mathbf{j}$. $\quad |\nabla\phi(1, 1, 1)| = |3\mathbf{j}| = 3|\mathbf{j}| = 3$. Thus, at the point $P(1, 1, 1)\dfrac{3\mathbf{j}}{3} = \mathbf{j}$ is a unit normal to $-x^2yz^2 + 2xy^2z = 1$.

4.7. Find an equation for the tangent plane to the surface $x^2yz - 4xyz^2 = -6$ at the point $P(1, 2, 1)$.

Solution

$$\nabla(x^2yz - 4xyz^2) = (2xyz - 4yz^2)\mathbf{i} + (x^2z - 4xz^2)\mathbf{j} + (x^2y - 8xyz)\mathbf{k}.$$

Evaluating the gradient at the point $P(1, 2, 1)$, we get $-4\mathbf{i} - 3\mathbf{j} - 14\mathbf{k}$. Then $4\mathbf{i} + 3\mathbf{j} + 14\mathbf{k}$ is normal to the surface at P. An equation of the plane with normal $\mathbf{N} = a\mathbf{i} + b\mathbf{j} + c\mathbf{k}$ has the form

$$ax + by + cz = \mathbf{k}$$

Thus the equation has the form $4x + 3y + 14z = \mathbf{k}$. Substituting P in the equation, we get $\mathbf{k} = 24$. Thus the required equation is $4x + 3y + 14z = 24$.

4.8. Let $\phi(x, y, z)$ and $\phi(x + \Delta x, y + \Delta y, z + \Delta z)$ be the temperatures at two neighboring points $P(x, y, z)$ and $Q(x + \Delta x, y + \Delta y, z + \Delta z)$ of a certain region.

(a) Interpret physically the quantity $\dfrac{\Delta\phi}{\Delta s} = \dfrac{\phi(x + \Delta x, y + \Delta y, z + \Delta z) - \phi(x, y, z)}{\Delta s}$ where Δs is the distance between points P and Q.

(b) Evaluate $\lim\limits_{\Delta s \to 0} \dfrac{\Delta \phi}{\Delta s} = \dfrac{d\phi}{ds}$ and interpret physically.

(c) Show that $\dfrac{d\phi}{ds} = \nabla \phi \cdot \dfrac{d\mathbf{r}}{ds}$.

Solution

(a) Since $\Delta \phi$ is the change in temperature between points P and Q and Δs is the distance between these points, $\Delta \phi / \Delta s$ represents the average rate of change in temperature per unit distance in the direction from P to Q.

(b) From the calculus,

$$\Delta \phi = \frac{\partial \phi}{\partial x} \Delta x + \frac{\partial \phi}{\partial y} \Delta y + \frac{\partial \phi}{\partial z} \Delta z + \text{infinitesimals of order higher than } \Delta x, \Delta y, \text{ and } \Delta z.$$

Then

$$\lim_{\Delta s \to 0} \frac{\Delta \phi}{\Delta s} = \lim_{\Delta s \to 0} \frac{\partial \phi}{\partial x} \frac{\Delta x}{\Delta s} + \frac{\partial \phi}{\partial y} \frac{\Delta y}{\Delta s} + \frac{\partial \phi}{\partial z} \frac{\Delta z}{\Delta s}$$

or

$$\frac{d\phi}{ds} = \frac{\partial \phi}{\partial x} \frac{dx}{ds} + \frac{\partial \phi}{\partial y} \frac{dy}{ds} + \frac{\partial \phi}{\partial z} \frac{dz}{ds}$$

where $\dfrac{d\phi}{ds}$ represents the rate of change of temperature with respect to distance at point P in a direction toward Q. This is also called the *directional derivative* of ϕ.

(c) $\dfrac{d\phi}{ds} = \dfrac{\partial \phi}{\partial x}\dfrac{dx}{ds} + \dfrac{\partial \phi}{\partial y}\dfrac{dy}{ds} + \dfrac{\partial \phi}{\partial z}\dfrac{dz}{ds} = \left(\dfrac{\partial \phi}{\partial x}\mathbf{i} + \dfrac{\partial \phi}{\partial y}\mathbf{j} + \dfrac{\partial \phi}{\partial z}\mathbf{k}\right) \cdot \left(\dfrac{dx}{ds}\mathbf{i} + \dfrac{dy}{ds}\mathbf{j} + \dfrac{dz}{ds}\mathbf{k}\right) = \nabla \phi \cdot \dfrac{d\mathbf{r}}{ds}.$

Note that since $\dfrac{d\mathbf{r}}{ds}$ is a unit vector, $\nabla \phi \cdot \dfrac{d\mathbf{r}}{ds}$ is the component of $\nabla \phi$ in the direction of this unit vector.

4.9. Show that the greatest rate of change of ϕ, i.e. the maximum directional derivative, takes place in the direction of, and has the magnitude of, the vector $\nabla \phi$.

Solution

By Problem 4.8(c), $\dfrac{d\phi}{ds} = \nabla \phi \cdot \dfrac{d\mathbf{r}}{ds}$ is the projection of $\nabla \phi$ in the direction $\dfrac{d\mathbf{r}}{ds}$. This projection will be a maximum when $\nabla \phi$ and $\dfrac{d\mathbf{r}}{ds}$ have the same direction. Then the maximum value of $\dfrac{d\phi}{ds}$ takes place in the direction of $\nabla \phi$ and its magnitude is $|\nabla \phi|$.

4.10. Let $\phi = x^2 yz - 4xyz^2$. Find the directional derivative of ϕ at $P(1, 3, 1)$ in the direction of $2\mathbf{i} - \mathbf{j} - 2\mathbf{k}$.

Solution

First find $\nabla \phi = (2xyz - 4yz^2)\mathbf{i} + (x^2 z - 4xz^2)\mathbf{j} + (x^2 y - 8xyz)\mathbf{k}$. Then $\nabla \phi(1, 3, 1) = -6\mathbf{i} - 3\mathbf{j} - 21\mathbf{k}$. The unit vector in the direction of $2\mathbf{i} - \mathbf{j} - 2\mathbf{k}$ is

$$\mathbf{a} = \frac{2\mathbf{i} - \mathbf{j} - 2\mathbf{k}}{\sqrt{(2)^2 + (-1)^2 + (-2)^2}} = \tfrac{2}{3}\mathbf{i} - \tfrac{1}{3}\mathbf{j} - \tfrac{2}{3}\mathbf{k}.$$

Thus the required directional derivative is

$$\nabla \phi(1, 3, 1) \cdot \mathbf{a} = (-6\mathbf{i} - 3\mathbf{j} - 21\mathbf{k}) \cdot \left(\tfrac{2}{3}\mathbf{i} - \tfrac{1}{3}\mathbf{j} - \tfrac{2}{3}\mathbf{k}\right) = -4 + 1 + 14 = 11.$$

4.11. Let $\phi = x^2 y^3 z^6$. (a) In what direction from the point $P(1, 1, 1)$ is the directional derivative of ϕ a maximum? (b) What is the magnitude of this maximum?

Solution

$\nabla \phi = \nabla(x^2 y^3 z^6) = 2xy^3 z^6 \mathbf{i} + 3x^2 y^2 z^6 \mathbf{j} + 6x^2 y^3 z^5 \mathbf{k}$. Then $\nabla \phi(1, 1, 1) = 2\mathbf{i} + 3\mathbf{j} + 6\mathbf{k}$. Then, by Problem 4.9:

(a)   The directional derivative is a maximum in the direction $\nabla \phi(1, 1, 1) = 2\mathbf{i} + 3\mathbf{j} + 6\mathbf{k}$.

(b)   The magnitude of this maximum is $|\nabla \phi(1, 1, 1)| = \sqrt{(2)^2 + (3)^2 + (6)^2} = 7$.

4.12. Find the angle between the surfaces $z = x^2 + y^2$ and $z = \left(x - \dfrac{\sqrt{6}}{6}\right)^2 + \left(y - \dfrac{\sqrt{6}}{6}\right)^2$ at the point $P = \left(\dfrac{\sqrt{6}}{12}, \dfrac{\sqrt{6}}{12}, \dfrac{1}{12}\right)$.

Solution

The angle between the surfaces at the point is the angle between the normals to the surfaces at the point.

Let $\phi_1 = x^2 + y^2 - z$ and $\phi_2 = \left(x - \dfrac{\sqrt{6}}{6}\right)^2 + \left(y - \dfrac{\sqrt{6}}{6}\right)^2 - z$.

A normal to $z = x^2 + y^2$ is

$$\nabla \phi_1 = 2x\mathbf{i} + 2y\mathbf{j} - \mathbf{k} \quad \text{and} \quad \nabla \phi_1(P) = \frac{\sqrt{6}}{6}\mathbf{i} + \frac{\sqrt{6}}{6}\mathbf{j} - \mathbf{k}.$$

A normal to $z = \left(x - \dfrac{\sqrt{6}}{6}\right)^2 + \left(y - \dfrac{\sqrt{6}}{6}\right)^2$ is

$$\nabla \phi_2 = 2\left(x - \frac{\sqrt{6}}{6}\right)\mathbf{i} + 2\left(y - \frac{\sqrt{6}}{6}\right)\mathbf{j} - \mathbf{k} \quad \text{and} \quad \nabla \phi_2(P) = -\frac{\sqrt{6}}{6}\mathbf{i} - \frac{\sqrt{6}}{6}\mathbf{j} - \mathbf{k}.$$

Now $(\nabla \phi_1(P)) \cdot (\nabla \phi_2(P)) = |\nabla \phi_1(P)||\nabla \phi_2(P)| \cos \theta$ where θ is the required angle.

$$\left(\frac{\sqrt{6}}{6}\mathbf{i} + \frac{\sqrt{6}}{6}\mathbf{j} - \mathbf{k}\right) \cdot \left(-\frac{\sqrt{6}}{6}\mathbf{i} - \frac{\sqrt{6}}{6}\mathbf{j} - \mathbf{k}\right) = \left|\frac{\sqrt{6}}{6}\mathbf{i} + \frac{\sqrt{6}}{6}\mathbf{j} - \mathbf{k}\right|\left|-\frac{\sqrt{6}}{6}\mathbf{i} - \frac{\sqrt{6}}{6}\mathbf{j} - \mathbf{k}\right| \cos \theta$$

$$-\frac{1}{6} - \frac{1}{6} + 1 = \sqrt{\frac{1}{6} + \frac{1}{6} + 1}\sqrt{\frac{1}{6} + \frac{1}{6} + 1} \cos \theta \quad \text{and} \quad \cos \theta = \frac{2/3}{4/3} = \frac{1}{2}.$$

Thus the acute angle is $\theta = \arccos\left(\dfrac{1}{2}\right) = 60°$.

4.13. Let R be the distance from a fixed point $A(a, b, c)$ to any point $P(x, y, z)$. Show that ∇R is a unit vector in the direction $\mathbf{AP} = \mathbf{R}$.

Solution

If $\mathbf{r}_A$ and $\mathbf{r}_P$ are the position vectors $a\mathbf{i} + b\mathbf{j} + c\mathbf{k}$ and $x\mathbf{i} + y\mathbf{j} + z\mathbf{k}$ of A and P, respectively, then $\mathbf{R} = \mathbf{r}_P - \mathbf{r}_A = (x - a)\mathbf{i} + (y - b)\mathbf{j} + (z - c)\mathbf{k}$, so that $R = \sqrt{(x - a)^2 + (y - b)^2 + (z - c)^2}$. Then

$$\nabla R = \nabla\left(\sqrt{(x - a)^2 + (y - b)^2 + (z - c)^2}\right) = \frac{(x - a)\mathbf{i} + (y - b)\mathbf{j} + (z - c)\mathbf{k}}{\sqrt{(x - a)^2 + (y - b)^2 + (z - c)^2}} = \frac{\mathbf{R}}{R}$$

is a unit vector in the direction $\mathbf{R}$.

4.14. Let P be any point on an ellipse whose foci are at points A and B, as shown in Fig. 4.2. Prove that lines AP and BP make equal angles with the tangent to the ellipse at P.

Solution

Let $\mathbf{R}_1 = \mathbf{AP}$ and $\mathbf{R}_2 = \mathbf{BP}$ denote vectors drawn respectively from foci A and B to point P on the ellipse, and let $\mathbf{T}$ be a unit tangent to the ellipse at P.

Since an ellipse is the locus of all points P, the sum of whose distances from two fixed points A and B is a constant p, it is seen that the equation of the ellipse is $R_1 + R_2 = p$.

By Problem 4.5, $\nabla(R_1 + R_2)$ is a normal to the ellipse; hence $[\nabla(R_1 + R_2)] \cdot \mathbf{T} = 0$ or $(\nabla R_2) \cdot \mathbf{T} = -(\nabla R_1) \cdot \mathbf{T}$.

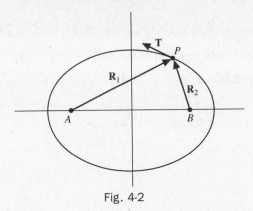

Fig. 4-2

Since ∇R_1 and ∇R_2 are unit vectors in direction $\mathbf{R}_1$ and $\mathbf{R}_2$ respectively (Problem 4.13), the cosine of the angle between ∇R_2 and $\mathbf{T}$ is equal to the cosine of the angle between ∇R_1 and $-\mathbf{T}$; hence the angles themselves are equal.

The problem has a physical interpretation. Light rays (or sound waves) originating at focus A, for example, will be reflected from the ellipse to focus B.

Divergence

4.15. Suppose $\mathbf{A} = x^2 z^2 \mathbf{i} - 2y^2 z^2 \mathbf{j} + xy^2 z \mathbf{k}$. Find $\nabla \cdot \mathbf{A}$ (or div $\mathbf{A}$) at the point $P(1, -1, 1)$.

Solution

$$\nabla \cdot \mathbf{A} = \left(\frac{\partial}{\partial x}\mathbf{i} + \frac{\partial}{\partial y}\mathbf{j} + \frac{\partial}{\partial z}\mathbf{k}\right) \cdot (x^2 z^2 \mathbf{i} - 2y^2 z^2 \mathbf{j} + xy^2 z \mathbf{k})$$

$$= \frac{\partial}{\partial x}(x^2 z^2) + \frac{\partial}{\partial y}(-2y^2 z^2) + \frac{\partial}{\partial z}(xy^2 z) = 2xz^2 - 4yz^2 + xy^2$$

$$\nabla \cdot \mathbf{A}(1, -1, 1) = 2(1)(1)^2 - 4(-1)(1)^2 + (1)(-1)^2 = 7$$

4.16. Given $\phi = 6x^3 y^2 z$. (a) Find $\nabla \cdot \nabla \phi$ (or div grad ϕ).

(b) Show that $\nabla \cdot \nabla \phi = \nabla^2 \phi$ where $\nabla^2 = \dfrac{\partial^2}{\partial x^2} + \dfrac{\partial^2}{\partial y^2} + \dfrac{\partial^2}{\partial z^2}$ denotes the Laplacian operator.

Solution

(a) $\nabla \phi = \dfrac{\partial}{\partial x}(6x^3 y^2 z)\mathbf{i} + \dfrac{\partial}{\partial y}(6x^3 y^2 z)\mathbf{j} + \dfrac{\partial}{\partial z}(6x^3 y^2 z)\mathbf{k} = 18x^2 y^2 z \mathbf{i} + 12x^3 yz \mathbf{j} + 6x^3 y^2 \mathbf{k}.$

Then

$$\nabla \cdot \nabla \phi = \left(\frac{\partial}{\partial x}\mathbf{i} + \frac{\partial}{\partial y}\mathbf{j} + \frac{\partial}{\partial z}\mathbf{k}\right) \cdot (18x^2 y^2 z \mathbf{i} + 12x^3 yz \mathbf{j} + 6x^3 y^2 \mathbf{k})$$

$$= \frac{\partial}{\partial x}(18x^2 y^2 z) + \frac{\partial}{\partial y}(12x^3 yz) + \frac{\partial}{\partial z}(6x^3 y^2) = 36xy^2 z + 12x^3 z.$$

(b) $\nabla \cdot \nabla \phi = \left(\dfrac{\partial}{\partial x}\mathbf{i} + \dfrac{\partial}{\partial y}\mathbf{j} + \dfrac{\partial}{\partial z}\mathbf{k}\right) \cdot \left(\dfrac{\partial \phi}{\partial x}\mathbf{i} + \dfrac{\partial \phi}{\partial y}\mathbf{j} + \dfrac{\partial \phi}{\partial z}\mathbf{k}\right)$

$$= \frac{\partial}{\partial x}\left(\frac{\partial \phi}{\partial x}\right) + \frac{\partial}{\partial y}\left(\frac{\partial \phi}{\partial y}\right) + \frac{\partial}{\partial z}\left(\frac{\partial \phi}{\partial z}\right) = \frac{\partial^2 \phi}{\partial x^2} + \frac{\partial^2 \phi}{\partial y^2} + \frac{\partial^2 \phi}{\partial z^2}$$

$$= \left(\frac{\partial^2}{\partial x^2} + \frac{\partial^2}{\partial y^2} + \frac{\partial^2}{\partial z^2}\right)\phi = \nabla^2 \phi$$

4.17. Prove that $\nabla^2\left(\dfrac{1}{r}\right) = 0$.

Solution

$$\nabla^2\left(\frac{1}{r}\right) = \left(\frac{\partial^2}{\partial x^2} + \frac{\partial^2}{\partial y^2} + \frac{\partial^2}{\partial z^2}\right)\left(\frac{1}{\sqrt{x^2 + y^2 + z^2}}\right)$$

$$\frac{\partial}{\partial x}\left(\frac{1}{\sqrt{x^2 + y^2 + z^2}}\right) = \frac{\partial}{\partial x}(x^2 + y^2 + z^2)^{-1/2} = -x(x^2 + y^2 + z^2)^{-3/2}$$

$$\frac{\partial^2}{\partial x^2}\left(\frac{1}{\sqrt{x^2 + y^2 + z^2}}\right) = \frac{\partial}{\partial x}\left[-x(x^2 + y^2 + z^2)^{-3/2}\right]$$

$$= 3x^2(x^2 + y^2 + z^2)^{-5/2} - (x^2 + y^2 + z^2)^{-3/2} = \frac{2x^2 - y^2 - z^2}{(x^2 + y^2 + z^2)^{5/2}}$$

Similarly,

$$\frac{\partial^2}{\partial y^2}\left(\frac{1}{\sqrt{x^2 + y^2 + z^2}}\right) = \frac{2y^2 - z^2 - x^2}{(x^2 + y^2 + z^2)^{5/2}} \quad \text{and} \quad \frac{\partial^2}{\partial z^2}\left(\frac{1}{\sqrt{x^2 + y^2 + z^2}}\right) = \frac{2z^2 - x^2 - y^2}{(x^2 + y^2 + z^2)^{5/2}}$$

Then, by addition,

$$\left(\frac{\partial^2}{\partial x^2} + \frac{\partial^2}{\partial y^2} + \frac{\partial^2}{\partial z^2}\right)\left(\frac{1}{\sqrt{x^2 + y^2 + z^2}}\right) = 0.$$

The equation $\nabla^2\phi = 0$ is called *Laplace's equation*. It follows that $\phi = 1/r$ is a solution of this equation.

4.18. Prove: (a) $\nabla\cdot(\mathbf{A} + \mathbf{B}) = \nabla\cdot\mathbf{A} + \nabla\cdot\mathbf{B}$,   (b) $\nabla\cdot(\phi\mathbf{A}) = (\nabla\phi)\cdot\mathbf{A} + \phi(\nabla\cdot\mathbf{A})$.

Solution

(a) Let $\mathbf{A} = A_1\mathbf{i} + A_2\mathbf{j} + A_3\mathbf{k}$, $\mathbf{B} = B_1\mathbf{i} + B_2\mathbf{j} + B_3\mathbf{k}$.
Then

$$\nabla\cdot(\mathbf{A} + \mathbf{B}) = \left(\frac{\partial}{\partial x}\mathbf{i} + \frac{\partial}{\partial y}\mathbf{j} + \frac{\partial}{\partial z}\mathbf{k}\right)\cdot[(A_1 + B_1)\mathbf{i} + (A_2 + B_2)\mathbf{j} + (A_3 + B_3)\mathbf{k}]$$

$$= \frac{\partial}{\partial x}(A_1 + B_1) + \frac{\partial}{\partial y}(A_2 + B_2) + \frac{\partial}{\partial z}(A_3 + B_3)$$

$$= \frac{\partial A_1}{\partial x} + \frac{\partial A_2}{\partial y} + \frac{\partial A_3}{\partial z} + \frac{\partial B_1}{\partial x} + \frac{\partial B_2}{\partial y} + \frac{\partial B_3}{\partial z}$$

$$= \left(\frac{\partial}{\partial x}\mathbf{i} + \frac{\partial}{\partial y}\mathbf{j} + \frac{\partial}{\partial z}\mathbf{k}\right)\cdot(A_1\mathbf{i} + A_2\mathbf{j} + A_3\mathbf{k}) + \left(\frac{\partial}{\partial x}\mathbf{i} + \frac{\partial}{\partial y}\mathbf{j} + \frac{\partial}{\partial z}\mathbf{k}\right)\cdot(B_1\mathbf{i} + B_2\mathbf{j} + B_3\mathbf{k})$$

$$= \nabla\cdot\mathbf{A} + \nabla\cdot\mathbf{B}$$

(b) $\nabla\cdot(\phi\mathbf{A}) = \nabla\cdot(\phi A_1\mathbf{i} + \phi A_2\mathbf{j} + \phi A_3\mathbf{k}) = \dfrac{\partial}{\partial x}(\phi A_1) + \dfrac{\partial}{\partial y}(\phi A_2) + \dfrac{\partial}{\partial z}(\phi A_3)$

$$= \frac{\partial\phi}{\partial x}A_1 + \phi\frac{\partial A_1}{\partial x} + \frac{\partial\phi}{\partial y}A_2 + \phi\frac{\partial A_2}{\partial y} + \frac{\partial\phi}{\partial z}A_3 + \phi\frac{\partial A_3}{\partial z}$$

$$= \frac{\partial\phi}{\partial x}A_1 + \frac{\partial\phi}{\partial y}A_2 + \frac{\partial\phi}{\partial z}A_3 + \phi\left(\frac{\partial A_1}{\partial x} + \frac{\partial A_2}{\partial y} + \frac{\partial A_3}{\partial z}\right)$$

$$= \left(\frac{\partial\phi}{\partial x}\mathbf{i} + \frac{\partial\phi}{\partial y}\mathbf{j} + \frac{\partial\phi}{\partial z}\mathbf{k}\right)\cdot(A_1\mathbf{i} + A_2\mathbf{j} + A_3\mathbf{k}) + \phi\left(\frac{\partial}{\partial x}\mathbf{i} + \frac{\partial}{\partial y}\mathbf{j} + \frac{\partial}{\partial z}\mathbf{k}\right)\cdot(A_1\mathbf{i} + A_2\mathbf{j} + A_3\mathbf{k})$$

$$= (\nabla\phi)\cdot\mathbf{A} + \phi(\nabla\cdot\mathbf{A})$$

4.19. Prove $\nabla\cdot\left(\dfrac{\mathbf{r}}{r^3}\right)=0$.

Solution

Let $\phi=r^{-3}$ and $\mathbf{A}=\mathbf{r}$ in the result of Problem 4.18(b).

Then $\nabla\cdot(r^{-3}\mathbf{r})=(\nabla r^{-3})\cdot\mathbf{r}+(r^{-3})\nabla\cdot\mathbf{r}$

$$=-3r^{-5}\mathbf{r}\cdot\mathbf{r}+3r^{-3}=0,\quad\text{using Problem 4.4.}$$

4.20. Prove $\nabla\cdot(U\nabla V-V\nabla U)=U\nabla^2 V-V\nabla^2 U$.

Solution

From Problem 4.18(b), with $\phi=U$ and $\mathbf{A}=\nabla V$,

$$\nabla\cdot(U\nabla V)=(\nabla U)\cdot(\nabla V)+U(\nabla\cdot\nabla V)=(\nabla U)\cdot(\nabla V)+U\nabla^2 V$$

Interchanging U and V yields

$$\nabla\cdot(V\nabla U)=(\nabla V)\cdot(\nabla U)+V\nabla^2 U.$$

Then subtracting,

$$\nabla\cdot(U\nabla V)-\nabla\cdot(V\nabla U)=\nabla\cdot(U\nabla V-V\nabla U)$$

$$=(\nabla U)\cdot(\nabla V)+U\nabla^2 V-[(\nabla V)\cdot(\nabla U)+V\nabla^2 U]$$

$$=U\nabla^2 V-V\nabla^2 U$$

4.21. A fluid moves so that its velocity at any point is $\mathbf{v}(x,y,z)$. Show that the loss of fluid per unit volume per unit time in a small parallelepiped having center at $P(x,y,z)$ and edges parallel to the coordinate axes and having magnitude $\triangle x$, $\triangle y$, $\triangle z$ respectively, is given approximately by div $\mathbf{v}=\nabla\cdot\mathbf{v}$.

Fig. 4-3

Solution

Referring to Fig. 4-3,

$$x\text{ component of velocity }\mathbf{v}\text{ at }P=v_1$$

$$x\text{ component of }\mathbf{v}\text{ at center of face }AFED=v_1-\frac{1}{2}\frac{\partial v_1}{\partial x}\triangle x\text{ approx.}$$

$$x\text{ component of }\mathbf{v}\text{ at center of face }GHCB=v_1+\frac{1}{2}\frac{\partial v_1}{\partial x}\triangle x\text{ approx.}$$

Then (1) volume of fluid crossing *AFED* per unit time $= \left(v_1 - \dfrac{1}{2}\dfrac{\partial v_1}{\partial x}\Delta x\right)\Delta y\Delta z,$

(2) volume of fluid crossing *GHCB* per unit time $= \left(v_1 + \dfrac{1}{2}\dfrac{\partial v_1}{\partial x}\Delta x\right)\Delta y\Delta z.$

Loss in volume per unit time in x direction $= (2) - (1) = \dfrac{\partial v_1}{\partial x}\Delta x\Delta y\Delta z.$

Similarly, loss in volume per unit time in y direction $= \dfrac{\partial v_2}{\partial y}\Delta x\Delta y\Delta z$

loss in volume per unit time in z direction $= \dfrac{\partial v_3}{\partial z}\Delta x\Delta y\Delta z.$

Then, total loss in volume per unit volume per unit time

$$= \dfrac{\left(\dfrac{\partial v_1}{\partial x} + \dfrac{\partial v_2}{\partial y} + \dfrac{\partial v_3}{\partial z}\right)\Delta x\Delta y\Delta z}{\Delta x\Delta y\Delta z} = \operatorname{div}\mathbf{v} = \nabla\cdot\mathbf{v}$$

This is true exactly only in the limit as the parallelepiped shrinks to P, i.e. as Δx, Δy, and Δz approach zero. If there is no loss of fluid anywhere, then $\nabla\cdot\mathbf{v} = 0$. This is called the *continuity equation* for an incompressible fluid. Since fluid is neither created nor destroyed at any point, it is said to have no sources or sinks. A vector such as $\mathbf{v}$ whose divergence is zero is sometimes called *solenoidal*.

4.22. Determine the constant a so that the following vector is solenoidal.

$$\mathbf{V} = (-4x - 6y + 3z)\mathbf{i} + (-2x + y - 5z)\mathbf{j} + (5x + 6y + az)\mathbf{k}$$

Solution

A vector $\mathbf{V}$ is *solenoidal* if its divergence is zero.

$$\nabla\cdot\mathbf{V} = \dfrac{\partial}{\partial x}(-4x - 6y + 3z) + \dfrac{\partial}{\partial y}(-2x + y - 5z) + \dfrac{\partial}{\partial z}(5x + 6y + az) = -4 + 1 + a = -3 + a.$$

Then $\nabla\cdot\mathbf{V} = -3 + a = 0$ when $a = 3$.

The Curl

4.23. Suppose $\mathbf{A} = x^2z^2\mathbf{i} - 2y^2z^2\mathbf{j} + xy^2z\mathbf{k}$. Find $\nabla\times\mathbf{A}$ (or curl $\mathbf{A}$) at the point $P = (1,\ -1,\ 1)$.

Solution

$$\nabla\times\mathbf{A} = \begin{vmatrix} \mathbf{i} & \mathbf{j} & \mathbf{k} \\ \dfrac{\partial}{\partial x} & \dfrac{\partial}{\partial y} & \dfrac{\partial}{\partial z} \\ x^2z^2 & -2y^2z^2 & xy^2z \end{vmatrix}$$

$$= \left[\dfrac{\partial}{\partial y}(xy^2z) - \dfrac{\partial}{\partial z}(-2y^2z^2)\right]\mathbf{i} - \left[\dfrac{\partial}{\partial x}(xy^2z) - \dfrac{\partial}{\partial z}(x^2z^2)\right]\mathbf{j}\left[\dfrac{\partial}{\partial x}(-2y^2z^2) + \dfrac{\partial}{\partial y}(x^2z^2)\right]\mathbf{k}$$

$$= (2xyz + 4yz^2)\mathbf{i} - (y^2z - 2x^2z)\mathbf{j}$$

Thus $\nabla\times\mathbf{A}(P) = 2\mathbf{i} + \mathbf{j}.$

4.24. Suppose $A = x^2 z^2 i - 2y^2 z^2 j + xy^2 z k$. Find curl curl $A = \nabla \times (\nabla \times A)$.

Solution

By the previous problem, $\nabla \times A = (2xyz + 4yz^2)i - (y^2 z - 2x^2 z)j$. Then

$$\nabla \times (\nabla \times A) = \nabla \times [(2xyz + 4y^2 z)i - (y^2 z - 2x^2 z)j]$$

$$= \begin{vmatrix} i & j & k \\ \dfrac{\partial}{\partial x} & \dfrac{\partial}{\partial y} & \dfrac{\partial}{\partial z} \\ 2xyz + 4y^2 z & -y^2 z + 2x^2 z & 0 \end{vmatrix}$$

$$= \left[\frac{\partial}{\partial y}(0) - \frac{\partial}{\partial z}(-y^2 z + 2x^2 z) \right]i - \left[\frac{\partial}{\partial x}(0) - \frac{\partial}{\partial z}(2xyz + 4y^2 z) \right]j$$

$$+ \left[\frac{\partial}{\partial x}(-y^2 z + 2x^2 z) + \frac{\partial}{\partial y}(2xyz + 4y^2 z) \right]k$$

$$= (y^2 - 2x^2)i + (2xy + 4y^2)j + (2xz - 8yz)k.$$

4.25. Let $A = A_1 i + A_2 j + A_3 k$, $B = B_1 i + B_2 j + B_3 k$. Show (a) $\nabla \times (A + B) = \nabla \times A + \nabla \times B$,
(b) $\nabla \times (\phi A) = (\nabla \phi) \times A + \phi(\nabla \times A)$.

Solution

(a)
$$\nabla \times (A + B) = \left(\frac{\partial}{\partial x}i + \frac{\partial}{\partial y}j + \frac{\partial}{\partial z}k \right) \times [(A_1 + B_1)i + (A_2 + B_2)j + (A_3 + B_3)k]$$

$$= \begin{vmatrix} i & j & k \\ \dfrac{\partial}{\partial x} & \dfrac{\partial}{\partial y} & \dfrac{\partial}{\partial z} \\ A_1 + B_1 & A_2 + B_2 & A_3 + B_3 \end{vmatrix}$$

$$= \left[\frac{\partial}{\partial y}(A_3 + B_3) - \frac{\partial}{\partial z}(A_2 + B_2) \right]i + \left[\frac{\partial}{\partial z}(A_1 + B_1) - \frac{\partial}{\partial x}(A_3 + B_3) \right]j$$

$$+ \left[\frac{\partial}{\partial x}(A_2 + B_2) - \frac{\partial}{\partial y}(A_1 + B_1) \right]k$$

$$= \left[\frac{\partial A_3}{\partial y} - \frac{\partial A_2}{\partial z} \right]i + \left[\frac{\partial A_1}{\partial z} - \frac{\partial A_3}{\partial x} \right]j + \left[\frac{\partial A_2}{\partial x} - \frac{\partial A_1}{\partial y} \right]k$$

$$+ \left[\frac{\partial B_3}{\partial y} - \frac{\partial B_2}{\partial z} \right]i + \left[\frac{\partial B_1}{\partial z} - \frac{\partial B_3}{\partial x} \right]j + \left[\frac{\partial B_2}{\partial x} - \frac{\partial B_1}{\partial y} \right]k$$

$$= \nabla \times A + \nabla \times B$$

(b) $\nabla \times (\phi A) = \nabla \times (\phi A_1 i + \phi A_2 j + \phi A_3 k)$

$$= \begin{vmatrix} i & j & k \\ \dfrac{\partial}{\partial x} & \dfrac{\partial}{\partial y} & \dfrac{\partial}{\partial z} \\ \phi A_1 & \phi A_2 & \phi A_3 \end{vmatrix}$$

$$= \left[\frac{\partial}{\partial y}(\phi A_3) - \frac{\partial}{\partial z}(\phi A_2) \right]i + \left[\frac{\partial}{\partial z}(\phi A_1) - \frac{\partial}{\partial x}(\phi A_3) \right]j + \left[\frac{\partial}{\partial x}(\phi A_2) - \frac{\partial}{\partial y}(\phi A_1) \right]k$$

$$= \left[\phi \frac{\partial A_3}{\partial y} + \frac{\partial \phi}{\partial y} A_3 - \phi \frac{\partial A_2}{\partial z} - \frac{\partial \phi}{\partial z} A_2 \right] \mathbf{i}$$

$$+ \left[\phi \frac{\partial A_1}{\partial z} + \frac{\partial \phi}{\partial z} A_1 - \phi \frac{\partial A_3}{\partial x} - \frac{\partial \phi}{\partial x} A_3 \right] \mathbf{j} + \left[\phi \frac{\partial A_2}{\partial x} + \frac{\partial \phi}{\partial x} A_2 - \phi \frac{\partial A_1}{\partial y} - \frac{\partial \phi}{\partial y} A_1 \right] \mathbf{k}$$

$$= \phi \left[\left(\frac{\partial A_3}{\partial y} - \frac{\partial A_2}{\partial z} \right) \mathbf{i} + \left(\frac{\partial A_1}{\partial z} - \frac{\partial A_3}{\partial x} \right) \mathbf{j} + \left(\frac{\partial A_2}{\partial x} - \frac{\partial A_1}{\partial y} \right) \mathbf{k} \right]$$

$$+ \left[\left(\frac{\partial \phi}{\partial y} A_3 - \frac{\partial \phi}{\partial z} A_2 \right) \mathbf{i} + \left(\frac{\partial \phi}{\partial z} A_1 - \frac{\partial \phi}{\partial x} A_3 \right) \mathbf{j} + \left(\frac{\partial \phi}{\partial x} A_2 - \frac{\partial \phi}{\partial y} A_1 \right) \mathbf{k} \right]$$

$$= \phi(\nabla \times \mathbf{A}) + \begin{vmatrix} \mathbf{i} & \mathbf{j} & \mathbf{k} \\ \frac{\partial \phi}{\partial x} & \frac{\partial \phi}{\partial y} & \frac{\partial \phi}{\partial z} \\ A_1 & A_2 & A_3 \end{vmatrix}$$

$$= \phi(\nabla \times \mathbf{A}) + (\nabla \phi) \times \mathbf{A}.$$

4.26. Suppose $\nabla \times \mathbf{A} = \mathbf{0}$. Evaluate $\nabla \cdot (\mathbf{A} \times \mathbf{r})$.

Solution

Let $\mathbf{A} = A_1 \mathbf{i} + A_2 \mathbf{j} + A_3 \mathbf{k}$, $r = x\mathbf{i} + y\mathbf{j} + z\mathbf{k}$. Then

$$\mathbf{A} \times \mathbf{r} = \begin{vmatrix} \mathbf{i} & \mathbf{j} & \mathbf{k} \\ A_1 & A_2 & A_3 \\ x & y & z \end{vmatrix}$$

$$= (zA_2 - yA_3)\mathbf{i} + (xA_3 - zA_1)\mathbf{j} + (yA_1 - xA_2)\mathbf{k}$$

and

$$\nabla \cdot (\mathbf{A} \times \mathbf{r}) = \frac{\partial}{\partial x}(zA_2 - yA_3) + \frac{\partial}{\partial y}(xA_3 - zA_1) + \frac{\partial}{\partial z}(yA_1 - xA_2)$$

$$= z\frac{\partial A_2}{\partial x} - y\frac{\partial A_3}{\partial x} + x\frac{\partial A_3}{\partial y} - z\frac{\partial A_1}{\partial y} + y\frac{\partial A_1}{\partial z} - x\frac{\partial A_2}{\partial z}$$

$$= x\left(\frac{\partial A_3}{\partial y} - \frac{\partial A_2}{\partial z} \right) + y\left(\frac{\partial A_1}{\partial z} - \frac{\partial A_3}{\partial x} \right) + z\left(\frac{\partial A_2}{\partial x} - \frac{\partial A_1}{\partial y} \right)$$

$$= [x\mathbf{i} + y\mathbf{j} + z\mathbf{k}] \cdot \left[\left(\frac{\partial A_3}{\partial y} - \frac{\partial A_2}{\partial z} \right)\mathbf{i} + \left(\frac{\partial A_1}{\partial z} - \frac{\partial A_3}{\partial x} \right)\mathbf{j} + \left(\frac{\partial A_2}{\partial x} - \frac{\partial A_1}{\partial y} \right)\mathbf{k} \right]$$

$$= \mathbf{r} \cdot (\nabla \times \mathbf{A}) = \mathbf{r} \cdot \text{curl } \mathbf{A}.$$

If $\nabla \times \mathbf{A} = \mathbf{0}$, this reduces to zero.

4.27. Prove: (a) $\nabla \times (\nabla \phi) = \mathbf{0}$ (curl grad $\phi = \mathbf{0}$), (b) $\nabla \cdot (\nabla \times \mathbf{A}) = \mathbf{0}$ (div curl $\mathbf{A} = 0$).

Solution

(a) $\quad \nabla \times (\nabla \phi) = \nabla \times \left(\frac{\partial \phi}{\partial x}\mathbf{i} + \frac{\partial \phi}{\partial y}\mathbf{j} + \frac{\partial \phi}{\partial z}\mathbf{k} \right) = \begin{vmatrix} \mathbf{i} & \mathbf{j} & \mathbf{k} \\ \frac{\partial}{\partial x} & \frac{\partial}{\partial y} & \frac{\partial}{\partial z} \\ \frac{\partial \phi}{\partial x} & \frac{\partial \phi}{\partial y} & \frac{\partial \phi}{\partial z} \end{vmatrix}$

$$= \left[\frac{\partial}{\partial y}\left(\frac{\partial \phi}{\partial z} \right) - \frac{\partial}{\partial z}\left(\frac{\partial \phi}{\partial y} \right) \right]\mathbf{i} + \left[\frac{\partial}{\partial z}\left(\frac{\partial \phi}{\partial x} \right) - \frac{\partial}{\partial x}\left(\frac{\partial \phi}{\partial z} \right) \right]\mathbf{j} + \left[\frac{\partial}{\partial x}\left(\frac{\partial \phi}{\partial y} \right) - \frac{\partial}{\partial y}\left(\frac{\partial \phi}{\partial x} \right) \right]\mathbf{k}$$

$$= \left(\frac{\partial^2 \phi}{\partial y \partial z} - \frac{\partial^2 \phi}{\partial z \partial y} \right)\mathbf{i} + \left(\frac{\partial^2 \phi}{\partial z \partial x} - \frac{\partial^2 \phi}{\partial x \partial z} \right)\mathbf{j} + \left(\frac{\partial^2 \phi}{\partial x \partial y} - \frac{\partial^2 \phi}{\partial y \partial x} \right)\mathbf{k} = \mathbf{0}$$

provided we assume that ϕ has continuous second partial derivatives so that the order of differentiation is immaterial.

(b) $\nabla \cdot (\nabla \times \mathbf{A}) = \nabla \cdot \begin{vmatrix} \mathbf{i} & \mathbf{j} & \mathbf{k} \\ \dfrac{\partial}{\partial x} & \dfrac{\partial}{\partial y} & \dfrac{\partial}{\partial z} \\ A_1 & A_2 & A_3 \end{vmatrix}$

$= \nabla \cdot \left[\left(\dfrac{\partial A_3}{\partial y} - \dfrac{\partial A_2}{\partial z} \right) \mathbf{i} + \left(\dfrac{\partial A_1}{\partial z} - \dfrac{\partial A_3}{\partial x} \right) \mathbf{j} + \left(\dfrac{\partial A_2}{\partial x} - \dfrac{\partial A_1}{\partial y} \right) \mathbf{k} \right]$

$= \dfrac{\partial}{\partial x} \left(\dfrac{\partial A_3}{\partial y} - \dfrac{\partial A_2}{\partial z} \right) + \dfrac{\partial}{\partial y} \left(\dfrac{\partial A_1}{\partial z} - \dfrac{\partial A_3}{\partial x} \right) + \dfrac{\partial}{\partial z} \left(\dfrac{\partial A_2}{\partial x} - \dfrac{\partial A_1}{\partial y} \right)$

$= \dfrac{\partial^2 A_3}{\partial x \partial y} - \dfrac{\partial^2 A_2}{\partial x \partial z} + \dfrac{\partial^2 A_1}{\partial y \partial z} - \dfrac{\partial^2 A_3}{\partial y \partial x} + \dfrac{\partial^2 A_2}{\partial z \partial x} - \dfrac{\partial^2 A_1}{\partial z \partial y} = 0$

assuming that **A** has continuous second partial derivatives.

Note the similarity between the above results and the results $(\mathbf{C} \times \mathbf{C}m) = (\mathbf{C} \times \mathbf{C})m = \mathbf{0}$, where m is a scalar and $\mathbf{C} \cdot (\mathbf{C} \times \mathbf{A}) = (\mathbf{C} \times \mathbf{C}) \cdot \mathbf{A} = 0$.

4.28. Find curl $(\mathbf{r}f(r))$ where $f(r)$ is differentiable.

Solution

$$\mathrm{curl}(\mathbf{r}f(r)) = \nabla \times (\mathbf{r}f(r))$$
$$= \nabla \times (xf(r)\mathbf{i} + yf(r)\mathbf{j} + zf(r)\mathbf{k})$$

$$= \begin{vmatrix} \mathbf{i} & \mathbf{j} & \mathbf{k} \\ \dfrac{\partial}{\partial x} & \dfrac{\partial}{\partial y} & \dfrac{\partial}{\partial z} \\ xf(r) & yf(r) & zf(r) \end{vmatrix}$$

$$= \left(z\dfrac{\partial f}{\partial y} - y\dfrac{\partial f}{\partial z} \right)\mathbf{i} + \left(x\dfrac{\partial f}{\partial z} - z\dfrac{\partial f}{\partial x} \right)\mathbf{j} + \left(y\dfrac{\partial f}{\partial x} - x\dfrac{\partial f}{\partial y} \right)\mathbf{k}$$

But $\dfrac{\partial f}{\partial x} = \left(\dfrac{\partial f}{\partial r} \right)\left(\dfrac{\partial r}{\partial x} \right) = \dfrac{\partial f}{\partial r}\dfrac{\partial}{\partial x}\left(\sqrt{x^2 + y^2 + z^2} \right) = \dfrac{f'(r)x}{\sqrt{x^2 + y^2 + x^2}} = \dfrac{f'x}{r}$.

Similarly, $\dfrac{\partial f}{\partial y} = \dfrac{f'y}{r}$ and $\dfrac{\partial f}{\partial z} = \dfrac{f'z}{r}$.

Then, the result $= \left(z\dfrac{f'y}{r} - y\dfrac{f'z}{r} \right)\mathbf{i} + \left(x\dfrac{f'z}{r} - z\dfrac{f'x}{r} \right)\mathbf{j} + \left(y\dfrac{f'x}{r} - x\dfrac{f'y}{r} \right)\mathbf{k} = \mathbf{0}$.

4.29. Prove $\nabla \times (\nabla \times \mathbf{A}) = -\nabla^2 \mathbf{A} + \nabla(\nabla \cdot \mathbf{A})$.

Solution

$$\nabla \times (\nabla \times \mathbf{A}) = \nabla \times \begin{vmatrix} \mathbf{i} & \mathbf{j} & \mathbf{k} \\ \dfrac{\partial}{\partial x} & \dfrac{\partial}{\partial y} & \dfrac{\partial}{\partial z} \\ A_1 & A_2 & A_3 \end{vmatrix}$$

$$= \nabla \times \left[\left(\dfrac{\partial A_3}{\partial y} - \dfrac{\partial A_2}{\partial z} \right)\mathbf{i} + \left(\dfrac{\partial A_1}{\partial z} - \dfrac{\partial A_3}{\partial x} \right)\mathbf{j} + \left(\dfrac{\partial A_2}{\partial x} - \dfrac{\partial A_1}{\partial y} \right)\mathbf{k} \right]$$

$$= \begin{vmatrix} \mathbf{i} & \mathbf{j} & \mathbf{k} \\ \dfrac{\partial}{\partial x} & \dfrac{\partial}{\partial y} & \dfrac{\partial}{\partial z} \\ \dfrac{\partial A_3}{\partial y} - \dfrac{\partial A_2}{\partial z} & \dfrac{\partial A_1}{\partial z} - \dfrac{\partial A_3}{\partial x} & \dfrac{\partial A_2}{\partial x} - \dfrac{\partial A_1}{\partial y} \end{vmatrix}$$

$$= \left[\dfrac{\partial}{\partial y}\left(\dfrac{\partial A_2}{\partial x} - \dfrac{\partial A_1}{\partial y} \right) - \dfrac{\partial}{\partial z}\left(\dfrac{\partial A_1}{\partial z} - \dfrac{\partial A_3}{\partial x} \right) \right]\mathbf{i}$$

$$+\left[\frac{\partial}{\partial z}\left(\frac{\partial A_3}{\partial y}-\frac{\partial A_2}{\partial z}\right)-\frac{\partial}{\partial x}\left(\frac{\partial A_2}{\partial x}-\frac{\partial A_1}{\partial y}\right)\right]\mathbf{j}$$

$$+\left[\frac{\partial}{\partial x}\left(\frac{\partial A_1}{\partial z}-\frac{\partial A_3}{\partial x}\right)-\frac{\partial}{\partial y}\left(\frac{\partial A_3}{\partial y}-\frac{\partial A_2}{\partial z}\right)\right]\mathbf{k}$$

$$=\left(-\frac{\partial^2 A_1}{\partial y^2}-\frac{\partial^2 A_1}{\partial z^2}\right)\mathbf{i}+\left(-\frac{\partial^2 A_2}{\partial z^2}-\frac{\partial^2 A_2}{\partial x^2}\right)\mathbf{j}+\left(-\frac{\partial^2 A_3}{\partial x^2}-\frac{\partial^2 A_3}{\partial y^2}\right)\mathbf{k}$$

$$+\left(\frac{\partial^2 A_2}{\partial y\partial x}+\frac{\partial^2 A_3}{\partial z\partial x}\right)\mathbf{i}+\left(\frac{\partial^2 A_3}{\partial z\partial y}+\frac{\partial^2 A_1}{\partial x\partial y}\right)\mathbf{j}+\left(\frac{\partial^2 A_1}{\partial x\partial z}+\frac{\partial^2 A_2}{\partial y\partial z}\right)\mathbf{k}$$

$$=\left(-\frac{\partial^2 A_1}{\partial x^2}-\frac{\partial^2 A_1}{\partial y^2}-\frac{\partial^2 A_1}{\partial z^2}\right)\mathbf{i}+\left(-\frac{\partial^2 A_2}{\partial x^2}-\frac{\partial^2 A_2}{\partial y^2}-\frac{\partial^2 A_2}{\partial z^2}\right)\mathbf{j}+\left(-\frac{\partial^2 A_3}{\partial x^2}-\frac{\partial^2 A_3}{\partial y^2}-\frac{\partial^2 A_3}{\partial z^2}\right)\mathbf{k}$$

$$+\left(\frac{\partial^2 A_1}{\partial x^2}+\frac{\partial^2 A_2}{\partial y\partial x}+\frac{\partial^2 A_3}{\partial z\partial x}\right)\mathbf{i}+\left(\frac{\partial^2 A_1}{\partial x\partial y}+\frac{\partial^2 A_2}{\partial y^2}+\frac{\partial^2 A_3}{\partial z\partial y}\right)\mathbf{j}+\left(\frac{\partial^2 A_1}{\partial x\partial z}+\frac{\partial^2 A_2}{\partial y\partial z}+\frac{\partial^2 A_3}{\partial z^2}\right)\mathbf{k}$$

$$=-\left(\frac{\partial^2}{\partial x^2}+\frac{\partial^2}{\partial y^2}+\frac{\partial^2}{\partial z^2}\right)(A_1\mathbf{i}+A_2\mathbf{j}+A_3\mathbf{k})$$

$$+\mathbf{i}\frac{\partial}{\partial x}\left(\frac{\partial A_1}{\partial x}+\frac{\partial A_2}{\partial y}+\frac{\partial A_3}{\partial z}\right)+\mathbf{j}\frac{\partial}{\partial y}\left(\frac{\partial A_1}{\partial x}+\frac{\partial A_2}{\partial y}+\frac{\partial A_3}{\partial z}\right)+\mathbf{k}\frac{\partial}{\partial z}\left(\frac{\partial A_1}{\partial x}+\frac{\partial A_2}{\partial y}+\frac{\partial A_3}{\partial z}\right)$$

$$=-\nabla^2\mathbf{A}+\nabla\left(\frac{\partial A_1}{\partial x}+\frac{\partial A_2}{\partial y}+\frac{\partial A_3}{\partial z}\right)=-\nabla^2\mathbf{A}+\nabla(\nabla\cdot\mathbf{A})$$

If desired, the labor of writing can be shortened in this as well as other derivations by writing only the **i** components since the others can be obtained by symmetry.

The result can also be established formally as follows. From Problem 47(a), Chapter 2,

$$\mathbf{A}\times(\mathbf{B}\times\mathbf{C})=\mathbf{B}(\mathbf{A}\cdot\mathbf{C})-(\mathbf{A}\cdot\mathbf{B})\mathbf{C} \tag{1}$$

Placing $\mathbf{A}=\mathbf{B}=\nabla$ and $\mathbf{C}=\mathbf{F}$,

$$\nabla\times(\nabla\times\mathbf{F})=\nabla(\nabla\cdot\mathbf{F})-(\nabla\cdot\nabla)\mathbf{F}=\nabla(\nabla\cdot\mathbf{F})-\nabla^2\mathbf{F}$$

Note that the formula (1) must be written so that the operators **A** and **B** precede the operand **C**, otherwise the formalism fails to apply.

4.30. Suppose $\mathbf{v}=\boldsymbol{\omega}\times\mathbf{r}$. Prove $\boldsymbol{\omega}=\frac{1}{2}$ curl **v** where $\boldsymbol{\omega}$ is a constant vector.

Solution

$$\text{curl } \mathbf{v}=\nabla\times\mathbf{v}=\nabla\times(\boldsymbol{\omega}\times\mathbf{r})=\nabla\times\begin{vmatrix}\mathbf{i}&\mathbf{j}&\mathbf{k}\\\omega_1&\omega_2&\omega_3\\x&y&z\end{vmatrix}$$

$$=\nabla\times[(\omega_2 z-\omega_3 y)\mathbf{i}+(\omega_3 x-\omega_1 z)\mathbf{j}+(\omega_1 y-\omega_2 x)\mathbf{k}]$$

$$=\begin{vmatrix}\mathbf{i}&\mathbf{j}&\mathbf{k}\\\dfrac{\partial}{\partial x}&\dfrac{\partial}{\partial y}&\dfrac{\partial}{\partial z}\\\omega_2 z-\omega_3 y&\omega_3 x-\omega_1 z&\omega_1 y-\omega_2 x\end{vmatrix}=2(\omega_1\mathbf{i}+\omega_2\mathbf{j}+\omega_3\mathbf{k})=2\boldsymbol{\omega}.$$

Then $\boldsymbol{\omega}=\frac{1}{2}\nabla\times\mathbf{v}=\frac{1}{2}$ curl **v**.

This problem indicates that the curl of a vector field has something to do with rotational properties of the field. This is confirmed in Chapter 6. If the field **F** is that due to a moving fluid, for example, then a paddle wheel placed at various points in the field would tend to rotate in regions where curl $\mathbf{F}\neq\mathbf{0}$, while if curl $\mathbf{F}=0$ in the region, there would be no rotation and the field **F** is then called *irrotational*. A field that is not irrotational is sometimes called a *vortex field*.

4.31. Suppose $\nabla \cdot \mathbf{E} = 0$, $\nabla \cdot \mathbf{H} = 0$, $\nabla \times \mathbf{E} = -\dfrac{\partial \mathbf{H}}{\partial t}$, $\nabla \times \mathbf{H} = \dfrac{\partial \mathbf{E}}{\partial t}$. Show that $\mathbf{E}$ and $\mathbf{H}$ satisfy $\nabla^2 u = \dfrac{\partial^2 u}{\partial t^2}$.

Solution

$$\nabla \times (\nabla \times \mathbf{E}) = \nabla \times \left(-\frac{\partial \mathbf{H}}{\partial t}\right) = -\frac{\partial}{\partial t}(\nabla \times \mathbf{H}) = -\frac{\partial}{\partial t}\left(\frac{\partial \mathbf{E}}{\partial t}\right) = -\frac{\partial^2 \mathbf{E}}{\partial t^2}$$

By Problem 4.29, $\nabla \times (\nabla \times \mathbf{E}) = -\nabla^2 \mathbf{E} + \nabla(\nabla \cdot \mathbf{E}) = -\nabla^2 \mathbf{E}$.　Then $\nabla^2 \mathbf{E} = \dfrac{\partial^2 \mathbf{E}}{\partial t^2}$.

Similarly, $\nabla \times (\nabla \times \mathbf{H}) = \nabla \times \left(\dfrac{\partial \mathbf{E}}{\partial t}\right) = \dfrac{\partial}{\partial t}(\nabla \times \mathbf{E}) = \dfrac{\partial}{\partial t}\left(-\dfrac{\partial \mathbf{H}}{\partial t}\right) = -\dfrac{\partial^2 \mathbf{H}}{\partial t^2}$.

But $\nabla \times (\nabla \times \mathbf{H}) = -\nabla^2 \mathbf{H} + \nabla(\nabla \cdot \mathbf{H}) = -\nabla^2 \mathbf{H}$.　Then $\nabla^2 \mathbf{H} = \dfrac{\partial^2 \mathbf{H}}{\partial t^2}$.

The given equations are related to *Maxwell's equations of electromagnetic theory*. The equation

$$\frac{\partial^2 u}{\partial x^2} + \frac{\partial^2 u}{\partial y^2} + \frac{\partial^2 u}{\partial z^2} = \frac{\partial^2 u}{\partial t^2}$$

is called the *wave equation*.

Miscellaneous Problems

4.32. A vector $\mathbf{V}$ is called *irrotational* if curl $\mathbf{V} = \mathbf{0}$. (a) Find constants a, b, and c so that

$$\mathbf{V} = (-4x - 3y + az)\mathbf{i} + (bx + 3y + 5z)\mathbf{j} + (4x + cy + 3z)\mathbf{k}$$

is irrotational. (b) Show that $\mathbf{V}$ can be expressed as the gradient of a scalar function.

Solution

(a)　curl $\mathbf{V} = \nabla \times \mathbf{V}$

$$\nabla \times \mathbf{V} = \begin{vmatrix} \mathbf{i} & \mathbf{j} & \mathbf{k} \\ \dfrac{\partial}{\partial x} & \dfrac{\partial}{\partial y} & \dfrac{\partial}{\partial z} \\ -4x-3y+az & bx+3y+5z & 4x+cy+3z \end{vmatrix}$$

$$= \begin{vmatrix} \dfrac{\partial}{\partial y} & \dfrac{\partial}{\partial z} \\ b+3y+5z & 4x+cy+3z \end{vmatrix}\mathbf{i} - \begin{vmatrix} \dfrac{\partial}{\partial x} & \dfrac{\partial}{\partial z} \\ -4x-3y+az & 4x+cy+3z \end{vmatrix}\mathbf{j}$$

$$+ \begin{vmatrix} \dfrac{\partial}{\partial x} & \dfrac{\partial}{\partial y} \\ -4x-3y+az & bx+3y+5z \end{vmatrix}\mathbf{k}$$

$$= (c-5)\mathbf{i} - (4-a)\mathbf{j} + (b+3)\mathbf{k}.$$

This equals the zero vector when $a = 4$, $b = -3$, and $c = 5$. So

$$\mathbf{V} = (-4x - 3y + 4z)\mathbf{i} + (-3x + 3y + 5z)\mathbf{j} + (4x + 5y + 3z)\mathbf{k}.$$

(b)　Assume $\mathbf{V} = \nabla\phi = \dfrac{\partial \phi}{\partial x}\mathbf{i} + \dfrac{\partial \phi}{\partial y}\mathbf{j} + \dfrac{\partial \phi}{\partial z}\mathbf{k}$. Then

$$\frac{\partial \phi}{\partial x} = -4x - 3y + 4z \tag{1}$$

$$\frac{\partial \phi}{\partial y} = -3x + 3y + 5z \tag{2}$$

$$\frac{\partial \phi}{\partial z} = 4x + 5y + 3z \tag{3}$$

Integrating (1) partially with respect to x keeping y and z constant, we obtain

$$\phi = -2x^2 - 3xy + 4xz + f(y, z) \tag{4}$$

where $f(y, z)$ is an arbitrary function of y and z. Similarly, we obtain from (2) and (3)

$$\phi = -3xy + \frac{3}{2}y^2 + 5yz + g(x, z) \tag{5}$$

and

$$\phi = 4xz + 5yz + \frac{3}{2}z^2 + h(x, y). \tag{6}$$

Comparison of (4), (5), and (6) shows that there will be a common value of ϕ if we choose

$$f(y, z) = \frac{3}{2}y^2 + 5yz + \frac{3}{2}z^2, \quad g(x, z) = -2x^2 + 4xz + \frac{3}{2}z^2, \quad h(x, y) = -2x^2 - 3xy + \frac{3}{2}y^2$$

so that

$$\phi = -2x^2 + \frac{3}{2}y^2 + \frac{3}{2}z^2 - 3xy + 4xz + 5yz$$

Note that we can add any constant to ϕ. In general, if $\nabla \times \mathbf{V} = \mathbf{0}$, then we can find ϕ so that $\mathbf{V} = \nabla\phi$.

A vector field $\mathbf{V}$, which can be obtained from a scalar field ϕ, so that $\mathbf{V} = \nabla\phi$ is called a *conservative vector field* and ϕ is called the *scalar potential*. Note conversely that, if $\mathbf{V} = \nabla\phi$, then $\nabla \times \mathbf{V} = \mathbf{0}$ (see Problem 4.27a).

4.33. Show that if $\phi(x, y, z)$ is any solution of Laplace's equation, then $\nabla\phi$ is a vector that is both solenoidal and irrotational.

Solution

By hypothesis, ϕ satisfies Laplace's equation $\nabla^2\phi = 0$, that is, $\nabla \cdot (\nabla\phi) = 0$. Then $\nabla\phi$ is solenoidal (see Problems 4.21 and 4.22).

From Problem 4.27a, $\nabla \times (\nabla\phi) = \mathbf{0}$, so that $\nabla\phi$ is also irrotational.

4.34. Give a possible definition of grad $\mathbf{B}$.

Solution

Assume $\mathbf{B} = B_1\mathbf{i} + B_2\mathbf{j} + B_3\mathbf{k}$. Formally, we can define grad $\mathbf{B}$ as

$$\begin{aligned}
\nabla\mathbf{B} &= \left(\frac{\partial}{\partial x}\mathbf{i} + \frac{\partial}{\partial y}\mathbf{j} + \frac{\partial}{\partial z}\mathbf{k}\right)(B_1\mathbf{i} + B_2\mathbf{j} + B_3\mathbf{k}) \\
&= \frac{\partial B_1}{\partial x}\mathbf{ii} + \frac{\partial B_2}{\partial x}\mathbf{ij} + \frac{\partial B_3}{\partial x}\mathbf{ik} \\
&\quad + \frac{\partial B_1}{\partial y}\mathbf{ji} + \frac{\partial B_2}{\partial y}\mathbf{jj} + \frac{\partial B_3}{\partial y}\mathbf{jk} \\
&\quad + \frac{\partial B_1}{\partial z}\mathbf{ki} + \frac{\partial B_2}{\partial z}\mathbf{kj} + \frac{\partial B_3}{\partial z}\mathbf{kk}
\end{aligned}$$

The quantities $\mathbf{ii}$, $\mathbf{ij}$, and so on, are called *unit dyads*. (Note that $\mathbf{ij}$, for example, is not the same as $\mathbf{ji}$.) A quantity of the form

$$a_{11}\mathbf{ii} + a_{12}\mathbf{ij} + a_{13}\mathbf{ik} + a_{21}\mathbf{ji} + a_{22}\mathbf{jj} + a_{23}\mathbf{jk} + a_{31}\mathbf{ki} + a_{32}\mathbf{kj} + a_{33}\mathbf{kk}$$

is called a *dyadic* and the coefficients $a_{11}, a_{12}, \ldots$ are its *components*. An array of these nine components in the form

$$\begin{bmatrix} a_{11} & a_{12} & a_{13} \\ a_{21} & a_{22} & a_{23} \\ a_{31} & a_{32} & a_{33} \end{bmatrix}$$

is called a 3 by 3 *matrix*. A dyadic is a generalization of a vector. Still further generalization leads to *triadics*, which are quantities consisting of 27 terms of the form $a_{111}\mathbf{iii} + a_{211}\mathbf{jii} + \cdots$. A study of how the components of a dyadic or triadic transform from one system of coordinates to another leads to the subject of *tensor analysis*, which is taken up in Chapter 8.

4.35. Let a vector $\mathbf{A}$ be defined by $\mathbf{A} = A_1\mathbf{i} + A_2\mathbf{j} + A_3\mathbf{k}$ and a dyadic $\boldsymbol{\Phi}$ by

$$\boldsymbol{\Phi} = a_{11}\mathbf{ii} + a_{12}\mathbf{ij} + a_{13}\mathbf{ik} + a_{21}\mathbf{ji} + a_{22}\mathbf{jj} + a_{23}\mathbf{jk} + a_{31}\mathbf{ki} + a_{32}\mathbf{kj} + a_{33}\mathbf{kk}$$

Give a possible definition of $\mathbf{A} \cdot \boldsymbol{\Phi}$.

Solution

Formally, assuming the distributive law to hold,

$$\mathbf{A} \cdot \boldsymbol{\Phi} = (A_1\mathbf{i} + A_2\mathbf{j} + A_3\mathbf{k}) \cdot \boldsymbol{\Phi} = A_1\mathbf{i} \cdot \boldsymbol{\Phi} + A_2\mathbf{j} \cdot \boldsymbol{\Phi} + A_3\mathbf{k} \cdot \boldsymbol{\Phi}$$

As an example, consider $\mathbf{i} \cdot \boldsymbol{\Phi}$. This product is formed by taking the dot product of $\mathbf{i}$ with each term of $\boldsymbol{\Phi}$ and adding results. Typical examples are $\mathbf{i} \cdot a_{11}\mathbf{ii}$, $\mathbf{i} \cdot a_{12}\mathbf{ij}$, $\mathbf{i} \cdot a_{21}\mathbf{ji}$, $\mathbf{i} \cdot a_{32}\mathbf{kj}$, and so on. If we give meaning to these as follows

$$\mathbf{i} \cdot a_{11}\mathbf{ii} = a_{11}(\mathbf{i} \cdot \mathbf{i})\mathbf{i} = a_{11}\mathbf{i} \quad \text{since } \mathbf{i} \cdot \mathbf{i} = 1$$
$$\mathbf{i} \cdot a_{12}\mathbf{ij} = a_{12}(\mathbf{i} \cdot \mathbf{i})\mathbf{j} = a_{12}\mathbf{j} \quad \text{since } \mathbf{i} \cdot \mathbf{i} = 1$$
$$\mathbf{i} \cdot a_{21}\mathbf{ji} = a_{21}(\mathbf{i} \cdot \mathbf{j})\mathbf{i} = \mathbf{0} \quad \text{since } \mathbf{i} \cdot \mathbf{j} = 0$$
$$\mathbf{i} \cdot a_{32}\mathbf{kj} = a_{32}(\mathbf{i} \cdot \mathbf{k})\mathbf{j} = \mathbf{0} \quad \text{since } \mathbf{i} \cdot \mathbf{k} = 0$$

and give analogous interpretation to the terms of $\mathbf{j} \cdot \boldsymbol{\Phi}$ and $\mathbf{k} \cdot \boldsymbol{\Phi}$, then

$$\mathbf{A} \cdot \boldsymbol{\Phi} = A_1(a_{11}\mathbf{i} + a_{12}\mathbf{j} + a_{13}\mathbf{k}) + A_2(a_{21}\mathbf{i} + a_{22}\mathbf{j} + a_{23}\mathbf{k}) + A_3(a_{31}\mathbf{i} + a_{32}\mathbf{j} + a_{33}\mathbf{k})$$
$$= (A_1a_{11} + A_2a_{21} + A_3a_{31})\mathbf{i} + (A_1a_{12} + A_2a_{22} + A_3a_{32})\mathbf{j} + (A_1a_{13} + A_2a_{23} + A_3a_{33})\mathbf{k}$$

which is a vector.

4.36. (a) Interpret the symbol $\mathbf{A} \cdot \nabla$. (b) Give a possible meaning to $(\mathbf{A} \cdot \nabla)\mathbf{B}$. (c) Is it possible to write this as $\mathbf{A} \cdot \nabla\mathbf{B}$ without ambiguity?

Solution

(a)　Let $\mathbf{A} = A_1\mathbf{i} + A_2\mathbf{j} + A_3\mathbf{k}$. Then, formally,

$$\mathbf{A} \cdot \nabla = (A_1\mathbf{i} + A_2\mathbf{j} + A_3\mathbf{k}) \cdot \left(\frac{\partial}{\partial x}\mathbf{i} + \frac{\partial}{\partial y}\mathbf{j} + \frac{\partial}{\partial z}\mathbf{k}\right)$$

$$= A_1\frac{\partial}{\partial x} + A_2\frac{\partial}{\partial y} + A_3\frac{\partial}{\partial z}$$

is an operator. For example,

$$(\mathbf{A} \cdot \nabla)\phi = \left(A_1\frac{\partial}{\partial x} + A_2\frac{\partial}{\partial y} + A_3\frac{\partial}{\partial z}\right)\phi = A_1\frac{\partial\phi}{\partial x} + A_2\frac{\partial\phi}{\partial y} + A_3\frac{\partial\phi}{\partial z}$$

Note that this is the same as $\mathbf{A} \cdot \nabla\phi$.

(b)　Formally, using (a) with ϕ replaced by $\mathbf{B} = B_1\mathbf{i} + B_2\mathbf{j} + B_3\mathbf{k}$,

$$(\mathbf{A} \cdot \nabla)\mathbf{B} = \left(A_1\frac{\partial}{\partial x} + A_2\frac{\partial}{\partial y} + A_3\frac{\partial}{\partial z}\right)\mathbf{B} = A_1\frac{\partial\mathbf{B}}{\partial x} + A_2\frac{\partial\mathbf{B}}{\partial y} + A_3\frac{\partial\mathbf{B}}{\partial z}$$

$$= \left(A_1\frac{\partial B_1}{\partial x} + A_2\frac{\partial B_1}{\partial y} + A_3\frac{\partial B_1}{\partial z}\right)\mathbf{i} + \left(A_1\frac{\partial B_2}{\partial x} + A_2\frac{\partial B_2}{\partial y} + A_3\frac{\partial B_2}{\partial z}\right)\mathbf{j}$$

$$+ \left(A_1\frac{\partial B_3}{\partial x} + A_2\frac{\partial B_3}{\partial y} + A_3\frac{\partial B_3}{\partial z}\right)\mathbf{k}$$

(c)   Use the interpretation of $\nabla \mathbf{B}$ as given in Problem 4.34. Then, according to the symbolism established in Problem 4.35,

$$\mathbf{A} \cdot \nabla \mathbf{B} = (A_1 \mathbf{i} + A_2 \mathbf{j} + A_3 \mathbf{k}) \cdot \nabla \mathbf{B} = A_1 \mathbf{i} \cdot \nabla \mathbf{B} + A_2 \mathbf{j} \cdot \nabla \mathbf{B} + A_3 \mathbf{k} \cdot \nabla \mathbf{B}$$

$$= A_1 \left(\frac{\partial B_1}{\partial x} \mathbf{i} + \frac{\partial B_2}{\partial x} \mathbf{j} + \frac{\partial B_3}{\partial x} \mathbf{k} \right) + A_2 \left(\frac{\partial B_1}{\partial y} \mathbf{i} + \frac{\partial B_2}{\partial y} \mathbf{j} + \frac{\partial B_3}{\partial y} \mathbf{k} \right) + A_3 \left(\frac{\partial B_1}{\partial z} \mathbf{i} + \frac{\partial B_2}{\partial z} \mathbf{j} + \frac{\partial B_3}{\partial z} \mathbf{k} \right)$$

which gives the same result as that given in part (b). It follows that $(\mathbf{A} \cdot \nabla)\mathbf{B} = \mathbf{A} \cdot \nabla \mathbf{B}$ without ambiguity provided the concept of dyadics is introduced with properties as indicated.

4.37.   Suppose $\mathbf{A} = 2yz\mathbf{i} - x^2 y\mathbf{j} + xz^2 \mathbf{k}$, $\mathbf{B} = x^2 \mathbf{i} + yz\mathbf{j} - xy\mathbf{k}$,   and $\phi = 2x^2 yz^3$.   Find   (a) $(\mathbf{A} \cdot \nabla)\phi$, (b) $\mathbf{A} \cdot \nabla\phi$,   (c) $(\mathbf{B} \cdot \nabla)\mathbf{A}$,   (d) $(\mathbf{A} \times \nabla)\phi$,   (e) $\mathbf{A} \times \nabla\phi$.

Solution

(a) $(\mathbf{A} \cdot \nabla)\phi = \left[(2yz\mathbf{i} - x^2 y\mathbf{j} + xz^2 \mathbf{k}) \cdot \left(\frac{\partial}{\partial x}\mathbf{i} + \frac{\partial}{\partial y}\mathbf{j} + \frac{\partial}{\partial z}\mathbf{k} \right) \right]\phi$

$$= \left(2yz\frac{\partial}{\partial x} - x^2 y\frac{\partial}{\partial y} + xz^2\frac{\partial}{\partial z} \right)(2x^2 yz^3)$$

$$= 2yz\frac{\partial}{\partial x}(2x^2 yz^3) - x^2 y\frac{\partial}{\partial y}(2x^2 yz^3) + xz^2\frac{\partial}{\partial z}(2x^2 yz^3)$$

$$= (2yz)(4xyz^3) - (x^2 y)(2x^2 z^3) + (xz^2)(6x^2 yz^2)$$

$$= 8xy^2 z^4 - 2x^4 yz^3 + 6x^3 yz^4$$

(b) $\mathbf{A} \cdot \nabla\phi = (2yz\mathbf{i} - x^2 y\mathbf{j} + xz^2 \mathbf{k}) \cdot \left(\frac{\partial\phi}{\partial x}\mathbf{i} + \frac{\partial\phi}{\partial y}\mathbf{j} + \frac{\partial\phi}{\partial z}\mathbf{k} \right)$

$$= (2yz\mathbf{i} - x^2 y\mathbf{j} + xz^2 \mathbf{k}) \cdot (4xyz^3 \mathbf{i} + 2x^2 z^3 \mathbf{j} + 6x^2 yz^2 \mathbf{k})$$

$$= 8xy^2 z^4 - 2x^4 yz^3 + 6x^3 yz^4$$

Comparison with (a) illustrates the result $(\mathbf{A} \cdot \nabla)\phi = \mathbf{A} \cdot \nabla\phi$.

(c) $(\mathbf{B} \cdot \nabla)\mathbf{A} = \left[(x^2 \mathbf{i} + yz\mathbf{j} - xy\mathbf{k}) \cdot \left(\frac{\partial}{\partial x}\mathbf{i} + \frac{\partial}{\partial y}\mathbf{j} + \frac{\partial}{\partial z}\mathbf{k} \right) \right]\mathbf{A}$

$$= \left(x^2\frac{\partial}{\partial x} + yz\frac{\partial}{\partial y} - xy\frac{\partial}{\partial z} \right)\mathbf{A} = x^2\frac{\partial \mathbf{A}}{\partial x} + yz\frac{\partial \mathbf{A}}{\partial y} - xy\frac{\partial \mathbf{A}}{\partial z}$$

$$= x^2(-2xy\mathbf{j} + z^2 \mathbf{k}) + yz(2z\mathbf{i} - x^2 \mathbf{j}) - xy(2yi + 2xz\mathbf{k})$$

$$= (2yz^2 - 2xy^2)\mathbf{i} - (2x^3 y + x^2 yz)\mathbf{j} + (x^2 z^2 - 2x^2 yz)\mathbf{k}$$

For comparison of this with $\mathbf{B} \cdot \nabla\mathbf{A}$, see Problem 4.36(c).

(d) $(\mathbf{A} \times \nabla)\phi = \left[(2yz\mathbf{i} - x^2 y\mathbf{j} + xz^2 \mathbf{k}) \times \left(\frac{\partial}{\partial x}\mathbf{i} + \frac{\partial}{\partial y}\mathbf{j} + \frac{\partial}{\partial z}\mathbf{k} \right) \right]\phi$

$$= \begin{vmatrix} \mathbf{i} & \mathbf{j} & \mathbf{k} \\ 2yz & -x^2 y & xz^2 \\ \dfrac{\partial}{\partial x} & \dfrac{\partial}{\partial y} & \dfrac{\partial}{\partial z} \end{vmatrix} \phi$$

$$= \left[\mathbf{i}\left(-x^2 y\frac{\partial}{\partial z} - xz^2\frac{\partial}{\partial y} \right) + \mathbf{j}\left(xz^2\frac{\partial}{\partial x} - 2yz\frac{\partial}{\partial z} \right) + \mathbf{k}\left(2yz\frac{\partial}{\partial y} + x^2 y\frac{\partial}{\partial x} \right) \right]\phi$$

$$= -\left(x^2 y\frac{\partial\phi}{\partial z} + xz^2\frac{\partial\phi}{\partial y} \right)\mathbf{i} + \left(xz^2\frac{\partial\phi}{\partial x} - 2yz\frac{\partial\phi}{\partial z} \right)\mathbf{j} + \left(2yz\frac{\partial\phi}{\partial y} + x^2 y\frac{\partial\phi}{\partial x} \right)\mathbf{k}$$

$$= -(6x^4 y^2 z^2 + 2x^3 z^5)\mathbf{i} + (4x^2 yz^5 - 12x^2 y^2 z^3)\mathbf{j} + (4x^2 yz^4 + 4x^3 y^2 z^3)\mathbf{k}$$

(e) $\mathbf{A} \times \nabla\phi = (2yz\mathbf{i} - x^2y\mathbf{j} + xz^2\mathbf{k}) \times \left(\dfrac{\partial\phi}{\partial x}\mathbf{i} + \dfrac{\partial\phi}{\partial y}\mathbf{j} + \dfrac{\partial\phi}{\partial z}\mathbf{k}\right)$

$$= \begin{vmatrix} \mathbf{i} & \mathbf{j} & \mathbf{k} \\ 2yz & -x^2y & xz^2 \\ \dfrac{\partial\phi}{\partial x} & \dfrac{\partial\phi}{\partial y} & \dfrac{\partial\phi}{\partial z} \end{vmatrix}.$$

$$= \left(-x^2y\frac{\partial\phi}{\partial z} - xz^2\frac{\partial\phi}{\partial y}\right)\mathbf{i} + \left(xz^2\frac{\partial\phi}{\partial x} - 2yz\frac{\partial\phi}{\partial z}\right)\mathbf{j} + \left(2yz\frac{\partial\phi}{\partial y} + x^2y\frac{\partial\phi}{\partial x}\right)\mathbf{k}$$

$$= -(6x^4y^2z^2 + 2x^3z^5)\mathbf{i} + (4x^2yz^5 - 12x^2y^2z^3)\mathbf{j} + (4x^2yz^4 + 4x^3y^2z^3)\mathbf{k}$$

Comparison with (d) illustrates the result $(\mathbf{A} \times \nabla)\phi = \mathbf{A} \times \nabla\phi$.

Invariance

4.38. Two rectangular xyz and $x'y'z'$ coordinate systems having the same origin are rotated with respect to each other. Derive the transformation equations between the coordinates of a point in the two systems.

Solution

Let $\mathbf{r}$ and $\mathbf{r}'$ be the position vectors of any point P in the two systems (see Fig. 4-1 on page 72). Then, since $\mathbf{r} = \mathbf{r}'$,

$$x'\mathbf{i}' + y'\mathbf{j}' + z'\mathbf{k}' = x\mathbf{i} + y\mathbf{j} + z\mathbf{k} \tag{1}$$

Now, for any vector $\mathbf{A}$, we have (Problem 4.20, Chapter 2),

$$\mathbf{A} = (\mathbf{A}\cdot\mathbf{i}')\mathbf{i}' + (\mathbf{A}\cdot\mathbf{j}')\mathbf{j}' + (\mathbf{A}\cdot\mathbf{k}')\mathbf{k}'$$

Then, letting $\mathbf{A} = \mathbf{i}$, $\mathbf{j}$, and $\mathbf{k}$ in succession,

$$\begin{cases} \mathbf{i} = (\mathbf{i}\cdot\mathbf{i}')\mathbf{i}' + (\mathbf{i}\cdot\mathbf{j}')\mathbf{j}' + (\mathbf{i}\cdot\mathbf{k}')\mathbf{k}' = l_{11}\mathbf{i}' + l_{21}\mathbf{j}' + l_{31}\mathbf{k}' \\ \mathbf{j} = (\mathbf{j}\cdot\mathbf{i}')\mathbf{i}' + (\mathbf{j}\cdot\mathbf{j}')\mathbf{j}' + (\mathbf{j}\cdot\mathbf{k}')\mathbf{k}' = l_{12}\mathbf{i}' + l_{22}\mathbf{j}' + l_{32}\mathbf{k}' \\ \mathbf{k} = (\mathbf{k}\cdot\mathbf{i}')\mathbf{i}' + (\mathbf{k}\cdot\mathbf{j}')\mathbf{j}' + (\mathbf{k}\cdot\mathbf{k}')\mathbf{k}' = l_{13}\mathbf{i}' + l_{23}\mathbf{j}' + l_{33}\mathbf{k}' \end{cases} \tag{2}$$

Substituting equations (2) into (1) and equating coefficients of $\mathbf{i}'$, $\mathbf{j}'$, and $\mathbf{k}'$, we find

$$x' = l_{11}x + l_{12}y + l_{13}z, \quad y' = l_{21}x + l_{22}y + l_{23}z, \quad z' = l_{31}x + l_{32}y + l_{33}z \tag{3}$$

the required transformation equations.

4.39. Prove

$$\mathbf{i}' = l_{11}\mathbf{i} + l_{12}\mathbf{j} + l_{13}\mathbf{k}$$
$$\mathbf{j}' = l_{21}\mathbf{i} + l_{22}\mathbf{j} + l_{23}\mathbf{k}$$
$$\mathbf{k}' = l_{31}\mathbf{i} + l_{32}\mathbf{j} + l_{33}\mathbf{k}$$

Solution

For any vector $\mathbf{A}$, we have $\mathbf{A} = (\mathbf{A}\cdot\mathbf{i})\mathbf{i} + (\mathbf{A}\cdot\mathbf{j})\mathbf{j} + (\mathbf{A}\cdot\mathbf{k})\mathbf{k}$.
Then, letting $\mathbf{A} = \mathbf{i}'$, $\mathbf{j}'$, and $\mathbf{k}'$ in succession,

$$\mathbf{i}' = (\mathbf{i}'\cdot\mathbf{i})\mathbf{i} + (\mathbf{i}'\cdot\mathbf{j})\mathbf{j} + (\mathbf{i}'\cdot\mathbf{k})\mathbf{k} = l_{11}\mathbf{i} + l_{12}\mathbf{j} + l_{13}\mathbf{k}$$
$$\mathbf{j}' = (\mathbf{j}'\cdot\mathbf{i})\mathbf{i} + (\mathbf{j}'\cdot\mathbf{j})\mathbf{j} + (\mathbf{j}'\cdot\mathbf{k})\mathbf{k} = l_{21}\mathbf{i} + l_{22}\mathbf{j} + l_{23}\mathbf{k}$$
$$\mathbf{k}' = (\mathbf{k}'\cdot\mathbf{i})\mathbf{i} + (\mathbf{k}'\cdot\mathbf{j})\mathbf{j} + (\mathbf{k}'\cdot\mathbf{k})\mathbf{k} = l_{31}\mathbf{i} + l_{32}\mathbf{j} + l_{33}\mathbf{k}$$

4.40. Prove that $\sum_{p=1}^{3} l_{pm}l_{pn} = 1$ if $m = n$, and 0 if $m \neq n$, where m and n can assume any of the values 1, 2, or 3.

Solution

From equation (2) of Problem 4.38,

$$\mathbf{i} \cdot \mathbf{i} = 1 = (l_{11}\mathbf{i'} + l_{21}\mathbf{j'} + l_{31}\mathbf{k'}) \cdot (l_{11}\mathbf{i'} + l_{21}\mathbf{j'} + l_{31}\mathbf{k'}) = l_{11}^2 + l_{21}^2 + l_{31}^2$$
$$\mathbf{i} \cdot \mathbf{j} = 0 = (l_{11}\mathbf{i'} + l_{21}\mathbf{j'} + l_{31}\mathbf{k'}) \cdot (l_{12}\mathbf{i'} + l_{22}\mathbf{j'} + l_{32}\mathbf{k'}) = l_{11}l_{12} + l_{21}l_{22} + l_{31}l_{32}$$
$$\mathbf{i} \cdot \mathbf{k} = 0 = (l_{11}\mathbf{i'} + l_{21}\mathbf{j'} + l_{31}\mathbf{k'}) \cdot (l_{13}\mathbf{i'} + l_{23}\mathbf{j'} + l_{33}\mathbf{k'}) = l_{11}l_{13} + l_{21}l_{23} + l_{31}l_{33}$$

These establish the required result where $m = 1$. By considering $\mathbf{j} \cdot \mathbf{i}$, $\mathbf{j} \cdot \mathbf{j}$, $\mathbf{j} \cdot \mathbf{k}$, $\mathbf{k} \cdot \mathbf{i}$, $\mathbf{k} \cdot \mathbf{j}$, and $\mathbf{k} \cdot \mathbf{k}$, the result can be proved for $m = 2$ and $m = 3$.

By writing

$$\delta_{mn} = \begin{cases} 1 & \text{if } m = n \\ 0 & \text{if } m \neq n \end{cases} \quad \text{the result can be written} \sum_{p=1}^{3} l_{pm}l_{pn} = \delta_{mn}.$$

The symbol δ_{mn} is called *Kronecker's symbol*.

4.41. Suppose $\phi(x, y, z)$ is a scalar invariant with respect to a rotation of axes. Prove that grad ϕ is a vector invariant under this transformation.

Solution

By hypothesis, $\phi(x, y, z) = \phi'(x', y', z')$. To establish the desired result, we must prove that

$$\frac{\partial \phi}{\partial x}\mathbf{i} + \frac{\partial \phi}{\partial y}\mathbf{j} + \frac{\partial \phi}{\partial z}\mathbf{k} = \frac{\partial \phi'}{\partial x'}\mathbf{i'} + \frac{\partial \phi'}{\partial y'}\mathbf{j'} + \frac{\partial \phi'}{\partial z'}\mathbf{k'}$$

Using the chain rule and the transformation equations (3) of Problem 4.38, we have

$$\frac{\partial \phi}{\partial x} = \frac{\partial \phi'}{\partial x'}\frac{\partial x'}{\partial x} + \frac{\partial \phi'}{\partial y'}\frac{\partial y'}{\partial x} + \frac{\partial \phi'}{\partial z'}\frac{\partial z'}{\partial x} = \frac{\partial \phi'}{\partial x'}l_{11} + \frac{\partial \phi'}{\partial y'}l_{21} + \frac{\partial \phi'}{\partial z'}l_{31}$$

$$\frac{\partial \phi}{\partial y} = \frac{\partial \phi'}{\partial x'}\frac{\partial x'}{\partial y} + \frac{\partial \phi'}{\partial y'}\frac{\partial y'}{\partial y} + \frac{\partial \phi'}{\partial z'}\frac{\partial z'}{\partial y} = \frac{\partial \phi'}{\partial x'}l_{12} + \frac{\partial \phi'}{\partial y'}l_{22} + \frac{\partial \phi'}{\partial z'}l_{32}$$

$$\frac{\partial \phi}{\partial z} = \frac{\partial \phi'}{\partial x'}\frac{\partial x'}{\partial z} + \frac{\partial \phi'}{\partial y'}\frac{\partial y'}{\partial z} + \frac{\partial \phi'}{\partial z'}\frac{\partial z'}{\partial z} = \frac{\partial \phi'}{\partial x'}l_{13} + \frac{\partial \phi'}{\partial y'}l_{23} + \frac{\partial \phi'}{\partial z'}l_{33}$$

Multiplying these equations by $\mathbf{i}$, $\mathbf{j}$, and $\mathbf{k}$, respectively, adding, and using Problem 4.39, the required result follows.

SUPPLEMENTARY PROBLEMS

4.42. Suppose $\phi = 2xz^4 - x^2y$. Find $\nabla\phi$ and $|\nabla\phi|$ at the point $(2, -2, -1)$.

4.43. Suppose $\mathbf{A} = 2x^2\mathbf{i} - 3yz\mathbf{j} + xz^2\mathbf{k}$ and $\phi = 2z - x^3y$. Find $\mathbf{A} \cdot \nabla\phi$ and $\mathbf{A} \times \nabla\phi$ at the point $(1, -1, 1)$.

4.44. Suppose $F = x^2z + e^{y/x}$ and $G = 2z^2y - xy^2$. Find (a) $\nabla(F + G)$ and (b) $\nabla(FG)$ at the point $(1, 0, -2)$.

4.45. Find $\nabla|\mathbf{r}|^3$.

4.46. Prove $\nabla f(r) = \dfrac{f'(r)\mathbf{r}}{r}$.

4.47. Evaluate $\nabla\left(3r^2 - 4\sqrt{r} + \dfrac{6}{\sqrt[3]{r}}\right)$.

4.48. Let $\nabla U = 2r^4\mathbf{r}$. Find U.

4.49. Find $\phi(r)$ such that $\nabla\phi = \mathbf{r}/r^5$ and $\phi(1) = 0$.

4.50. Find $\nabla \psi$ where $\psi = (x^2 + y^2 + z^2)e^{-\sqrt{x^2+y^2+z^2}}$.

4.51. Let $\nabla \phi = 2xyz^3\mathbf{i} + x^2z^3\mathbf{j} + 3x^2yz^2\mathbf{k}$. Find $\phi(x, y, z)$ if $\phi(1, -2, 2) = 4$.

4.52. Suppose $\nabla \psi = (y^2 - 2xyz^3)\mathbf{i} + (3 + 2xy - x^2z^3)\mathbf{j} + (6z^3 - 3x^2yz^2)\mathbf{k}$. Find ψ.

4.53. Let U be a differentiable function of x, y, and z. Prove $\nabla U \cdot d\mathbf{r} = dU$.

4.54. Suppose F is a differentiable function of x, y, z, t where x, y, z are differentiable functions of t. Prove that

$$\frac{dF}{dt} = \frac{\partial F}{\partial t} + \nabla F \cdot \frac{d\mathbf{r}}{dt}$$

4.55. Let $\mathbf{A}$ be a constant vector. Prove $\nabla(\mathbf{r} \cdot \mathbf{A}) = \mathbf{A}$.

4.56. Suppose $\mathbf{A}(x, y, z) = A_1\mathbf{i} + A_2\mathbf{j} + A_3\mathbf{k}$. Show that $d\mathbf{A} = (\nabla A_1 \cdot d\mathbf{r})\mathbf{i} + (\nabla A_2 \cdot d\mathbf{r})\mathbf{j} + (\nabla A_3 \cdot d\mathbf{r})\mathbf{k}$.

4.57. Prove $\nabla\left(\dfrac{F}{G}\right) = \dfrac{G\nabla F - F\nabla G}{G^2}$ if $G \neq 0$.

4.58. Find a unit vector that is perpendicular to the surface of the paraboloid of revolution $z = x^2 + y^2$ at the point $(1, 2, 5)$.

4.59. Find the unit outward drawn normal to the surface $(x - 1)^2 + y^2 + (z + 2)^2 = 9$ at the point $(3, 1, -4)$.

4.60. Find an equation for the tangent plane to the surface $xz^2 + x^2y = z - 1$ at the point $(1, -3, 2)$.

4.61. Find equations for the tangent plane and normal line to the surface $z = x^2 + y^2$ at the point $(2, -1, 5)$.

4.62. Find the directional derivative of $\phi = 4xz^3 - 3x^2y^2z$ at $(2, -1, 2)$ in the direction $2\mathbf{i} - 3\mathbf{j} + 6\mathbf{k}$.

4.63. Find the directional derivative of $P = 4e^{2x-y+z}$ at the point $(1, 1, -1)$ in a direction toward the point $(-3, 5, 6)$.

4.64. In what direction from the point $(1, 3, 2)$ is the directional derivative of $\phi = 2xz - y^2$ a maximum? What is the magnitude of this maximum?

4.65. Find the values of the constants a, b, and c so that the directional derivative of $\phi = axy^2 + byz + cz^2x^3$ at $(1, 2, -1)$ has a maximum of magnitude 64 in a direction parallel to the z axis.

4.66. Find the acute angle between the surfaces $xy^2z = 3x + z^2$ and $3x^2 - y^2 + 2z = 1$ at the point $(1, -2, 1)$.

4.67. Find the constants a and b so that the surface $ax^2 - byz = (a + 2)x$ will be orthogonal to the surface $4x^2y + z^3 = 4$ at the point $(1, -1, 2)$.

4.68. (a) Let u and v be differentiable functions of x, y, and z. Show that a necessary and sufficient condition that u and v are functionally related by the equation $F(u, v) = 0$ is that $\nabla u \times \nabla v = \mathbf{0}$.

(b) Determine whether $u = \arctan x + \arctan y$ and $v = \dfrac{x + y}{1 - xy}$ are functionally related.

4.69. (a) Show that $\nabla u \cdot \nabla v \times \nabla w = 0$ a necessary and sufficient condition that $u(x, y, z)$, $v(x, y, z)$, and $w(x, y, z)$ be functionally related through the equation $F(u, v, w) = 0$.

(b) Express $\nabla u \cdot \nabla v \times \nabla w$ in determinant form. This determinant is called the Jacobian of u, v, and w with respect to x, y, and z, and is written $\dfrac{\partial(u, v, w)}{\partial(x, y, z)}$ or $J\left(\dfrac{u, v, w}{x, y, z}\right)$.

(c) Determine whether $u = x + y + z$, $v = x^2 + y^2 + z^2$ and $w = xy + yz + zx$ are functionally related.

4.70. Let $\mathbf{A} = 3xyz^2\mathbf{i} + 2xy^3\mathbf{j} - x^2yz\mathbf{k}$ and $\phi = 3x^2 - yz$. Find, at the point $(1, -1, 1)$, (a) $\nabla \cdot \mathbf{A}$, (b) $\mathbf{A} \cdot \nabla \phi$, (c) $\nabla \cdot (\phi \mathbf{A})$, (d) $\nabla \cdot (\nabla \phi)$, .

4.71. Evaluate $\operatorname{div}(2x^2z\mathbf{i} - xy^2z\mathbf{j} + 3yz^2\mathbf{k})$.

4.72. Let $\phi = 3x^2z - y^2z^3 + 4x^3y + 2x - 3y - 5$. Find $\nabla^2\phi$.

4.73. Evaluate $\nabla^2(\ln r)$.

4.74. Prove $\nabla^2 r^n = n(n+1)r^{n-2}$ where n is a constant.

4.75. Let $\mathbf{F} = (3x^2y - z)\mathbf{i} + (xz^3 + y^4)\mathbf{j} - 2x^3z^2\mathbf{k}$. Find $\nabla(\nabla \cdot \mathbf{F})$ at the point $(2, -1, 0)$.

4.76. Suppose $\boldsymbol{\omega}$ is a constant vector and $\mathbf{v} = \boldsymbol{\omega} \times \mathbf{r}$. Prove that div $\mathbf{v} = 0$.

4.77. Prove $\nabla^2(\phi\psi) = \phi\nabla^2\psi + 2\nabla\phi \cdot \nabla\psi + \psi\nabla^2\phi$.

4.78. Let $U = 3x^2y$ and $V = xz^2 - 2y$. Evaluate grad$[(\text{grad } U) \cdot (\text{grad } V)]$.

4.79. Evaluate $\nabla \cdot (r^3\mathbf{r})$.

4.80. Evaluate $\nabla \cdot [r\nabla(1/r^3)]$.

4.81. Evaluate $\nabla^2[\nabla \cdot (\mathbf{r}/r^2)]$.

4.82. If $\mathbf{A} = \mathbf{r}/r$, find grad div $\mathbf{A}$.

4.83. (a) Prove $\nabla^2 f(r) = \dfrac{d^2f}{dr^2} + \dfrac{2}{r}\dfrac{df}{dr}$. (b) Find $f(r)$ such that $\nabla^2 f(r) = 0$.

4.84. Prove that the vector $\mathbf{A} = 3y^4z^2\mathbf{i} + 4x^3z^2\mathbf{j} - 3x^2y^2\mathbf{k}$ is solenoidal.

4.85. Show that $\mathbf{A} = (2x^2 + 8xy^2z)\mathbf{i} + (3x^3y - 3xy)\mathbf{j} - (4y^2z^2 + 2x^3z)\mathbf{k}$ is not solenoidal but $\mathbf{B} = xyz^2\mathbf{A}$ is solenoidal.

4.86. Find the most general differentiable function $f(r)$ so that $f(r)\mathbf{r}$ is solenoidal.

4.87. Show that the vector field $\mathbf{V} = \dfrac{-x\mathbf{i} - y\mathbf{j}}{\sqrt{x^2 + y^2}}$ is a "sink field". Plot and give a physical interpretation.

4.88. Suppose U and V are differentiable scalar fields. Prove that $\nabla U \times \nabla V$ is solenoidal.

4.89. Let $\mathbf{A} = 2xz^2\mathbf{i} - yz\mathbf{j} + 3xz^3\mathbf{k}$ and $\phi = x^2yz$. Find, at the point $(1, 1, 1)$:
(a) $\nabla \times \mathbf{A}$, (b) curl$(\phi\mathbf{A})$, (c) $\nabla \times (\nabla \times \mathbf{A})$, (d) $\nabla[\mathbf{A} \cdot \text{curl } \mathbf{A}]$, (e) curl grad$(\phi\mathbf{A})$.

4.90. Let $F = x^2yz$, $G = xy - 3z^2$. Find (a) $\nabla[(\nabla F) \cdot (\nabla G)]$, (b) $\nabla \cdot [(\nabla F) \times (\nabla G)]$, (c) $\nabla \times [(\nabla F) \times (\nabla G)]$.

4.91. Evaluate $\nabla \times (\mathbf{r}/r^2)$.

4.92. For what value of the constant a will the vector $\mathbf{A} = (axy - z^3)\mathbf{i} + (a - 2)x^2\mathbf{j} + (1 - a)xz^2\mathbf{k}$ have its curl identically equal to zero?

4.93. Prove curl$(\phi \text{ grad } \phi) = \mathbf{0}$.

4.94. Graph the vector fields $\mathbf{A} = x\mathbf{i} + y\mathbf{j}$ and $\mathbf{B} = y\mathbf{i} - x\mathbf{j}$. Compute the divergence and curl of each vector field and explain the physical significance of the results obtained.

4.95. Given $\mathbf{A} = x^2z\mathbf{i} + yz^3\mathbf{j} - 3xy\mathbf{k}$, $\mathbf{B} = y^2\mathbf{i} - yz\mathbf{j} + 2x\mathbf{k}$ and $\phi = 2x^2 + yz$. Find:
(a) $\mathbf{A} \cdot (\nabla\phi)$, (b) $(\mathbf{A} \cdot \nabla)\phi$, (c) $(\mathbf{A} \cdot \nabla)\mathbf{B}$, (d) $\mathbf{B}(\mathbf{A} \cdot \nabla)$, (e) $(\nabla \cdot \mathbf{A})\mathbf{B}$.

4.96. Suppose $\mathbf{A} = yz^2\mathbf{i} - 3xz^2\mathbf{j} + 2xyz\mathbf{k}$, $\mathbf{B} = 3x\mathbf{i} + 4z\mathbf{j} - xy\mathbf{k}$, and $\phi = xyz$. Find (a) $\mathbf{A} \times (\nabla\phi)$, (b) $(\mathbf{A} \times \nabla)\phi$, (c) $(\nabla \times \mathbf{A}) \times \mathbf{B}$, (d) $\mathbf{B} \cdot \nabla \times \mathbf{A}$.

4.97. Given $\mathbf{A} = xz^2\mathbf{i} + 2y\mathbf{j} - 3xz\mathbf{k}$ and $\mathbf{B} = 3xz\mathbf{i} + 2yz\mathbf{j} - z^2\mathbf{k}$. Find $\mathbf{A} \times (\nabla \times \mathbf{B})$ and $(\mathbf{A} \times \nabla) \times \mathbf{B}$ at the point $(1, -1, 2)$.

4.98. Prove $(\mathbf{v} \cdot \nabla)\mathbf{v} = \frac{1}{2}\nabla v^2 - \mathbf{v} \times (\nabla \times \mathbf{v})$.

4.99. Prove $\nabla \cdot (\mathbf{A} \times \mathbf{B}) = \mathbf{B} \cdot (\nabla \times \mathbf{A}) - \mathbf{A} \cdot (\nabla \times \mathbf{B})$.

4.100. Prove $\nabla \times (\mathbf{A} \times \mathbf{B}) = (\mathbf{B} \cdot \nabla)\mathbf{A} - \mathbf{B}(\nabla \cdot \mathbf{A}) - (\mathbf{A} \cdot \nabla)\mathbf{B} + \mathbf{A}(\nabla \cdot \mathbf{B})$.

4.101. Prove $\nabla(\mathbf{A} \cdot \mathbf{B}) = (\mathbf{B} \cdot \nabla)\mathbf{A} + (\mathbf{A} \cdot \nabla)\mathbf{B} + \mathbf{B} \times (\nabla \times \mathbf{A}) + \mathbf{A} \times (\nabla \times \mathbf{B})$.

4.102. Show that $\mathbf{A} = (6xy + z^3)\mathbf{i} + (3x^2 - z)\mathbf{j} + (3xz^2 - y)\mathbf{k}$ is irrotational. Find ϕ such that $\mathbf{A} = \nabla\phi$.

4.103. Show that $\mathbf{E} = \mathbf{r}/r^2$ is irrotational. Find ϕ such that $\mathbf{E} = -\nabla\phi$ and such that $\phi(a) = 0$ where $a > 0$.

4.104. Suppose $\mathbf{A}$ and $\mathbf{B}$ are irrotational. Prove that $\mathbf{A} \times \mathbf{B}$ is solenoidal.

4.105. Suppose $f(r)$ is differentiable. Prove that $f(r)\mathbf{r}$ is irrotational.

4.106. Is there a differentiable vector function $\mathbf{V}$ such that (a) curl $\mathbf{V} = \mathbf{r}$, (b) curl $\mathbf{V} = 2\mathbf{i} + \mathbf{j} + 3\mathbf{k}$? If so, find $\mathbf{V}$.

4.107. Show that solutions to Maxwell's equations

$$\nabla \times \mathbf{H} = \frac{1}{c}\frac{\partial \mathbf{E}}{\partial t}, \quad \nabla \times \mathbf{E} = -\frac{1}{c}\frac{\partial \mathbf{H}}{\partial t}, \quad \nabla \cdot \mathbf{H} = 0, \quad \nabla \cdot \mathbf{E} = 4\pi\rho$$

where ρ is a function of x, y, and z, and c is the velocity of light, assumed constant, are given by

$$\mathbf{E} = -\nabla\phi - \frac{1}{c}\frac{\partial \mathbf{A}}{\partial t}, \quad \mathbf{H} = \nabla \times \mathbf{A}$$

where $\mathbf{A}$ and ϕ, called the *vector* and *scalar potentials*, respectively, satisfy the equations

$$\nabla \cdot \mathbf{A} + \frac{1}{c}\frac{\partial \phi}{\partial t} = 0, \tag{1}$$

$$\nabla^2\phi - \frac{1}{c^2}\frac{\partial^2 \phi}{\partial t^2} = -4\pi\rho, \tag{2}$$

$$\nabla^2\mathbf{A} = \frac{1}{c^2}\frac{\partial^2 \mathbf{A}}{\partial t^2} \tag{3}$$

4.108. (a) Given the dyadic $\mathbf{\Phi} = \mathbf{ii} + \mathbf{jj} + \mathbf{kk}$, evaluate $\mathbf{r} \cdot (\mathbf{\Phi} \cdot \mathbf{r})$ and $(\mathbf{r} \cdot \mathbf{\Phi}) \cdot \mathbf{r}$. (b) Is there any ambiguity in writing $\mathbf{r} \cdot \mathbf{\Phi} \cdot \mathbf{r}$? (c) What does $\mathbf{r} \cdot \mathbf{\Phi} \cdot \mathbf{r} = 1$ represent geometrically?

4.109. (a) Suppose $\mathbf{A} = xz\mathbf{i} - y^2\mathbf{j} + yz^2\mathbf{k}$ and $\mathbf{B} = 2z^2\mathbf{i} - xy\mathbf{j} + y^3\mathbf{k}$. Give a possible significance to $(\mathbf{A} \times \nabla)\mathbf{B}$ at the point $(1, -1, 1)$.

(b) Is it possible to write the result as $\mathbf{A} \times (\nabla\mathbf{B})$ by use of dyadics?

4.110. Prove that $\phi(x, y, z) = x^2 + y^2 + z^2$ is a scalar invariant under a rotation of axes.

4.111. Let $\mathbf{A}(x, y, z)$ be an invariant differentiable vector field with respect to a rotation of axes. Prove that (a) div $\mathbf{A}$ and (b) curl $\mathbf{A}$ are invariant scalar and vector fields, respectively.

4.112. Solve equation (3) of Solved Problem 4.38 for x, y, and z in terms of x', y', and z'.

4.113. Suppose $\mathbf{A}$ and $\mathbf{B}$ are invariant under rotation. Show that $\mathbf{A} \cdot \mathbf{B}$ and $\mathbf{A} \times \mathbf{B}$ are also invariant.

4.114. Show that under a rotation

$$\nabla = \mathbf{i}\frac{\partial}{\partial x} + \mathbf{j}\frac{\partial}{\partial y} + \mathbf{k}\frac{\partial}{\partial z} = \mathbf{i}'\frac{\partial}{\partial x'} + \mathbf{j}'\frac{\partial}{\partial y'} + \mathbf{k}'\frac{\partial}{\partial z'} = \nabla'$$

4.115. Show that the Laplacian operator is invariant under a rotation.

4.116. Suppose $\mathbf{A} = x^2z^2\mathbf{i} - 2y^2z^2\mathbf{j} + xy^2z\mathbf{k}$, $\mathbf{B} = x^2\mathbf{i} + yz\mathbf{j} - xy\mathbf{k}$, and $\phi = 2x^2yz^3$. Find:
(a) $(\mathbf{A} \cdot \nabla)\phi$, (b) $\mathbf{A} \cdot \nabla\phi$, (c) $(\mathbf{B} \cdot \nabla)\phi$, (d) $(\mathbf{A} \times \nabla)\phi$, and (e) $\mathbf{A} \times \nabla\phi$.

4.117. Prove: (a) $\nabla \times (\mathbf{A} + \mathbf{B}) = \nabla \times \mathbf{A} + \nabla \times \mathbf{B}$, (b) $\nabla \times (\phi\mathbf{A}) = (\nabla\phi) \times \mathbf{A} + \phi(\nabla \times \mathbf{A})$.

ANSWERS TO SUPPLEMENTARY PROBLEMS

4.42. $10\mathbf{i} - 4\mathbf{j} - 16\mathbf{k}$, $2\sqrt{93}$

4.43. 5, $7\mathbf{i} - \mathbf{j} - 11\mathbf{k}$

4.44. (a) $-4\mathbf{i} + 9\mathbf{j} + \mathbf{k}$, (b) $-8\mathbf{j}$

4.45. $3r\mathbf{r}$

4.47. $(6 - 2r^{-3/2} - 2r^{-7/3})\mathbf{r}$

4.48. $\mathbf{r}^6/3 + \text{constant}$

4.49. $\phi(r) = \dfrac{1}{3}\left(1 - \dfrac{1}{r^3}\right)$

4.50. $(2 - r)e^{-r}\mathbf{r}$

4.51. $\phi = x^2 yz^3 + 20$

4.52. $\omega = xy^2 - x^2 yz^3 + 3y + (3/2)z^4 + \text{constant}$

4.58. $(2\mathbf{i} + 4\mathbf{j} - \mathbf{k})/\pm\sqrt{21}$

4.59. $(2\mathbf{i} + \mathbf{j} - 2\mathbf{k})/3$

4.60. $2x - y - 3z + 1$

4.61. $4x - 2y - z = 5$, $\dfrac{x-2}{4} = \dfrac{y+1}{-2} = \dfrac{z-5}{-1}$

or $x = 4t + 2$, $y = -2t - 1$, $z = -t + 5$

4.62. $376/7$

4.63. $-20/9$

4.81. $2r^{-4}$

4.64. In the direction of $4\mathbf{i} - 6\mathbf{j} + 2\mathbf{k}$, $2\sqrt{14}$

4.65. $a = 6$, $b = 24$, $c = -8$

4.66. $\arccos\dfrac{3}{\sqrt{14}\sqrt{21}} = \arccos\dfrac{\sqrt{6}}{14} = 79°55'$

4.67. $a = 5/2$, $b = 1$

4.68. (b) Yes $(v = \tan u)$

4.69. (b) $\begin{vmatrix} \dfrac{\partial u}{\partial x} & \dfrac{\partial u}{\partial y} & \dfrac{\partial u}{\partial z} \\[2mm] \dfrac{\partial v}{\partial x} & \dfrac{\partial v}{\partial y} & \dfrac{\partial v}{\partial z} \\[2mm] \dfrac{\partial w}{\partial x} & \dfrac{\partial w}{\partial y} & \dfrac{\partial w}{\partial z} \end{vmatrix}$ (c) Yes $(u^2 - v - 2w = 0)$

4.70. (a) 4, (b) -15, (c) 1, (d) 6

4.71. $4xz - 2xyz + 6y^2 z$

4.72. $6z + 24xy - 2z^3 - 6y^2 z$

4.73. $1/r^2$

4.75. $-6\mathbf{i} + 24\mathbf{j} - 32\mathbf{k}$

4.78. $(6yz^2 - 12x)\mathbf{i} + 6xz^2\mathbf{j} + 12xyz\mathbf{k}$

4.79. $6r^3$

4.80. $3r^{-4}$

4.82. $-2r^{-3}\mathbf{r}$

4.83. $f(r) = A + B/r$ where A and B are arbitrary constants.

4.86. $f(r) = C/r^3$ where C is an arbitrary constant.

4.89. (a) $\mathbf{i} + \mathbf{j}$, (b) $5\mathbf{i} - 3\mathbf{j} - 4\mathbf{k}$, (c) $5\mathbf{i} + 3\mathbf{k}$, (d) $-2\mathbf{i} + \mathbf{j} + 8\mathbf{k}$, (e) $\mathbf{0}$

4.90. (a) $(2y^2 z + 3x^2 z - 12xyz)\mathbf{i} + (4xyz - 6x^2 z)\mathbf{j} + (2xy^2 + x^3 - 6x^2 y)\mathbf{k}$

(b) $\mathbf{0}$

(c) $(x^2 z - 24xyz)\mathbf{i} - (12x^2 z + 2xyz)\mathbf{j} + (2xy^2 + 12yz^2 + x^3)\mathbf{k}$

4.91. $\mathbf{0}$

4.92. $a = 4$

4.95. (a) $4x^3 z + yz^4 - 3xy^2$

(b) $4x^3 z + yz^4 - 3xy^2$ (same as (a))

(c) $2y^2 z^3\mathbf{i} + (3xy^2 - yz^4)\mathbf{j} + 2x^2 z\mathbf{k}$

(d) the operator $(x^2 y^2 z\mathbf{i} - x^2 yz^2\mathbf{j} + 2x^3 z\mathbf{k})\dfrac{\partial}{\partial x} + (y^3 z^3\mathbf{i} - y^2 z^4\mathbf{j} + 2xyz^3\mathbf{k})\dfrac{\partial}{\partial y} + (-3xy^3\mathbf{i} + 3xy^2 z\mathbf{j} - 6x^2 y\mathbf{k})\dfrac{\partial}{\partial z}$

(e) $(2xy^2 z + y^2 z^3)\mathbf{i} - (2xyz^2 + yz^4)\mathbf{j} + (4x^2 z + 2xz^3)\mathbf{k}$

4.96.   (a) $-5x^2yz^2\mathbf{i} + xy^2z^2\mathbf{j} + 4xyz^3\mathbf{k}$

     (b) $-5x^2yz^2\mathbf{i} + xy^2z^2\mathbf{j} + 4xyz^3\mathbf{k}$ (same as (a))

     (c) $16z^3\mathbf{i} + (8x^2yz - 12xz^2)\mathbf{j} + 32xz^2\mathbf{k}$ (d) $24x^2z + 4xyz^2$

4.97. $\mathbf{A} \times (\nabla \times \mathbf{B}) = 18\mathbf{i} - 12\mathbf{j} + 16\mathbf{k}$, $(\mathbf{A} \times \nabla) \times \mathbf{B} = 4\mathbf{j} + 76\mathbf{k}$

4.102. $\phi = 3x^2 + xz^3 - yz +$ constant   **4.103.** $\phi = \ln(a/r)$

4.106.   (a) No, (b) $\mathbf{V} = 3x\mathbf{j} + (2y - x)\mathbf{k} + \nabla\phi$, where ϕ is an arbitrary twice differentiable function.

4.108.   (a) $\mathbf{r}\cdot(\mathbf{\Phi}\cdot\mathbf{r}) = (\mathbf{r}\cdot\mathbf{\Phi})\cdot\mathbf{r} = x^2 + y^2 + z^2$, (b) No, (c) Sphere of radius one with center at the origin.

4.109.   (a) $-4\mathbf{ii} - \mathbf{ij} + 3\mathbf{ik} - \mathbf{jj} - 4\mathbf{ji} + 3\mathbf{kk}$

     (b) Yes, if the operations are suitably performed.

4.112. $x = l_{11}x' + l_{21}y' + l_{31}z'$, $y = l_{12}x' + l_{22}y' + l_{32}z'$, $z = l_{13}x' + l_{23}y' + l_{33}z'$

4.116.   (a) = (b) $4x^3yz^5 - 4x^2y^2z^5 + 6x^3y^3z^3$

     (c) $(2x^3z^2 - 2x^3yz)\mathbf{i} + (-4y^2z^3 + 4xy^3z)\mathbf{j} + (x^2y^2z + 2xy^2z^2 - x^3y^3)\mathbf{k}$

     (d) = (e) $(-12x^2y^3z^4 - 2x^3y^2z^4)\mathbf{i} + (-6x^4yz^4 + 4x^2y^3z^4)\mathbf{j} + (2x^4z^5 + 8xy^3z^5)\mathbf{k}$

CHAPTER 5

Vector Integration

5.1 Introduction

The reader is familiar with the integration of real-valued functions $f(x)$ of one variable. Specifically, we have the indefinite integral or anti-derivative, denoted by

$$\int f(x)\, dx$$

and the definite integral on a closed interval, say $[a, b]$, denoted by

$$\int_a^b f(x)\, dx$$

Here we extend these definitions to vector value functions of a single variable.

5.2 Ordinary Integrals of Vector Valued Functions

Let $\mathbf{R}(u) = R_1(u)\mathbf{i} + R_2(u)\mathbf{j} + R_3(u)\mathbf{k}$ be a vector depending on a single scalar variable u, where $R_1(u)$, $R_2(u)$, $R_3(u)$ are assumed to be continuous in a specific interval. Then

$$\int \mathbf{R}(u)\, du = \mathbf{i} \int R_1(u)\, du + \mathbf{j} \int R_2(u)\, du + \mathbf{k} \int R_3(u)\, du$$

is called an *indefinite integral* of $\mathbf{R}(u)$. If there exists a vector $\mathbf{S}(u)$ such that

$$\mathbf{R}(u) = \frac{d}{du}(\mathbf{S}(u)),$$

then

$$\int \mathbf{R}(u)\, du = \int \frac{d}{du}(\mathbf{S}(u))\, du = \mathbf{S}(u) + \mathbf{c}$$

where $\mathbf{c}$ is an *arbitrary constant vector* independent of u. The *definite integral* between limits $u = a$ and $u = b$ can in such case be written

$$\int_a^b \mathbf{R}(u)\, du = \int_a^b \frac{d}{du}(\mathbf{S}(u))\, du = \mathbf{S}(u) + \mathbf{c} \Big|_a^b = \mathbf{S}(b) - \mathbf{S}(a)$$

This integral can also be defined as a limit of a sum in a manner analogous to that of elementary integral calculus.

EXAMPLE 5.1   Suppose $\mathbf{R}(u) = u^2\mathbf{i} + 2u^3\mathbf{j} - 5\mathbf{k}$. Find: (a) $\int \mathbf{R}(u)\,du$, (b) $\int_1^2 \mathbf{R}(u)\,du$.

(a) $\displaystyle \int \mathbf{R}(u)\,du = \int [u^2\mathbf{i} + 2u^3\mathbf{j} - 5\mathbf{k}]\,du = \mathbf{i}\int u^2\,du + \mathbf{j}\int 2u^3\,du + \mathbf{k}\int -5\,du$

$\displaystyle = \left(\frac{u^3}{3} + c_1\right)\mathbf{i} + \left(\frac{u^4}{2} + c_2\right)\mathbf{j} + (-5u + c_3)\mathbf{k}$

$\displaystyle = \frac{u^3}{3}\mathbf{i} + \frac{u^4}{2}\mathbf{j} - 5u\mathbf{k} + \mathbf{c}$

where $\mathbf{c}$ is the constant vector $c_1\mathbf{i} + c_2\mathbf{j} + c_3\mathbf{k}$.

(b)   From (a):

$$\int_1^2 \mathbf{R}(u)\,du = \frac{u^3}{3}\mathbf{i} + \frac{u^4}{2}\mathbf{j} - 5u\mathbf{k} + \mathbf{c}\,\Big|_1^2 = [(8/3)\mathbf{i} + 4\mathbf{j} - 10\mathbf{k}] - [-(1/3)\mathbf{i} + (1/2)\mathbf{j} - 5\mathbf{k}]$$

$$= (7/3)\mathbf{i} + (7/2)\mathbf{j} - 5\mathbf{k}$$

5.3   Line Integrals

Suppose $\mathbf{r}(u) = x(u)\mathbf{i} + y(u)\mathbf{j} + z(u)\mathbf{k}$ is the position vector of points $P(x, y, z)$ and suppose $\mathbf{r}(u)$ defines a curve C joining points P_1 and P_2 where $u = u_1$ and $u = u_2$, respectively.

We assume that C is composed of a finite number of curves for each of which $\mathbf{r}(u)$ has a continuous derivative. Let $\mathbf{A}(x, y, z) = A_1\mathbf{i} + A_2\mathbf{j} + A_3\mathbf{k}$ be a vector function of position defined and continuous along C. Then the integral of the tangential component of $\mathbf{A}$ along C from P_1 to P_2, written as

$$\int_{P_1}^{P_2} \mathbf{A} \cdot d\mathbf{r} = \int_C \mathbf{A} \cdot d\mathbf{r} = \int_C A_1\,dx + A_2\,dy + A_3\,dz$$

is an example of a *line integral*. If $\mathbf{A}$ is the force $\mathbf{F}$ on a particle moving along C, this line integral represents the work done by the force. If C is a closed curve (which we shall suppose is a *simple closed curve*, that is, a curve that does not intersect itself anywhere), the integral around C is often denoted by

$$\oint \mathbf{A} \cdot d\mathbf{r} = \oint A_1\,dx + A_2\,dy + A_3\,dz$$

In aerodynamics and fluid mechanics, this integral is called the *circulation* of $\mathbf{A}$ about C, where $\mathbf{A}$ represents the velocity of a fluid.

In general, any integral that is to be evaluated along a curve is called a line integral. Such integrals can be defined in terms of limits of sums as are the integrals of elementary calculus.

EXAMPLE 5.2   Suppose $\mathbf{F} = -3x^2\mathbf{i} + 5xy\mathbf{j}$ and let C be the curve $y = 2x^2$ in the xy-plane. Evaluate the line integral $\int_C \mathbf{F} \cdot d\mathbf{r}$ from $P_1(0,0)$ to $P_2(1, 2)$.

Since the integration is performed in the xy-plane ($z = 0$), we may take $\mathbf{r} = x\mathbf{i} + y\mathbf{j}$. Then:

$$\int_C \mathbf{F} \cdot d\mathbf{r} = \int_C (-3x^2\mathbf{i} + 5xy\mathbf{j}) \cdot (dx\mathbf{i} + dy\mathbf{j}) = \int_C (-3x^2\,dx + 5xy\,dy).$$

First Method. Let $x = t$ in $y = 2x^2$. Then the parametric equations of C are $x = t$, $y = 2t^2$. Points $(0, 0)$ and $(1, 2)$ correspond to $t = 0$ and $t = 1$, respectively. Then:

$$\int_C \mathbf{F} \cdot d\mathbf{r} = \int_{t=0}^{1} [-3t^2\, dt + 5t(2t^2)\, d(2t^2)] = \int_{t=0}^{1} (-3t^2 + 40t^4)\, dt = [-t^3 + 8t^5]_0^1 = 7.$$

Second Method. Substitute $y = 2x^2$ directly where x goes from 0 to 1. Then:

$$\int_C \mathbf{F} \cdot d\mathbf{r} = \int_{x=0}^{1} [-3x^2\, dx + 5x(2x^2)\, d(2x^2)] = \int_{x=0}^{1} (-3x^2 + 40x^4)\, dx = [-x^3 + 8x^5]_0^1 = 7.$$

Conservative Fields

The following theorem applies.

THEOREM 5.1. Suppose $\mathbf{A} = \nabla\phi$ everywhere in a region R of space, where R is defined by $a_1 \leq x \leq a_2$, $b_1 \leq y \leq b_2$, $c_1 \leq z \leq c_2$, and where $\phi(x, y, z)$ is single-valued and has continuous derivatives in R. Then:

    (i) $\int_{P_1}^{P_2} \mathbf{A} \cdot d\mathbf{r}$ is independent of the path C in R joining P_1 and P_2.

    (ii) $\oint_C \mathbf{A} \cdot d\mathbf{r} = 0$ around any closed curve C in R.

In such a case, $\mathbf{A}$ is called a *conservative vector field* and ϕ is its *scalar potential*.

5.4 Surface Integrals

Let S be a two-sided surface, such as shown in Fig. 5-1. Let one side of S be considered arbitrarily as the positive side. (If S is a closed surface, such as a sphere, then the outer side is considered the positive side.) A unit normal $\mathbf{n}$ to any point of the positive side of S is called a *positive* or *outward drawn unit normal*.

Fig. 5-1

Associate with the differential of surface area dS a vector $d\mathbf{S}$ whose magnitude is dS and whose direction is that of $\mathbf{n}$. Then $d\mathbf{S} = \mathbf{n}\, dS$. The integral

$$\iint_S \mathbf{A} \cdot d\mathbf{S} = \iint_S \mathbf{A} \cdot \mathbf{n}\, dS$$

is an example of a surface integral called the *flux* of **A** over *S*. Other surface integrals are

$$\iint_S \phi \, dS, \quad \iint_S \phi \mathbf{n} \, dS, \quad \iint_S \mathbf{A} \times d\mathbf{S}$$

where ϕ is a scalar function. Such integrals can be defined in terms of limits of sums as in elementary calculus (see Problem 5.17).

The notation $\iint\!\!\!\!\oint_S$ is sometimes used to indicate integration over the closed surface S. Where no confusion can arise the notation $\oint_S$ may also be used.

To evaluate surface integrals, it is convenient to express them as double integrals taken over the projected area of the surface S on one of the coordinate planes. This is possible if any line perpendicular to the coordinate plane chosen meets the surface in no more than one point. However, this does not pose any real problem since we can generally subdivide S into surfaces that do satisfy this restriction.

5.5   Volume Integrals

Consider a closed surface in space enclosing a volume V. Then the following denote *volume integrals* or *space integrals* as they are sometimes called:

$$\iiint_V \mathbf{A} \, dV \quad \text{and} \quad \iiint_V \phi \, dV$$

The Solved Problems evaluate some such integrals.

SOLVED PROBLEMS

5.1. Suppose $\mathbf{R}(u) = 3\mathbf{i} + (u^3 + 4u^7)\mathbf{j} + u\mathbf{k}$. Find: (a) $\int \mathbf{R}(u) \, du$, (b) $\int_1^2 \mathbf{R}(u) \, du$.

Solution

(a) $\displaystyle \int \mathbf{R}(u) \, du = \int \left[3\mathbf{i} + (u^3 + 4u^7)\mathbf{j} + u\mathbf{k} \right] du$

$\displaystyle \qquad = \mathbf{i} \int 3 \, du + \mathbf{j} \int (u^3 + 4u^7) \, du + \mathbf{k} \int u \, du$

$\displaystyle \qquad = (3u + c_1)\mathbf{i} + \left(\frac{1}{4} u^4 + \frac{1}{2} u^8 c_2 \right)\mathbf{j} + \left(\frac{1}{2} u^2 + c_3 \right)\mathbf{k}$

$\displaystyle \qquad = (3u)\mathbf{i} + \left(\frac{1}{4} u^4 + \frac{1}{2} u^8 \right)\mathbf{j} + \left(\frac{1}{2} u^2 \right)\mathbf{k} + \mathbf{c}$

where **c** is the constant vector $c_1\mathbf{i} + c_2\mathbf{j} + c_3\mathbf{k}$.

(b) From (a),

$$\int_1^2 \mathbf{R}(u) \, du = \left[(3u)\mathbf{i} + \left(\frac{1}{4} u^4 + \frac{1}{2} u^8 \right)\mathbf{j} + \left(\frac{1}{2} u^2 \right)\mathbf{k} + \mathbf{c} \right]_1^2 = 3\mathbf{i} + \frac{525}{4}\mathbf{j} + \frac{3}{2}\mathbf{k}.$$

Another Method

$$\int_1^2 \mathbf{R}(u)\, du = \mathbf{i} \int_1^2 3\, du + \mathbf{j} \int_1^2 (u^3 + 4u^7)\, du + \mathbf{k} \int_1^2 u\, du$$

$$= [3u]_1^2 \mathbf{i} + \left[\frac{1}{4}u^4 + \frac{1}{2}u^8\right]_1^2 + \left[\frac{1}{2}u^2\right]_1^2 \mathbf{k} = 3\mathbf{i} + \frac{525}{4}\mathbf{j} + \frac{3}{2}\mathbf{k}.$$

5.2. The acceleration of a particle at any time $t \geq 0$ is given by

$$\mathbf{a} = \frac{d\mathbf{r}}{dt} = (25 \cos 2t)\mathbf{i} + (16 \sin 2t)\mathbf{j} + (9t)\mathbf{k}.$$

Solution

Suppose the velocity $\mathbf{v}$ and the displacement $\mathbf{r}$ are the zero vector at $t = 0$. Find $\mathbf{v}$ and $\mathbf{r}$ at any time.
Integrating:

$$\mathbf{v} = \mathbf{i} \int (25 \cos 2t)\, dt + \mathbf{j} \int (16 \sin 2t)\, dt + \mathbf{k} \int (9t)\, dt$$

$$= \left(\frac{25}{2} \sin 2t\right)\mathbf{i} + (-8 \cos 2t)\mathbf{j} + \left(\frac{9}{2}t^2\right)\mathbf{k} + \mathbf{c}_1.$$

Putting $\mathbf{v} = 0$ when $t = 0$, we find $\mathbf{0} = 0\mathbf{i} - 8\mathbf{j} + 0\mathbf{k} + \mathbf{c}_1$ and $\mathbf{c}_1 = 8\mathbf{j}$. Then

$$\mathbf{v} = \frac{d\mathbf{r}}{dt} = \left(\frac{25}{2} \sin 2t\right)\mathbf{i} + (8 - 8 \cos 2t)\mathbf{j} + \left(\frac{9}{2}t^2\right)\mathbf{k}.$$

Integrating,

$$\mathbf{r} = \mathbf{i} \int \left(\frac{25}{2} \sin 2t\right) dt + \mathbf{j} \int (8 - 8 \cos 2t)\, dt + \mathbf{k} \int \left(\frac{9}{2}t^2\right) dt$$

$$= \left(-\frac{25}{4} \cos 2t\right)\mathbf{i} + (8t + 4 \sin 2t)\mathbf{j} + \left(\frac{3}{2}t^3\right)\mathbf{j} + \mathbf{c}_2.$$

Putting $\mathbf{r} = 0$ when $t = 0$, we get

$$\mathbf{0} = -\frac{25}{4}\mathbf{i} + \mathbf{c}_2 \quad \text{and} \quad \mathbf{c}_2 = \frac{25}{4}\mathbf{i}.$$

Then

$$\mathbf{r} = \left(\frac{25}{4} - \frac{25}{4} \cos 2t\right)\mathbf{i} + (8 + 4 \sin 2t)\mathbf{j} + \left(\frac{3}{2}t^3\right)\mathbf{k}.$$

5.3. Evaluate $\displaystyle\int \mathbf{A} \times \frac{d^2\mathbf{A}}{dt^2}\, dt$.

Solution

$$\frac{d}{dt}\left(\mathbf{A} \times \frac{d\mathbf{A}}{dt}\right) = \mathbf{A} \times \frac{d^2\mathbf{A}}{dt^2} + \frac{d\mathbf{A}}{dt} \times \frac{d\mathbf{A}}{dt} = \mathbf{A} \times \frac{d^2\mathbf{A}}{dt^2}$$

Integrating,

$$\int \mathbf{A} \times \frac{d^2\mathbf{A}}{dt^2}\, dt = \int \frac{d}{dt}\left(\mathbf{A} \times \frac{d\mathbf{A}}{dt}\right) dt = \mathbf{A} \times \frac{d\mathbf{A}}{dt} + \mathbf{c}.$$

5.4. The equation of motion of a particle P of mass m is given by

$$m\frac{d^2\mathbf{r}}{dt^2} = f(r)\mathbf{r}_1$$

where $\mathbf{r}$ is the position vector of P measured from an origin O, $\mathbf{r}_1$ is a unit vector in the direction $\mathbf{r}$, and $f(r)$ is a function of the distance of P from O.

(a) Show that $\mathbf{r} \times (d\mathbf{r}/dt) = \mathbf{c}$ where $\mathbf{c}$ is a constant vector.

(b) Interpret physically the cases $f(r) < 0$ and $f(r) > 0$.

(c) Interpret the result in (a) geometrically.

(d) Describe how the results obtained relate to the motion of the planets in our solar system.

Solution

(a) Multiply both sides of $m(d^2\mathbf{r}/dt^2) = f(r)\mathbf{r}_1$ by $\mathbf{r}\times$. Then

$$m\mathbf{r} \times \frac{d^2\mathbf{r}}{dt^2} = f(r)\mathbf{r} \times \mathbf{r}_1 = 0$$

since $\mathbf{r}$ and $\mathbf{r}_1$ are collinear and so $\mathbf{r} \times \mathbf{r}_1 = 0$. Thus

$$\mathbf{r} \times \frac{d^2\mathbf{r}}{dt^2} = 0 \quad \text{and} \quad \frac{d}{dt}\left(\mathbf{r} \times \frac{d\mathbf{r}}{dt}\right) = 0$$

Integrating, $\mathbf{r} \times \dfrac{d\mathbf{r}}{dt} = \mathbf{c}$, where $\mathbf{c}$ is a constant vector. (Compare with Problem 5.3.)

(b) If $f(r) < 0$, the acceleration $d^2\mathbf{r}/dt^2$ has a direction opposite to $\mathbf{r}_1$; hence, the force is directed toward O and the particle is always *attracted* toward O.

   If $f(r) > 0$, the force is directed away from O and the particle is under the influence of a *repulsive* force at O.

   A force directed toward or away from a fixed point O and having magnitude depending only on the distance r from O is called a *central force*.

(c) In time Δt, the particle moves from M to N (see Fig. 5-2). The area swept out by the position vector in this time is approximately half the area of a parallelogram with sides $\mathbf{r}$ and $\Delta\mathbf{r}$, or $\frac{1}{2}\mathbf{r} \times \Delta\mathbf{r}$. Then the approximate area swept out by the radius vector per unit time is $\frac{1}{2}\mathbf{r} \times \Delta\mathbf{r}/\Delta t$; hence, the instantaneous time rate of change in area is

$$\lim_{\Delta t \to 0} \tfrac{1}{2}\mathbf{r} \times \frac{\Delta\mathbf{r}}{\Delta t} = \tfrac{1}{2}\mathbf{r} \times \frac{d\mathbf{r}}{dt} = \tfrac{1}{2}\mathbf{r} \times \mathbf{v}$$

where $\mathbf{v}$ is the instantaneous velocity of the particle. The quantity $\mathbf{H} = \frac{1}{2}\mathbf{r} \times (d\mathbf{r}/dt) = \frac{1}{2}\mathbf{r} \times \mathbf{v}$ is called the *areal velocity*. From part (a),

$$\text{areal velocity} = \mathbf{H} = \tfrac{1}{2}\mathbf{r} \times \frac{d\mathbf{r}}{dt} = \text{constant}$$

Since $\mathbf{r} \cdot \mathbf{H} = 0$, the motion takes place in a plane, which we take as the xy-plane in Fig. 5-2.

(d) A planet (such as Earth) is attracted toward the Sun according to Newton's universal law of gravitation, which states that any two objects of mass m and M, respectively, are attracted toward each other with a force of magnitude $F = GMm/r^2$, where r is the distance between objects and G is a universal constant.

Let m and M be the masses of the planet and sun, respectively, and choose a set of coordinate axes with the origin O at the Sun. Then, the equation of motion of the planet is

$$m\frac{d^2\mathbf{r}}{dt^2} = -\frac{GMm}{r^2}\mathbf{r}_1 \quad \text{or} \quad \frac{d^2\mathbf{r}}{dt^2} = -\frac{GM}{r^2}\mathbf{r}_1$$

assuming the influence of the other planets to be negligible.

According to part (c), a planet moves around the Sun so that its position vector sweeps out equal areas in equal times. This result and that of Problem 5.5 are two of Kepler's three famous laws that he deduced empirically from volumes of data compiled by the astronomer Tycho Brahe. These laws enabled Newton to formulate his universal law of gravitation. For Kepler's third law, see Problem 5.36.

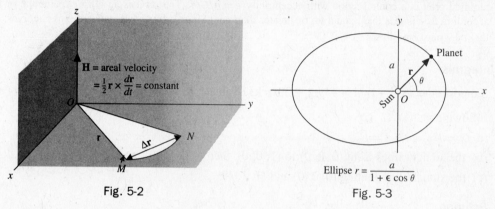

Fig. 5-2   Fig. 5-3

5.5. Show that the path of a planet around the Sun is an ellipse with the Sun at one focus.

Solution

From Problems 5.4(c) and 5.4(d),

$$\frac{d\mathbf{v}}{dt} = -\frac{GM}{r^2}\mathbf{r}_1 \tag{1}$$

$$\mathbf{r} \times \mathbf{v} = 2\mathbf{H} = \mathbf{h} \tag{2}$$

Now $\mathbf{r} = r\mathbf{r}_1, d\mathbf{r}/dt = r(d\mathbf{r}_1/dt) + (dr/dt)\mathbf{r}_1$ so that

$$\mathbf{h} = \mathbf{r} \times \mathbf{v} = r\mathbf{r}_1 \times \left(r\frac{d\mathbf{r}_1}{dt} + \frac{dr}{dt}\mathbf{r}_1 \right) = r^2\mathbf{r}_1 \times \frac{d\mathbf{r}_1}{dt} \tag{3}$$

From (1),

$$\frac{d\mathbf{v}}{dt} \times \mathbf{h} = -\frac{GM}{r^2}\mathbf{r}_1 \times \mathbf{h} = -GM\mathbf{r}_1 \times \left(\mathbf{r}_1 \times \frac{d\mathbf{r}_1}{dt} \right)$$

$$= -GM\left[\left(\mathbf{r}_1 \cdot \frac{d\mathbf{r}_1}{dt} \right)\mathbf{r}_1 - (\mathbf{r}_1 \cdot \mathbf{r}_1)\frac{d\mathbf{r}_1}{dt} \right] = GM\frac{d\mathbf{r}_1}{dt}$$

using equation (3) and the fact that $\mathbf{r}_1 \cdot (d\mathbf{r}_1/dt) = 0$ (Problem 3.9). But since $\mathbf{h}$ is a constant vector,

$$\frac{d\mathbf{v}}{dt} \times \mathbf{h} = \frac{d}{dt}(\mathbf{v} \times \mathbf{h})$$

so that

$$\frac{d}{dt}(\mathbf{v} \times \mathbf{h}) = GM\frac{d\mathbf{r}_1}{dt}$$

Integrating,

$$\mathbf{v} \times \mathbf{h} = GM\mathbf{r}_1 + \mathbf{p}$$

from which

$$\mathbf{r} \cdot (\mathbf{v} \times \mathbf{h}) = GM\mathbf{r} \cdot \mathbf{r}_1 + \mathbf{r} \cdot \mathbf{p}$$
$$= GMr + r\mathbf{r}_1 \cdot \mathbf{p} = GMr + rp\cos\theta$$

where $\mathbf{p}$ is an arbitrary constant vector with magnitude p, and θ is the angle between $\mathbf{p}$ and $\mathbf{r}_1$.

Since $\mathbf{r} \cdot (\mathbf{v} \times \mathbf{h}) = (\mathbf{r} \times \mathbf{v}) \cdot \mathbf{h} = \mathbf{h} \cdot \mathbf{h} = h^2$, we have $h^2 = GMr + rp\cos\theta$ and

$$r = \frac{h^2}{GM + p\cos\theta} = \frac{h^2/GM}{1 + (p/GM)\cos\theta}$$

From analytic geometry, the polar equation of a conic section with focus at the origin and eccentricity ϵ is $r = a/(1 + \epsilon\cos\theta)$ where a is a constant. See Fig. 5-3. Comparing this with the equation derived, it is seen that the required orbit is a conic section with eccentricity $\epsilon = p/GM$. The orbit is an ellipse, parabola, or hyperbola according as ϵ is less than, equal to, or greater than one. Since orbits of planets are closed curves, it follows that they must be ellipses.

Line Integrals

5.6. Suppose $\mathbf{A} = (3x^2 + 6y)\mathbf{i} - 14yz\mathbf{j} + 20xz^2\mathbf{k}$. Evaluate $\int_C \mathbf{A} \cdot d\mathbf{r}$ from $(0, 0, 0)$ to $(1, 1, 1)$ along the following paths C:

(a) $x = t$, $y = t^2$, $z = t^3$.

(b) the straight lines from $(0, 0, 0)$ to $(1, 0, 0)$, then to $(1, 1, 0)$, and then to $(1, 1, 1)$.

(c) the straight line joining $(0, 0, 0)$ and $(1, 1, 1)$.

Solution

$$\int_C \mathbf{A} \cdot d\mathbf{r} = \int_C \left[(3x^2 + 6y)\mathbf{i} - 14yz\mathbf{j} + 20xz^2\mathbf{k} \right] \cdot (dx\mathbf{i} + dy\mathbf{j} + dz\mathbf{k})$$

$$= \int_C (3x^2 + 6y)\, dx - 14yz\, dy + 20xz^2\, dz$$

(a) If $x = t, y = t^2, z = t^3$, points $(0, 0, 0)$ and $(1, 1, 1)$ correspond to $t = 0$ and $t = 1$, respectively. Then

$$\int_C \mathbf{A} \cdot d\mathbf{r} = \int_{t=0}^{1} (3t^2 + 6t^2)\, dt - 14(t^2)(t^3)\, d(t^2) + 20(t)(t^3)^2\, d(t^3)$$

$$= \int_{t=0}^{1} 9t^2\, dt - 28t^6\, dt + 60t^9\, dt$$

$$= \int_{t=0}^{1} (9t^2 - 28t^6 + 60t^9)\, dt = 3t^3 - 4t^7 + 6t^{10}\Big|_0^1 = 5$$

Another Method

Along C, $\mathbf{A} = 9t^2\mathbf{i} - 14t^5\mathbf{j} + 20t^7\mathbf{k}$ and $\mathbf{r} = x\mathbf{i} + y\mathbf{j} + z\mathbf{k} = t\mathbf{i} + t^2\mathbf{j} + t^3\mathbf{k}$ and $d\mathbf{r} = (\mathbf{i} + 2t\mathbf{j} + 3t^2\mathbf{k})\, dt$. Then

$$\int_C \mathbf{A} \cdot d\mathbf{r} = \int_{t=0}^{1} (9t^2\mathbf{i} - 14t^5\mathbf{j} + 20t^7\mathbf{k}) \cdot (\mathbf{i} + 2t\mathbf{j} + 3t^2\mathbf{k})\, dt$$

$$= \int_0^1 (9t^2 - 28t^6 + 60t^9)\, dt = 5$$

(b) Along the straight line from $(0, 0, 0)$ to $(1, 0, 0)$, $y = 0$, $z = 0$, $dy = 0$, $dz = 0$ while x varies from 0 to 1. Then the integral over this part of the path is

$$\int_{x=0}^{1} (3x^2 + 6(0))\, dx - 14(0)(0)(0) + 20x(0)^2(0) = \int_{x=0}^{1} 3x^2 \, dx = x^3 \Big|_0^1 = 1$$

Along the straight line from $(1, 0, 0)$ to $(1, 1, 0)$, $x = 1$, $z = 0$, $dx = 0$, $dz = 0$ while y varies from 0 to 1. Then the integral over this part of the path is

$$\int_{y=0}^{1} (3(1)^2 + 6y)0 - 14y(0)\, dy + 20(1)(0)^2 0 = 0$$

Along the straight line from $(1, 1, 0)$ to $(1, 1, 1)$, $x = 1$, $y = 1$, $dx = 0$, $dy = 0$ while z varies from 0 to 1. Then the integral over this part of the path is

$$\int_{z=0}^{1} (3(1)^2 + 6(1))0 - 14(1)z(0) + 20(1)z^2 \, dz = \int_{z=0}^{1} 20z^2 \, dz = \frac{20z^3}{3} \Big|_0^1 = \frac{20}{3}$$

Adding,

$$\int_C \mathbf{A} \cdot d\mathbf{r} = 1 + 0 + \frac{20}{3} = \frac{23}{3}$$

(c) The straight line joining $(0, 0, 0)$ and $(1, 1, 1)$ is given in parametric form by $x = t$, $y = t$, $z = t$. Then

$$\int_C \mathbf{A} \cdot d\mathbf{r} = \int_{t=0}^{1} (3t^2 + 6t)\, dt - 14(t)(t)\, dt + 20(t)(t)^2 \, dt$$

$$= \int_{t=0}^{1} (3t^2 + 6t - 14t^2 + 20t^3) \, dt = \int_{t=0}^{1} (6t - 11t^2 + 20t^3) \, dt = \frac{13}{3}$$

5.7. Find the total work done in moving a particle in the force field given by $\mathbf{F} = z\mathbf{i} + z\mathbf{j} + x\mathbf{k}$ along the helix C given by $x = \cos t$, $y = \sin t$, $z = t$ from $t = 0$ to $t = \pi/2$.

Solution

$$\text{Total work} = \int_C \mathbf{F} \cdot d\mathbf{r} = \int_C (z\mathbf{i} + z\mathbf{j} + x\mathbf{k}) \cdot (dx\mathbf{i} + dy\mathbf{j} + dz\mathbf{k}) = \int_C (z\,dx + z\,dy + x\,dz)$$

$$= \int_0^{\pi/2} (t\, d(\cos t) + t\, d(\sin t) + \cos t\, dt) = \int_0^{\pi/2} (-t \sin t)\, dt + \int_0^{\pi/2} (t + 1) \cos t\, dt$$

Evaluating $\int_0^{\pi/2} (-t \sin t)\, dt$ by parts we get

$$[t \cos t]_0^{\pi/2} - \int_0^{\pi/2} \cos t\, dt = 0 - [\sin t]_0^{\pi/2} = -1.$$

Evaluating $\int_0^{\pi/2} (t + 1) \cos t\, dt$ by parts we get

$$[(t + 1) \sin t]_0^{\pi/2} - \int_0^{\pi/2} \sin t\, dt = \frac{\pi}{2} + 1 + [\cos t]_0^{\pi/2} = \frac{\pi}{2}.$$

Thus the total work is $(\pi/2) - 1$.

5.8. Suppose $\mathbf{F} = -3x^2\mathbf{i} + 5xy\mathbf{j}$. Evaluate $\int_C \mathbf{F} \cdot d\mathbf{r}$ where C is the curve in the xy-plane, $y = 2x^2$, from $(0, 0)$ to $(1, 2)$.

Solution

Since the integration is performed in the xy-plane ($z = 0$), we may take $\mathbf{r} = x\mathbf{i} + y\mathbf{j}$. Then:

$$\int_C \mathbf{F} \cdot d\mathbf{r} = \int_C (-3x^2\mathbf{i} + 5xy\mathbf{j}) \cdot (dx\mathbf{i} + dy\mathbf{j}) = \int_C (-3x^2\,dx + 5xy\,dy).$$

First Method. Let $x = t$ in $y = 2x^2$. Then the parametric equations of C are $x = t$, $y = 2t^2$. Points $(0, 0)$ and $(1, 2)$ correspond to $t = 0$ and $t = 1$, respectively. Then:

$$\int_C \mathbf{F} \cdot d\mathbf{r} = \int_{t=0}^{1} [-3t^2\,dt + 5t(2t^2)\,d(2t^2)] = \int_{t=0}^{1} (-3t^2 + 40t^4)\,dt = \left[-t^3 + 8t^5\right]_0^1 = 7.$$

Second Method. Substitute $y = 2x^2$ directly where x goes from 0 to 1. Then:

$$\int_C \mathbf{F} \cdot d\mathbf{r} = \int_{x=0}^{1} [-3x^2\,dx + 5x(2x^2)\,d(2x^2)] = \int_{x=0}^{1} (-3x^2 + 40x^4)\,dx = \left[-x^3 + 8x^5\right]_0^1 = 7.$$

5.9. Suppose a force field is given by

$$\mathbf{F} = (2x - y + z)\mathbf{i} + (x + y - z^2)\mathbf{j} + (3x - 2y + 4z)\mathbf{k}$$

Find the work done in moving a particle once around a circle C in the xy-plane with its center at the origin and a radius of 3.

Solution

In the plane $z = 0$, $\mathbf{F} = (2x - y)\mathbf{i} + (x + y)\mathbf{j} + (3x - 2y)\mathbf{k}$ and $d\mathbf{r} = dx\mathbf{i} + dy\mathbf{j}$ so that the work done is

$$\int_C \mathbf{F} \cdot d\mathbf{r} = \int_C [(2x - y)\mathbf{i} + (x + y)\mathbf{j} + (3x - 2y)\mathbf{k}] \cdot (dx\mathbf{i} + dy\mathbf{j})$$

$$= \int_C (2x - y)\,dx + (x + y)\,dy$$

$\mathbf{r} = x\mathbf{i} + y\mathbf{j}$
$= 3\cos t\,\mathbf{i} + 3\sin t\,\mathbf{j}$

Fig. 5-4

Choose the parametric equations of the circle as $x = 3\cos t$, $y = 3\sin t$ where t varies from 0 to 2π (as in Fig. 5-4). Then the line integral equals

$$\int_{t=0}^{2\pi} [2(3\cos t) - 3\sin t](-3\sin t)\, dt + (3\cos t + 3\sin t)(3\cos t)\, dt$$

$$= \int_0^{2\pi} (9 - 9\sin t \cos t)\, dt = 9t - \frac{9}{2}\sin^2 t \Big|_0^{2\pi} = 18\pi$$

In traversing C, we have chosen the counterclockwise direction indicated in the adjoining figure. We call this the *positive* direction, or say that C has been traversed in the positive sense. If C were traversed in the clockwise (negative) direction the value of the integral would be -18π.

5.10. (a) Suppose $\mathbf{F} = \nabla\phi$, where ϕ is single-valued and has continuous partial derivatives. Show that the work done in moving a particle from one point $P_1 \equiv (x_1, y_1, z_1)$ in this field to another point $P_2 \equiv (x_2, y_2, z_2)$ is independent of the path joining the two points.

(b) Conversely, suppose $\int_C \mathbf{F} \cdot d\mathbf{r}$ is independent of the path C joining any two points. Show that there exists a function ϕ such that $\mathbf{F} = \nabla\phi$.

Solution

(a)

$$\text{Work done} = \int_{P_1}^{P_2} \mathbf{F} \cdot d\mathbf{r} = \int_{P_1}^{P_2} \nabla\phi \cdot d\mathbf{r}$$

$$= \int_{P_1}^{P_2} \left(\frac{\partial\phi}{\partial x}\mathbf{i} + \frac{\partial\phi}{\partial y}\mathbf{j} + \frac{\partial\phi}{\partial z}\mathbf{k} \right) \cdot (dx\mathbf{i} + dy\mathbf{j} + dz\mathbf{k})$$

$$= \int_{P_1}^{P_2} \frac{\partial\phi}{\partial x} dx + \frac{\partial\phi}{\partial y} dy + \frac{\partial\phi}{\partial z} dz$$

$$= \int_{P_1}^{P_2} d\phi = \phi(P_2) - \phi(P_1) = \phi(x_2, y_2, z_2) - \phi(x_1, y_1, z_1)$$

Then the integral depends only on points P_1 and P_2 and not on the path joining them. This is true of course only if $\phi(x, y, z)$ is single-valued at all points P_1 and P_2.

(b) Let $\mathbf{F} = F_1\mathbf{i} + F_2\mathbf{j} + F_3\mathbf{k}$. By hypothesis, $\int_C \mathbf{F} \cdot d\mathbf{r}$ is independent of the path C joining any two points, which we take as (x_1, y_1, z_1) and (x, y, z), respectively. Then

$$\phi(x, y, z) = \int_{(x_1, y_1, z_1)}^{(x, y, z)} \mathbf{F} \cdot d\mathbf{r} = \int_{(x_1, y_1, z_1)}^{(x, y, z)} F_1\, dx + F_2\, dy + F_3\, dz$$

is independent of the path joining (x_1, y_1, z_1) and (x, y, z). Thus

$$\phi(x + \Delta x, y, z) - \phi(x, y, z) = \int_{(x_1, y_1, z_1)}^{(x+\Delta x, y, z)} \mathbf{F} \cdot d\mathbf{r} - \int_{(x_1, y_1, z_1)}^{(x, y, z)} \mathbf{F} \cdot d\mathbf{r}$$

$$= \int_{(x, y, z)}^{(x_1, y_1, z_1)} \mathbf{F} \cdot d\mathbf{r} + \int_{(x_1, y_1, z_1)}^{(x+\Delta x, y, z)} \mathbf{F} \cdot d\mathbf{r}$$

$$= \int_{(x, y, z)}^{(x+\Delta x, y, z)} \mathbf{F} \cdot d\mathbf{r} = \int_{(x, y, z)}^{(x+\Delta x, y, z)} F_1 \, dx + F_2 \, dy + F_3 \, dz$$

Since the last integral must be independent of the path joining (x, y, z) and $(x + \Delta x, y, z)$, we may choose the path to be a straight line joining these points so that dy and dz are zero. Then

$$\frac{\phi(x + \Delta x, y, z) - \phi(x, y, z)}{\Delta x} = \frac{1}{\Delta x} \int_{(x, y, z)}^{(x+\Delta x, y, z)} F_1 \, dx$$

Taking the limit of both sides as $\Delta x \to 0$, we have $\partial\phi/\partial x = F_1$. Similarly, we can show that $\partial\phi/\partial y = F_2$ and $\partial\phi/\partial z = F_3$. Then

$$\mathbf{F} = F_1 \mathbf{i} + F_2 \mathbf{j} + F_3 \mathbf{k} = \frac{\partial\phi}{\partial x} \mathbf{i} + \frac{\partial\phi}{\partial y} \mathbf{j} + \frac{\partial\phi}{\partial z} \mathbf{k} = \nabla\phi.$$

If $\int_{P_1}^{P_2} \mathbf{F} \cdot d\mathbf{r}$ is independent of the path C joining P_1 and P_2, then $\mathbf{F}$ is called a *conservative field*. It follows that if $\mathbf{F} = \nabla\phi$ then $\mathbf{F}$ is conservative, and conversely.

Proof using vectors. If the line integral is independent of the path, then

$$\phi(x, y, z) = \int_{(x_1, y_1, z_1)}^{(x, y, z)} \mathbf{F} \cdot d\mathbf{r} = \int_{(x_1, y_1, z_1)}^{(x, y, z)} \mathbf{F} \cdot \frac{d\mathbf{r}}{ds} \, ds$$

By differentiation, $d\phi/ds = \mathbf{F} \cdot (d\mathbf{r}/ds)$. But $d\phi/ds = \nabla\phi \cdot (d\mathbf{r}/ds)$ so that $(\nabla\phi - \mathbf{F}) \cdot (d\mathbf{r}/ds) = 0$. Since this must hold irrespective of $d\mathbf{r}/ds$, we have $\mathbf{F} = \nabla\phi$.

5.11. (a) Suppose $\mathbf{F}$ is a conservative field. Prove that curl $\mathbf{F} = \nabla \times \mathbf{F} = 0$ (i.e. $\mathbf{F}$ is irrotational).

    (b) Conversely, if $\nabla \times \mathbf{F} = 0$ (i.e. $\mathbf{F}$ is irrotational), prove that $\mathbf{F}$ is conservative.

Solution

(a) If $\mathbf{F}$ is a conservative field, then by Problem 5.10, $\mathbf{F} = \nabla\phi$.
    Thus curl $\mathbf{F} = \nabla \times \nabla\phi = 0$ (see Problem 4.27(a), Chapter 4).

(b) If $\nabla \times \mathbf{F} = \mathbf{0}$, then $\begin{vmatrix} \mathbf{i} & \mathbf{j} & \mathbf{k} \\ \dfrac{\partial}{\partial x} & \dfrac{\partial}{\partial y} & \dfrac{\partial}{\partial z} \\ F_1 & F_2 & F_3 \end{vmatrix} = \mathbf{0}$ and thus

$$\frac{\partial F_3}{\partial y} = \frac{\partial F_2}{\partial z}, \quad \frac{\partial F_1}{\partial z} = \frac{\partial F_3}{\partial x}, \quad \frac{\partial F_2}{\partial x} = \frac{\partial F_1}{\partial y}$$

We must prove that $\mathbf{F} = \nabla \phi$ follows as a consequence of this.

The work done in moving a particle from (x_1, y_1, z_1) to (x, y, z) in the force field $\mathbf{F}$ is

$$\int_C F_1(x, y, z)\, dx + F_2(x, y, z)\, dy + F_3(x, y, z)\, dz$$

where C is a path joining (x_1, y_1, z_1) and (x, y, z). Let us choose as a particular path the straight line segments from (x_1, y_1, z_1) to (x, y_1, z_1) to (x, y, z_1) to (x, y, z) and call $\phi(x, y, z)$ the work done along this particular path. Then

$$\phi(x, y, z) = \int_{x_1}^{x} F_1(x, y_1, z_1)\, dx + \int_{y_1}^{y} F_2(x, y, z_1)\, dy + \int_{z_1}^{z} F_3(x, y, z)\, dz$$

It follows that

$$\frac{\partial \phi}{\partial z} = F_3(x, y, z)$$

$$\frac{\partial \phi}{\partial y} = F_2(x, y, z_1) + \int_{z_1}^{z} \frac{\partial F_3}{\partial y}(x, y, z)\, dz$$

$$= F_2(x, y, z_1) + \int_{z_1}^{z} \frac{\partial F_2}{\partial z}(x, y, z)\, dz$$

$$= F_2(x, y, z_1) + F_2(x, y, z)\Big|_{z_1}^{z} = F_2(x, y, z_1) + F_2(x, y, z) - F_2(x, y, z_1) = F_2(x, y, z)$$

$$\frac{\partial \phi}{\partial x} = F_1(x, y_1, z_1) + \int_{y_1}^{y} \frac{\partial F_2}{\partial x}(x, y, z_1)\, dy + \int_{z_1}^{z} \frac{\partial F_3}{\partial x}(x, y, z)\, dz$$

$$= F_1(x, y_1, z_1) + \int_{y_1}^{y} \frac{\partial F_1}{\partial y}(x, y, z_1)\, dy + \int_{z_1}^{z} \frac{\partial F_1}{\partial z}(x, y, z)\, dz$$

$$= F_1(x, y_1, z_1) + F_1(x, y, z_1)\Big|_{y_1}^{y} + F_1(x, y, z)\Big|_{z_1}^{z}$$

$$= F_1(x, y_1, z_1) + F_1(x, y, z_1) - F_1(x, y_1, z_1) + F_1(x, y, z) - F_1(x, y, z_1) = F_1(x, y, z)$$

Then

$$\mathbf{F} = F_1 \mathbf{i} + F_2 \mathbf{j} + F_3 \mathbf{k} = \frac{\partial \phi}{\partial x}\mathbf{i} + \frac{\partial \phi}{\partial y}\mathbf{j} + \frac{\partial \phi}{\partial z}\mathbf{k} = \nabla \phi.$$

Thus a necessary and sufficient condition that a field $\mathbf{F}$ be conservative is that curl $\mathbf{F} = \nabla \times \mathbf{F} = \mathbf{0}$.

5.12. (a) Show that $\mathbf{F} = (2xy + z^3)\mathbf{i} + x^2\mathbf{j} + 3xz^2\mathbf{k}$ is a conservative force field. (b) Find the scalar potential. (c) Find the work done in moving an object in this field from $(1, -2, 1)$ to $(3, 1, 4)$.

Solution

(a) From Problem 5.11, a necessary and sufficient condition that a force will be conservative is that curl $\mathbf{F} = \nabla \times \mathbf{F} = \mathbf{0}$. Now

$$\nabla \times \mathbf{F} = \begin{vmatrix} \mathbf{i} & \mathbf{j} & \mathbf{k} \\ \dfrac{\partial}{\partial x} & \dfrac{\partial}{\partial y} & \dfrac{\partial}{\partial z} \\ 2xy + z^3 & x^2 & 3xz^2 \end{vmatrix} = \mathbf{0}.$$

Thus $\mathbf{F}$ is a conservative force field.

(b) *First Method.* By Problem 5.10,

$$\mathbf{F} = \nabla\phi \text{ or } \frac{\partial\phi}{\partial x}\mathbf{i} + \frac{\partial\phi}{\partial y}\mathbf{j} + \frac{\partial\phi}{\partial z}\mathbf{k} = (2xy + z^3)\mathbf{i} + x^2\mathbf{j} + 3xz^2\mathbf{k}.$$

Then

$$\frac{\partial\phi}{\partial x} = 2xy + z^3 \tag{1}$$

$$\frac{\partial\phi}{\partial y} = x^2 \tag{2}$$

$$\frac{\partial\phi}{\partial z} = 3xz^2 \tag{3}$$

Integrating, we find from (1), (2), and (3), respectively, that

$$\begin{aligned} \phi &= x^2y &+ xz^3 &+ f(y,z) \\ \phi &= x^2y & &+ g(x,z) \\ \phi &= &xz^3 &+ h(x,y) \end{aligned}$$

These agree if we choose $f(y,z) = 0, g(x,z) = xz^3, h(x,y) = x^2y$ so that $\phi = x^2y + xz^3$ to which may be added any constant.

Second Method. Since $\mathbf{F}$ is conservative, $\int_C \mathbf{F} \cdot d\mathbf{r}$ is independent of the path C joining (x_1, y_1, z_1) and (x, y, z). Using the method of Problem 5.11(b),

$$\phi(x, y, z) = \int_{x_1}^{x} (2xy_1 + z_1^3)\, dx + \int_{y_1}^{y} x^2\, dy + \int_{z_1}^{z} 3xz^2\, dz$$

$$= \left. (x^2y_1 + xz_1^3) \right|_{x_1}^{x} + \left. x^2y \right|_{y_1}^{y} + \left. xz^3 \right|_{z_1}^{z}$$

$$= x^2y_1 + xz_1^3 - x_1^2y_1 - x_1z_1^3 + x^2y - x^2y_1 + xz^3 - xz_1^3$$

$$= x^2y + xz^3 - x_1^2y_1 - x_1z_1^3 = x^2y + xz^3 + \text{constant}$$

Third Method.

$$\mathbf{F} \cdot d\mathbf{r} = \nabla\phi \cdot d\mathbf{r} = \frac{\partial\phi}{\partial x}\,dx + \frac{\partial\phi}{\partial y}\,dy + \frac{\partial\phi}{\partial z}\,dz = d\phi.$$

Then

$$d\phi = \mathbf{F} \cdot d\mathbf{r} = (2xy + z^3)\, dx + x^2\, dy + 3xz^2\, dz$$
$$= (2xy\, dx + x^2\, dy) + (z^3\, dx + 3xz^2\, dz)$$
$$= d(x^2 y) + d(xz^3) = d(x^2 y + xz^3)$$

and $\phi = x^2 y + xz^3 + \text{constant}$.

(c) Work done $= \displaystyle\int_{P_1}^{P_2} \mathbf{F} \cdot d\mathbf{r}$

$$= \int_{P_1}^{P_2} (2xy + z^3)\, dx + x^2\, dy + 3xz^2\, dz$$

$$= \int_{P_1}^{P_2} d(x^2 y + xz^3) = x^2 y + xz^3 \Big|_{P_1}^{P_2} = x^2 y + xz^3 \Big|_{(1,-2,1)}^{(3,1,4)} = 202$$

Another Method. From part (b), $\phi(x, y, z) = x^2 y + xz^3 + \text{constant}$.
Then work done $= \phi(3, 1, 4) - \phi(1, -2, 1) = 202$.

5.13. Prove that if $\int_{P_1}^{P_2} \mathbf{F} \cdot d\mathbf{r}$ is independent of the path joining any two points P_1 and P_2 in a given region, then $\oint \mathbf{F} \cdot d\mathbf{r} = 0$ for all closed paths in the region and conversely.

Solution

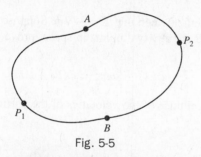

Fig. 5-5

Let $P_1 A P_2 B P_1$ (see Fig. 5-5) be a closed curve. Then

$$\oint \mathbf{F} \cdot d\mathbf{r} = \int_{P_1 A P_2 B P_1} \mathbf{F} \cdot d\mathbf{r} = \int_{P_1 A P_2} \mathbf{F} \cdot d\mathbf{r} + \int_{P_2 B P_1} \mathbf{F} \cdot d\mathbf{r}$$

$$= \int_{P_1 A P_2} \mathbf{F} \cdot d\mathbf{r} - \int_{P_1 B P_2} \mathbf{F} \cdot d\mathbf{r} = 0$$

since the integral from P_1 to P_2 along a path through A is the same as that along a path through B, by hypothesis. Conversely, if $\oint \mathbf{F} \cdot d\mathbf{r} = 0$, then

$$\int_{P_1 A P_2 B P_1} \mathbf{F} \cdot d\mathbf{r} = \int_{P_1 A P_2} \mathbf{F} \cdot d\mathbf{r} + \int_{P_2 B P_1} \mathbf{F} \cdot d\mathbf{r} = \int_{P_1 A P_2} \mathbf{F} \cdot d\mathbf{r} - \int_{P_1 B P_2} \mathbf{F} \cdot d\mathbf{r} = 0$$

so that,

$$\int_{P_1 A P_2} \mathbf{F} \cdot d\mathbf{r} = \int_{P_1 B P_2} \mathbf{F} \cdot d\mathbf{r}.$$

5.14. (a) Show that a necessary and sufficient condition that $F_1\, dx + F_2\, dy + F_3\, dz$ be an exact differential is that $\nabla \times \mathbf{F} = \mathbf{0}$ where $\mathbf{F} = F_1 \mathbf{i} + F_2 \mathbf{j} + F_3 \mathbf{k}$.

(b) Show that $(y^2 z^3 \cos x - 4x^3 z)\, dx + 2z^3 y \sin x\, dy + (3y^2 z^2 \sin x - x^4)\, dz$ is an exact differential of a function ϕ and find ϕ.

Solution

(a) Suppose

$$F_1 \, dx + F_2 \, dy + F_3 \, dz = d\phi = \frac{\partial \phi}{\partial x} \, dx + \frac{\partial \phi}{\partial y} \, dy + \frac{\partial \phi}{\partial z} \, dz,$$

an exact differential. Then, since x, y, and z are independent variables,

$$F_1 = \frac{\partial \phi}{\partial x}, \quad F_2 = \frac{\partial \phi}{\partial y}, \quad F_3 = \frac{\partial \phi}{\partial z}$$

and so $\mathbf{F} = F_1 \mathbf{i} + F_2 \mathbf{j} + F_3 \mathbf{k} = (\partial \phi / \partial x)\mathbf{i} + (\partial \phi / \partial y)\mathbf{j} + (\partial \phi / \partial z)\mathbf{k} = \nabla \phi$. Thus $\nabla \times \mathbf{F} = \nabla \times \nabla \phi = \mathbf{0}$.

Conversely, if $\nabla \times \mathbf{F} = \mathbf{0}$, then by Problem 5.11, $\mathbf{F} = \nabla \phi$ and so $\mathbf{F} \cdot d\mathbf{r} = \nabla \phi \cdot d\mathbf{r} = d\phi$, that is, $F_1 \, dx + F_2 \, dy + F_3 \, dz = d\phi$, an exact differential.

(b) $\mathbf{F} = (y^2 z^3 \cos x - 4x^3 z)\mathbf{i} + 2z^3 y \sin x \mathbf{j} + (3y^2 z^2 \sin x - x^4)\mathbf{k}$ and $\nabla \times \mathbf{F}$ is computed to be zero, so that by part (a)

$$(y^2 z^3 \cos x - 4x^3 z) \, dx + 2z^3 y \sin x \, dy + (3y^2 z^2 \sin x - x^4) \, dz = d\phi$$

By any of the methods of Problem 5.12, we find $\phi = y^2 z^3 \sin x - x^4 z + \text{constant}$.

5.15. Let $\mathbf{F}$ be a conservative force field such that $\mathbf{F} = -\nabla \phi$. Suppose a particle of constant mass m to move in this field. If A and B are any two points in space, prove that

$$\phi(A) + \tfrac{1}{2}mv_A^2 = \phi(B) + \tfrac{1}{2}mv_B^2$$

where v_A and v_B are the magnitudes of the velocities of the particle at A and B, respectively.

Solution

$$\mathbf{F} = m\mathbf{a} = m\frac{d^2\mathbf{r}}{dt^2}.$$

Then

$$\mathbf{F} \cdot \frac{d\mathbf{r}}{dt} = m\frac{d\mathbf{r}}{dt} \cdot \frac{d^2\mathbf{r}}{dt^2} = \frac{m}{2}\frac{d}{dt}\left(\frac{d\mathbf{r}}{dt}\right)^2.$$

Integrating,

$$\int_A^B \mathbf{F} \cdot d\mathbf{r} = \frac{m}{2}v^2 \Big|_A^B = \frac{1}{2}mv_B^2 - \frac{1}{2}mv_A^2.$$

If $\mathbf{F} = -\nabla \phi$,

$$\int_A^B \mathbf{F} \cdot d\mathbf{r} = -\int_A^B \nabla \phi \cdot d\mathbf{r} = -\int_A^B d\phi = \phi(A) - \phi(B).$$

Then $\phi(A) - \phi(B) = \frac{1}{2}mv_B^2 - \frac{1}{2}mv_A^2$ and the result follows.

$\phi(A)$ is called the *potential energy* at A and $\frac{1}{2}mv_A^2$ is the *kinetic energy* at A. The result states that the total energy at A equals the total energy at B (conservation of energy). Note the use of the minus sign in $\mathbf{F} = -\nabla \phi$.

5.16. Suppose $\phi = 2xyz^2$, $\mathbf{F} = xy\mathbf{i} - z\mathbf{j} + x^2\mathbf{k}$ and C is the curve $x = t^2, y = 2t, z = t^3$ from $t = 0$ to $t = 1$. Evaluate the line integrals (a) $\int_C \phi \, d\mathbf{r}$, (b) $\int_C \mathbf{F} \times d\mathbf{r}$.

Solution

(a) Along C,

$$\phi = 2xyz^2 = 2(t^2)(2t)(t^3)^2 = 4t^9,$$

$$\mathbf{r} = x\mathbf{i} + y\mathbf{j} + z\mathbf{k} = t^2\mathbf{i} + 2t\mathbf{j} + t^3\mathbf{k}, \quad \text{and}$$

$$d\mathbf{r} = (2t\mathbf{i} + 2\mathbf{j} + 3t^2\mathbf{k}) \, dt.$$

Then

$$\int_C \phi \, d\mathbf{r} = \int_{t=0}^{1} 4t^9(2t\mathbf{i} + 2\mathbf{j} + 3t^2\mathbf{k}) \, dt$$

$$= \mathbf{i} \int_0^1 8t^{10} \, dt + \mathbf{j} \int_0^1 8t^9 \, dt + \mathbf{k} \int_0^1 12t^{11} \, dt = \frac{8}{11}\mathbf{i} + \frac{4}{5}\mathbf{j} + \mathbf{k}$$

(b) Along C, we have $\mathbf{F} = xy\mathbf{i} - z\mathbf{j} + x^2\mathbf{k} = 2t^3\mathbf{i} - t^3\mathbf{j} + t^4\mathbf{k}$. Then

$$\mathbf{F} \times d\mathbf{r} = (2t^3\mathbf{i} - t^3\mathbf{j} + t^4\mathbf{k}) \times (2t\mathbf{i} + 2\mathbf{j} + 3t^2\mathbf{k}) \, dt$$

$$= \begin{vmatrix} \mathbf{i} & \mathbf{j} & \mathbf{k} \\ 2t^3 & -t^3 & t^4 \\ 2t & 2 & 3t^2 \end{vmatrix} \, dt = \left[(-3t^5 - 2t^4)\mathbf{i} + (2t^5 - 6t^5)\mathbf{j} + (4t^3 + 2t^4)\mathbf{k}\right] dt$$

and

$$\int_C \mathbf{F} \times d\mathbf{r} = \mathbf{i} \int_0^1 (-3t^5 - 2t^4) \, dt + \mathbf{j} \int_0^1 (-4t^5) \, dt + \mathbf{k} \int_0^1 (4t^3 + 2t^4) \, dt$$

$$= -\frac{9}{10}\mathbf{i} - \frac{2}{3}\mathbf{j} + \frac{7}{5}\mathbf{k}$$

Surface Integrals

5.17. Give a definition of $\iint_S \mathbf{A} \cdot \mathbf{n} \, dS$ over a surface S in terms of limit of a sum (see Fig. 5-6).

Solution

Subdivide the area S into M elements of area ΔS_p where $p = 1, 2, 3, \ldots, M$. Choose any point P_p within ΔS_p whose coordinates are (x_p, y_p, z_p). Define $\mathbf{A}(x_p, y_p, z_p) = \mathbf{A}_p$. Let $\mathbf{n}_p$ be the positive unit normal to ΔS_p at P. From the sum

$$\sum_{p=1}^{M} \mathbf{A}_p \cdot \mathbf{n}_p \, \Delta S_p$$

where $\mathbf{A}_p \cdot \mathbf{n}_p$ is the normal component of $\mathbf{A}_p$ at P_p.

Now take the limit of this sum as $M \to \infty$ in such a way that the largest dimension of each ΔS_p approaches zero. This limit, if it exists, is called the surface integral of the normal component of $\mathbf{A}$ over S and is denoted by

$$\iint_S \mathbf{A} \cdot \mathbf{n} \, dS$$

5.18. Suppose that the surface S has projection R on the xy-plane (see Fig. 5-6). Show that

$$\iint_S \mathbf{A} \cdot \mathbf{n} \, dS = \iint_R \mathbf{A} \cdot \mathbf{n} \frac{dx \, dy}{|\mathbf{n} \cdot \mathbf{k}|}$$

Solution

By Problem 5.17, the surface integral is the limit of the sum

$$\sum_{p=1}^{M} \mathbf{A}_p \cdot \mathbf{n}_p \, \Delta S_p \tag{1}$$

The projection of ΔS_p on the xy-plane is $|(\mathbf{n}_p \, \Delta S_p) \cdot \mathbf{k}|$ or $|\mathbf{n}_p \cdot \mathbf{k}| \Delta S_p$, which is equal to $\Delta x_p \Delta y_p$ so that $\Delta S_p = \Delta x_p \Delta y_p / |\mathbf{n}_p \cdot \mathbf{k}|$. Thus sum (1) becomes

$$\sum_{p=1}^{M} \mathbf{A}_p \cdot \mathbf{n}_p \frac{\Delta x_p \, \Delta y_p}{|\mathbf{n}_p \cdot \mathbf{k}|} \tag{2}$$

By the fundamental theorem of integral calculus, the limit of this sum as $M \to \infty$ in such a manner that the largest Δx_p and Δy_p approach zero is

$$\iint_R \mathbf{A} \cdot \mathbf{n} \frac{dx \, dy}{|\mathbf{n} \cdot \mathbf{k}|}$$

and so the required result follows.

Strictly speaking, the result $\Delta S_p = \Delta x_p \Delta y_p / |\mathbf{n}_p \cdot \mathbf{k}|$ is only approximately true but it can be shown on closer examination that they differ from each other by infinitesimals of order higher than $\Delta x_p \Delta y_p$, and using this the *limits* of (1) and (2) can in fact be shown equal.

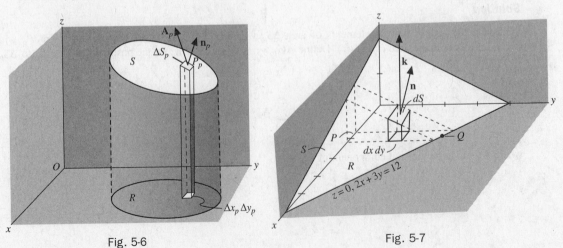

Fig. 5-6   Fig. 5-7

5.19. Evaluate $\iint_S \mathbf{A} \cdot \mathbf{n} \, dS$, where $\mathbf{A} = 18z\mathbf{i} - 12\mathbf{j} + 3y\mathbf{k}$ and S is that part of the plane $2x + 3y + 6z = 12$, which is located in the first octant.

Solution

The surface S and its projection R on the xy-plane are shown in Fig. 5-7.
From Problem 5.18,

$$\iint_S \mathbf{A} \cdot \mathbf{n} \, dS = \iint_R \mathbf{A} \cdot \mathbf{n} \frac{dx \, dy}{|\mathbf{n} \cdot \mathbf{k}|}$$

To obtain $\mathbf{n}$, note that a vector perpendicular to the surface $2x + 3y + 6z = 12$ is given by $\nabla(2x + 3y + 6z) = 2\mathbf{i} + 3\mathbf{j} + 6\mathbf{k}$ (see Problem 4.5 of Chapter 4). Then a unit normal to any point of S (see Fig. 5-7) is

$$\mathbf{n} = \frac{2\mathbf{i} + 3\mathbf{j} + 6\mathbf{k}}{\sqrt{2^2 + 3^2 + 6^2}} = \tfrac{2}{7}\mathbf{i} + \tfrac{3}{7}\mathbf{j} + \tfrac{6}{7}\mathbf{k}$$

Thus $\mathbf{n} \cdot \mathbf{k} = (\tfrac{2}{7}\mathbf{i} + \tfrac{3}{7}\mathbf{j} + \tfrac{6}{7}\mathbf{k}) \cdot \mathbf{k} = \tfrac{6}{7}$ and so

$$\frac{dx \, dy}{|\mathbf{n} \cdot \mathbf{k}|} = \frac{7}{6} dx \, dy.$$

Also

$$\mathbf{A} \cdot \mathbf{n} = (18z\mathbf{i} - 12\mathbf{j} + 3y\mathbf{k}) \cdot (\tfrac{2}{7}\mathbf{i} + \tfrac{3}{7}\mathbf{j} + \tfrac{6}{7}\mathbf{k}) = \frac{36z - 36 + 18y}{7} = \frac{36 - 12x}{7},$$

using the fact that $z = (12 - 2x - 3y)/6$ from the equation of S. Then

$$\iint_S \mathbf{A} \cdot \mathbf{n} \, dS = \iint_R \mathbf{A} \cdot \mathbf{n} \frac{dx \, dy}{|\mathbf{n} \cdot \mathbf{k}|} = \iint_R \left(\frac{36 - 12x}{7}\right) \frac{7}{6} \, dx \, dy = \iint_R (6 - 2x) \, dx \, dy$$

To evaluate this double integral over R, keep x fixed and integrate with respect to y from $y = 0$ (P in the figure above) to $y = (12 - 2x)/3$ (Q in the figure above); then integrate with respect to x from $x = 0$ to $x = 6$. In this manner, R is completely covered. The integral becomes

$$\int_{x=0}^{6} \int_{y=0}^{(12-2x)/3} (6 - 2x) \, dy \, dx = \int_{x=0}^{6} \left(24 - 12x + \frac{4x^2}{3}\right) dx = 24$$

If we had chosen the positive unit normal $\mathbf{n}$ opposite to that in Fig. 5-7, we would have obtained the result -24.

5.20. Evaluate $\iint_S \mathbf{A} \cdot \mathbf{n} \, dS$, where $\mathbf{A} = z\mathbf{i} + x\mathbf{j} - 3y^2z\mathbf{k}$ and S is the surface of the cylinder $x^2 + y^2 = 16$ included in the first octant between $z = 0$ and $z = 5$.

Solution

Project S on the xz-plane as in Fig. 5-8 and call the projection R. Note that the projection of S on the xy-plane cannot be used here. Then

$$\iint_S \mathbf{A} \cdot \mathbf{n} \, dS = \iint_R \mathbf{A} \cdot \mathbf{n} \frac{dx \, dz}{|\mathbf{n} \cdot \mathbf{j}|}$$

A normal to $x^2 + y^2 = 16$ is $\nabla(x^2 + y^2) = 2x\mathbf{i} + 2y\mathbf{j}$. Thus as shown in Fig. 5-8 the unit normal to S is

$$\mathbf{n} = \frac{2x\mathbf{i} + 2y\mathbf{j}}{\sqrt{(2x)^2 + (2y)^2}} = \frac{x\mathbf{i} + y\mathbf{j}}{4}$$

since $x^2 + y^2 = 16$ on S.

$$\mathbf{A} \cdot \mathbf{n} = (z\mathbf{i} + x\mathbf{j} - 3y^2 z\mathbf{k}) \cdot \left(\frac{x\mathbf{i} + y\mathbf{j}}{4}\right) = \frac{1}{4}(xz + xy)$$

$$\mathbf{n} \cdot \mathbf{j} = \frac{x\mathbf{i} + y\mathbf{j}}{4} \cdot \mathbf{j} = \frac{y}{4}.$$

Then the surface integral equals

$$\iint\limits_{R} \frac{xz + xy}{y}\, dx\, dz = \int\limits_{z=0}^{5} \int\limits_{x=0}^{4} \left(\frac{xz}{\sqrt{16 - x^2}} + x\right) dx\, dz = \int\limits_{z=0}^{5} (4z + 8)\, dz = 90$$

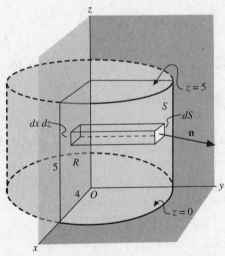

Fig. 5-8

5.21. Evaluate $\iint_S \phi \mathbf{n}\, dS$ where $\phi = \frac{3}{8}xyz$ and S is the surface of Problem 5.20.

Solution

We have

$$\iint\limits_{S} \phi \mathbf{n}\, dS = \iint\limits_{R} \phi \mathbf{n}\, \frac{dx\, dz}{|\mathbf{n} \cdot \mathbf{j}|}$$

Using $\mathbf{n} = (x\mathbf{i} + y\mathbf{j})/4, \mathbf{n} \cdot \mathbf{j} = y/4$ as in Problem 5.20, this last integral becomes

$$\iint_R \frac{3}{8} xz(x\mathbf{i} + y\mathbf{j}) \, dx \, dz = \frac{3}{8} \int_{z=0}^{5} \int_{x=0}^{4} (x^2 z\mathbf{i} + xz\sqrt{16 - x^2}\mathbf{j}) \, dx \, dz$$

$$= \frac{3}{8} \int_{z=0}^{5} \left(\frac{64}{3} z\mathbf{i} + \frac{64}{3} z\mathbf{j} \right) dz = 100\mathbf{i} + 100\mathbf{j}$$

5.22. Suppose $\mathbf{F} = y\mathbf{i} + (x - 2xz)\mathbf{j} - xy\mathbf{k}$. Evaluate $\iint_S (\nabla \times \mathbf{F}) \cdot \mathbf{n} \, dS$ where S is the surface of the sphere $x^2 + y^2 + z^2 = a^2$ above the xy-plane (see Fig. 5-9).

Solution

$$\nabla \times \mathbf{F} = \begin{vmatrix} \mathbf{i} & \mathbf{j} & \mathbf{k} \\ \dfrac{\partial}{\partial x} & \dfrac{\partial}{\partial y} & \dfrac{\partial}{\partial z} \\ y & x - 2xz & -xy \end{vmatrix} = x\mathbf{i} + y\mathbf{j} - 2z\mathbf{k}$$

A normal to $x^2 + y^2 + z^2 = a^2$ is

$$\nabla(x^2 + y^2 + z^2) = 2x\mathbf{i} + 2y\mathbf{j} + 2z\mathbf{k}$$

Then the unit normal $\mathbf{n}$ of Fig. 5-9 is given by

$$\mathbf{n} = \frac{2x\mathbf{i} + 2y\mathbf{j} + 2z\mathbf{k}}{\sqrt{4x^2 + 4y^2 + 4z^2}} = \frac{x\mathbf{i} + y\mathbf{j} + z\mathbf{k}}{a}$$

since $x^2 + y^2 + z^2 = a^2$.

The projection of S on the xy-plane is the region R bounded by the circle $x^2 + y^2 = a^2, z = 0$ (see Fig. 5-9). Then

$$\iint_S (\nabla \times \mathbf{F}) \cdot \mathbf{n} \, dS = \iint_R (\nabla \times \mathbf{F}) \cdot \mathbf{n} \frac{dx \, dy}{|\mathbf{n} \cdot \mathbf{k}|}$$

$$= \iint_R (x\mathbf{i} + y\mathbf{j} - 2z\mathbf{k}) \cdot \left(\frac{x\mathbf{i} + y\mathbf{j} + z\mathbf{k}}{a} \right) \frac{dx \, dy}{z/a}$$

$$= \int_{x=-a}^{a} \int_{y=-\sqrt{a^2-x^2}}^{\sqrt{a^2-x^2}} \frac{3(x^2 + y^2) - 2a^2}{\sqrt{a^2 - x^2 - y^2}} \, dy \, dx$$

using the fact that $z = \sqrt{a^2 - x^2 - y^2}$. To evaluate the double integral, transform to polar coordinates (ρ, ϕ) where $x = \rho \cos \phi, y = \rho \sin \phi$, and $dy\, dx$ is replaced by $\rho\, d\rho\, d\phi$. The double integral becomes

$$\int_{\phi=0}^{2\pi} \int_{\rho=0}^{a} \frac{3\rho^2 - 2a^2}{\sqrt{a^2 - \rho^2}} \rho\, d\rho\, d\phi = \int_{\phi=0}^{2\pi} \int_{\rho=0}^{a} \frac{3(\rho^2 - a^2) + a^2}{\sqrt{a^2 - \rho^2}} \rho\, d\rho\, d\phi$$

$$= \int_{\phi=0}^{2\pi} \int_{\rho=0}^{a} \left(-3\rho\sqrt{a^2 - \rho^2} + \frac{a^2 \rho}{\sqrt{a^2 - \rho^2}} \right) d\rho\, d\phi$$

$$= \int_{\phi=0}^{2\pi} \left[(a^2 - \rho^2)^{3/2} - a^2 \sqrt{a^2 - \rho^2} \Big|_{\rho=0}^{a} \right] d\phi$$

$$= \int_{\phi=0}^{2\pi} (a^3 - \rho^3)\, d\phi = 0$$

Fig. 5-9

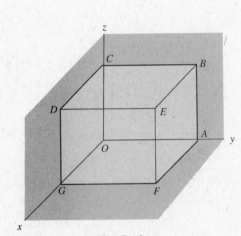

Fig. 5-10

5.23. Let $\mathbf{F} = 4xz\mathbf{i} - y^2\mathbf{j} + yz\mathbf{k}$. Evaluate $\iint_S \mathbf{F} \cdot \mathbf{n}\, dS$ where S is the surface of the cube bounded by $x = 0$, $x = 1$, $y = 0$, $y = 1$, $z = 0$, $z = 1$. (See Fig. 5-10).

Solution

Face DEFG: $\mathbf{n} = \mathbf{i}$, $x = 1$. Then

$$\iint_{DEFG} \mathbf{F} \cdot \mathbf{n}\, dS = \int_0^1 \int_0^1 (4z\mathbf{i} - y^2\mathbf{j} + yz\mathbf{k}) \cdot \mathbf{i}\, dy\, dz$$

$$= \int_0^1 \int_0^1 4z\, dy\, dz = 2$$

Face ABCO: $\mathbf{n} = -\mathbf{i}$, $x = 0$. Then

$$\iint_{ABCO} \mathbf{F} \cdot \mathbf{n}\, dS = \int_0^1 \int_0^1 (-y^2\mathbf{j} + yz\mathbf{k}) \cdot (-\mathbf{i})\, dy\, dz = 0$$

Face ABEF: $\mathbf{n} = \mathbf{j}$, $y = 1$. Then

$$\iint\limits_{ABEF} \mathbf{F} \cdot \mathbf{n} \, dS = \int_0^1 \int_0^1 (4xz\mathbf{i} - \mathbf{j} + z\mathbf{k}) \cdot \mathbf{j} \, dx \, dz = \int_0^1 \int_0^1 -dx \, dz = -1$$

Face OGDC: $\mathbf{n} = -\mathbf{j}$, $y = 0$. Then

$$\iint\limits_{OGDC} \mathbf{F} \cdot \mathbf{n} \, dS = \int_0^1 \int_0^1 (4xz\mathbf{i}) \cdot (-\mathbf{j}) \, dx \, dz = 0$$

Face BCDE: $\mathbf{n} = \mathbf{k}$, $z = 1$. Then

$$\iint\limits_{BCDE} \mathbf{F} \cdot \mathbf{n} \, dS = \int_0^1 \int_0^1 (4x\mathbf{i} - y^2\mathbf{j} + y\mathbf{k}) \cdot \mathbf{k} \, dx \, dy = \int_0^1 \int_0^1 y \, dx \, dy = \tfrac{1}{2}$$

Face AFGO: $\mathbf{n} = -\mathbf{k}$, $z = 0$. Then

$$\iint\limits_{AFGO} \mathbf{F} \cdot \mathbf{n} \, dS = \int_0^1 \int_0^1 (-y^2\mathbf{j}) \cdot (-\mathbf{k}) \, dx \, dy = 0$$

Adding, $\iint_S \mathbf{F} \cdot \mathbf{n} \, dS = 2 + 0 + (-1) + 0 + \tfrac{1}{2} + 0 = \tfrac{3}{2}$.

5.24. In dealing with surface integrals, we have restricted ourselves to surfaces that are two-sided. Give an example of a surface that is not two-sided.

Solution

Take a strip of paper such as *ABCD* as shown in Fig. 5-11. Twist the strip so that points *A* and *B* fall on *D* and *C*, respectively, as in Fig. 5-11. If **n** is the positive normal at point *P* of the surface, we find that as **n** moves around the surface, it reverses its original direction when it reaches *P* again. If we tried to color only one side of the surface, we would find the whole thing colored. This surface, called a *Moebius strip*, is an example of a one-sided surface. This is sometimes called a *non-orientable* surface. A two-sided surface is *orientable*.

Fig. 5-11

Volume Integrals

5.25. Let $\phi = 45x^2y$ and let V denote the closed region bounded by the planes $4x + 2y + z = 8$, $x = 0$, $y = 0$, $z = 0$. (a) Express $\iiint_V \phi \, dV$ as the limit of a sum. (b) Evaluate the integral in (a).

Solution

(a) Subdivide region V into M cubes having volume $\Delta V_k = \Delta x_k \Delta y_k \Delta z_k$, $k = 1, 2, \ldots, M$ as indicated in Fig. 5-12 and let (x_k, y_k, z_k) be a point within this cube. Define $\phi(x_k, y_k, z_k) = \phi_k$. Consider the sum

$$\sum_{k=1}^{M} \phi_k \Delta V_k \tag{1}$$

taken over all possible cubes in the region. The limit of this sum, when $M \to \infty$ in such a manner that the largest of the quantities ΔV_k will approach zero, if it exists, is denoted by $\iiint_V \phi \, dV$. It can be shown that this limit is independent of the method of subdivision if ϕ is continuous throughout V.

In forming the sum (1) over all possible cubes in the region, it is advisable to proceed in an orderly fashion. One possibility is to first add all terms in (1) corresponding to volume elements contained in a column such as PQ in the above figure. This amounts to keeping x_k and y_k fixed and adding over all z_ks. Next, keep x_k fixed but sum over all y_ks. This amounts to adding all columns such as PQ contained in a slab RS, and consequently amounts to summing over all cubes contained in such a slab. Finally, vary x_k. This amounts to addition of all slabs such as RS.

In the process outlined, the summation is taken first over z_ks, then over y_ks, and finally over x_ks. However, the summation can clearly be taken in any other order.

Fig. 5-12

(b) The ideas involved in the method of summation outlined in (a) can be used in evaluating the integral. Keeping x and y constant, integrate from $z = 0$ (base of column PQ) to $z = 8 - 4x - 2y$ (top of column PQ). Next keep x constant and integrate with respect to y. This amounts to addition of columns having bases in the xy-plane ($z = 0$) located anywhere from R (where $y = 0$) to S (where $4x + 2y = 8$ or $y = 4 - 2x$), and the integration is from $y = 0$ to $y = 4 - 2x$. Finally, we add all slabs parallel to the yz-plane, which amounts to integration from $x = 0$ to $x = 2$. The integration can be written

$$\int_{x=0}^{2} \int_{y=0}^{4-2x} \int_{z=0}^{8-4x-2y} 45x^2 y \, dz \, dy \, dx = 45 \int_{x=0}^{2} \int_{y=0}^{4-2x} x^2 y (8 - 4x - 2y) \, dy \, dx$$

$$= 45 \int_{x=0}^{2} \frac{1}{3} x^2 (4 - 2x)^3 \, dx = 128$$

Note: Physically, the result can be interpreted as the mass of the region V in which the density ϕ varies according to the formula $\phi = 45x^2 y$.

5.26. Let $\mathbf{F} = 2xz\mathbf{i} - x\mathbf{j} + y^2\mathbf{k}$. Evaluate $\iiint_V \mathbf{F}\,dV$ where V is the region bounded by the surfaces $x = 0$, $y = 0$, $y = 6$, $z = x^2$, $z = 4$, as pictured in Fig. 5-13.

Solution

The region V is covered (a) by keeping x and y fixed and integrating from $z = x^2$ to $z = 4$ (base to top of column PQ), (b) then by keeping x fixed and integrating from $y = 0$ to $y = 6$ (R to S in the slab), (c) finally integrating from $x = 0$ to $x = 2$ (where $z = x^2$ meets $z = 4$). Then the required integral is

$$\int_{x=0}^{2}\int_{y=0}^{6}\int_{z=x^2}^{4}(2xz\mathbf{i} - x\mathbf{j} + y^2\mathbf{k})\,dz\,dy\,dx$$

$$= \mathbf{i}\int_{0}^{2}\int_{0}^{6}\int_{x^2}^{4}2xz\,dz\,dy\,dx - \mathbf{j}\int_{0}^{2}\int_{0}^{6}\int_{x^2}^{4}x\,dz\,dy\,dx + \mathbf{k}\int_{0}^{2}\int_{0}^{6}\int_{x^2}^{4}y^2\,dz\,dy\,dx = 128\mathbf{i} - 24\mathbf{j} + 384\mathbf{k}$$

Fig. 5-13   Fig. 5-14

5.27. Find the volume of the region common to the intersecting cylinders $x^2 + y^2 = a^2$ and $x^2 + z^2 = a^2$.

Solution

Required volume = 8 times volume of region shown in Fig. 5-14

$$= 8\int_{x=0}^{a}\int_{y=0}^{\sqrt{a^2-x^2}}\int_{z=0}^{\sqrt{a^2-x^2}}dz\,dy\,dx$$

$$= 8\int_{x=0}^{a}\int_{y=0}^{\sqrt{a^2-x^2}}\sqrt{a^2-x^2}\,dy\,dx = 8\int_{x=0}^{a}(a^2-x^2)\,dx = \frac{16a^3}{3}$$

SUPPLEMENTARY PROBLEMS

5.28. Suppose $\mathbf{R}(t) = (3t^2 - t)\mathbf{i} + (2 - 6t)\mathbf{j} - 4t\mathbf{k}$. Find (a) $\int \mathbf{R}(t)\,dt$ and (b) $\int_2^4 \mathbf{R}(t)\,dt$.

5.29. Evaluate $\int_0^{\pi/2}(3\sin u\mathbf{i} + 2\cos u\mathbf{j})\,du$.

5.30. Let $\mathbf{A}(t) = t\mathbf{i} - t^2\mathbf{j} + (t - 1)\mathbf{k}$ and $\mathbf{B}(t) = 2t^2\mathbf{i} + 6t\mathbf{k}$. Evaluate (a) $\int_0^2 \mathbf{A}\cdot\mathbf{B}\,dt$, (b) $\int_0^2 \mathbf{A}\times\mathbf{B}\,dt$.

5.31. Let $\mathbf{A} = t\mathbf{i} - 3\mathbf{j} + 2t\mathbf{k}$, $\mathbf{B} = \mathbf{i} - 2\mathbf{j} + 2\mathbf{k}$, $\mathbf{C} = 3\mathbf{i} + t\mathbf{j} - \mathbf{k}$. Evaluate (a) $\int_1^2 \mathbf{A}\cdot\mathbf{B}\times\mathbf{C}\,dt$, (b) $\int_1^2 \mathbf{A}\times(\mathbf{B}\times\mathbf{C})\,dt$.

5.32. The acceleration **a** of a particle at any time $t \geq 0$ is given by $\mathbf{a} = e^{-t}\mathbf{i} - 6(t+1)\mathbf{j} + 3\sin t\mathbf{k}$. If the velocity **v** and displacement **r** are zero at $t = 0$, find **v** and **r** at any time.

5.33. The acceleration **a** of an object at any time t is given by $\mathbf{a} = -g\mathbf{j}$, where g is a constant. At $t = 0$, the velocity is given by $\mathbf{v} = v_0 \cos \theta_0 \mathbf{i} + v_0 \sin \theta_0 \mathbf{j}$ and the displacement $\mathbf{r} = \mathbf{0}$. Find **v** and **r** at any time $t > 0$. This describes the motion of a projectile fired from a cannon inclined at angle θ_0 with the positive x-axis with initial velocity of magnitude v_0.

5.34. Suppose $\mathbf{A}(2) = 2\mathbf{i} - \mathbf{j} + 2\mathbf{k}$ and $\mathbf{A}(3) = 4\mathbf{i} - 2\mathbf{j} + 3\mathbf{k}$. Evaluate $\int_2^3 \mathbf{A} \cdot (d\mathbf{A}/dt)\, dt$.

5.35. Find the areal velocity of a particle that moves along the path $\mathbf{r} = a \cos \omega t\mathbf{i} + b \sin \omega t\mathbf{j}$ where a, b, and ω are constants and t is time.

5.36. Prove that the squares of the periods of the planets in their motion around the Sun are proportional to the cubes of the major axes of their elliptical paths (Kepler's third law).

5.37. Let $\mathbf{A} = (2y+3)\mathbf{i} + xz\mathbf{j} + (yz-x)\mathbf{k}$. Evaluate $\int_C \mathbf{A} \cdot d\mathbf{r}$ along the following paths C:

(a) $x = 2t^2$, $y = t$, $z = t^3$ from $t = 0$ to $t = 1$,

(b) the straight lines from $(0, 0, 0)$ to $(0, 0, 1)$, then to $(0, 1, 1)$, and then to $(2, 1, 1)$,

(c) the straight line joining $(0, 0, 0)$ and $(2, 1, 1)$.

5.38. Suppose $\mathbf{F} = (5xy - 6x^2)\mathbf{i} + (2y - 4x)\mathbf{j}$. Evaluate $\int_C \mathbf{F} \cdot d\mathbf{r}$ along the curve C in the xy-plane, $y = x^3$ from the point $(1, 1)$ to $(2, 8)$.

5.39. Let $\mathbf{F} = (2x + y)\mathbf{i} + (3y - x)\mathbf{j}$. Evaluate $\int_C \mathbf{F} \cdot d\mathbf{r}$ where C is the curve in the xy-plane consisting of the straight lines from $(0, 0)$ to $(2, 0)$ and then to $(3, 2)$.

5.40. Find the work done in moving a particle in the force field $\mathbf{F} = 3x^2\mathbf{i} + (2xz - y)\mathbf{j} + z\mathbf{k}$ along

(a) the straight line from $(0, 0, 0)$ to $(2, 1, 3)$.

(b) the space curve $x = 2t^2$, $y = t$, $z = 4t^2 - t$ from $t = 0$ to $t = 1$.

(c) the curve defined by $x^2 = 4y$, $3x^3 = 8z$ from $x = 0$ to $x = 2$.

5.41. Evaluate $\oint_C \mathbf{F} \cdot d\mathbf{r}$ where $\mathbf{F} = (x - 3y)\mathbf{i} + (y - 2x)\mathbf{j}$ and C is the closed curve in the xy-plane, $x = 2 \cos t$, $y = 3 \sin t$ from $t = 0$ to $t = 2\pi$.

5.42. Suppose **T** is a unit tangent vector to the curve $C, \mathbf{r} = \mathbf{r}(u)$. Show that the work done in moving a particle in a force field **F** along C is given by $\int_C \mathbf{F} \cdot \mathbf{T}\, ds$ where s is the arc length.

5.43. Let $\mathbf{F} = (2x + y^2)\mathbf{i} + (3y - 4x)\mathbf{j}$. Evaluate $\oint_C \mathbf{F} \cdot d\mathbf{r}$ around the triangle C of Fig. 5-15 (a) in the indicated direction, (b) opposite to the indicated direction.

Fig. 5-15

Fig. 5-16

5.44. Let $\mathbf{A} = (x - y)\mathbf{i} + (x + y)\mathbf{j}$. Evaluate $\oint_C \mathbf{A} \cdot d\mathbf{r}$ around the closed curve C of Fig. 5-16.

5.45. Let $\mathbf{A} = (y - 2x)\mathbf{i} + (3x + 2y)\mathbf{j}$. Compute the circulation of $\mathbf{A}$ about a circle C in the xy-plane with center at the origin and radius 2, if C is traversed in the positive direction.

5.46. (a) Suppose $\mathbf{A} = (4xy - 3x^2z^2)\mathbf{i} + 2x^2\mathbf{j} - 2x^3z\mathbf{k}$. Prove that $\int_C \mathbf{A} \cdot d\mathbf{r}$ is independent of the curve C joining two given points. (b) Show that there is a differentiable function ϕ such that $\mathbf{A} = \nabla\phi$ and find it.

5.47. (a) Prove that $\mathbf{F} = (y^2 \cos x + z^3)\mathbf{i} + (2y \sin x - 4)\mathbf{j} + (3xz^2 + 2)\mathbf{k}$ is a conservative force field.

   (b) Find the scalar potential for $\mathbf{F}$.

   (c) Find the work done in moving an object in this field from $(0, 1, -1)$ to $(\pi/2, -1, 2)$.

5.48. Prove that $\mathbf{F} = r^2\mathbf{r}$ is conservative and find the scalar potential.

5.49. Determine whether the force field $\mathbf{F} = 2xz\mathbf{i} + (x^2 - y)\mathbf{j} + (2z - x^2)\mathbf{k}$ is conservative or non-conservative.

5.50. Show that the work done on a particle in moving it from A to B equals its change in kinetic energies at these points whether the force field is conservative or not.

5.51. Given $\mathbf{A} = (yz + 2x)\mathbf{i} + xz\mathbf{j} + (xy + 2z)\mathbf{k}$. Evaluate $\int_C \mathbf{A} \cdot d\mathbf{r}$ along the curve $x^2 + y^2 = 1$, $z = 1$ in the positive direction from $(0, 1, 1)$ to $(1, 0, 1)$.

5.52. (a) Let $\mathbf{E} = r\mathbf{r}_4$. Is there a function ϕ such that $\mathbf{E} = -\nabla\phi$? If so, find it. (b) Evaluate $\oint_C \mathbf{E} \cdot d\mathbf{r}$ if C is any simple closed curve.

5.53. Show that $(2x \cos y + z \sin y)\,dx + (xz \cos y - x^2 \sin y)\,dy + x \sin y\,dz$ is an exact differential. Hence, solve the differential equation $(2x \cos y + z \sin y)\,dx + (xz \cos y - x^2 \sin y)\,dy + x \sin y\,dz = 0$.

5.54. Solve (a) $(e^{-y} + 3x^2y^2)\,dx + (2x^3y - xe^{-y})\,dy = 0$,

   (b) $(z - e^{-x} \sin y)\,dx + (1 + e^{-x} \cos y)\,dy + (x - 8z)\,dz = 0$.

5.55. Given $\phi = 2xy^2z + x^2y$. Evaluate $\int_C \phi\,d\mathbf{r}$ where C

   (a) is the curve $x = t$, $y = t^2$, $z = t^3$ from $t = 0$ to $t = 1$,

   (b) consists of the straight lines from $(0, 0, 0)$ to $(1, 0, 0)$, then to $(1, 1, 0)$, and then to $(1, 1, 1)$.

5.56. Let $\mathbf{F} = 2y\mathbf{i} - z\mathbf{j} + x\mathbf{k}$. Evaluate $\int_C \mathbf{F} \times d\mathbf{r}$ along the curve $x = \cos t$, $y = \sin t$, $z = 2 \cos t$ from $t = 0$ to $t = \pi/2$.

5.57. Suppose $\mathbf{A} = (3x + y)\mathbf{i} - x\mathbf{j} + (y - 2)\mathbf{k}$ and $\mathbf{B} = 2\mathbf{i} - 3\mathbf{j} + \mathbf{k}$. Evaluate $\oint_C (\mathbf{A} \times \mathbf{B}) \times d\mathbf{r}$ around the circle in the xy-plane having center at the origin and radius 2 traversed in the positive direction.

5.58. Evaluate $\iint_S \mathbf{A} \cdot \mathbf{n}\,dS$ for each of the following cases.

   (a) $\mathbf{A} = y\mathbf{i} + 2x\mathbf{j} - z\mathbf{k}$ and S is the surface of the plane $2x + y = 6$ in the first octant cut off by the plane $z = 4$.

   (b) $\mathbf{A} = (x + y^2)\mathbf{i} - 2x\mathbf{j} + 2yz\mathbf{k}$ and S is the surface of the plane $2x + y + 2z = 6$ in the first octant.

5.59. Suppose $\mathbf{F} = 2y\mathbf{i} - z\mathbf{j} + x^2\mathbf{k}$ and S is the surface of the parabolic cylinder $y^2 = 8x$ in the first octant bounded by the planes $y = 4$ and $z = 6$. Evaluate $\iint_S \mathbf{F} \cdot \mathbf{n}\,dS$.

5.60. Suppose $\mathbf{A} = 6z\mathbf{i} + (2x + y)\mathbf{j} - x\mathbf{k}$. Evaluate $\iint_S \mathbf{A} \cdot \mathbf{n}\,dS$ over the entire surface S of the region bounded by the cylinder $x^2 + z^2 = 9$, $x = 0$, $y = 0$, $z = 0$, and $y = 8$.

5.61. Evaluate $\iint_S \mathbf{r} \cdot \mathbf{n}\, dS$ over: (a) the surface S of the unit cube bounded by the coordinate planes and the planes $x = 1$, $y = 1$, $z = 1$; (b) the surface of a sphere of radius a with center at $(0, 0, 0)$.

5.62. Suppose $\mathbf{A} = 4xz\mathbf{i} + xyz^2\mathbf{j} + 3z\mathbf{k}$. Evaluate $\iint_S \mathbf{A} \cdot \mathbf{n}\, dS$ over the entire surface of the region above the xy-plane bounded by the cone $z^2 = x^2 + y^2$ and the plane $z = 4$.

5.63. (a) Let R be the projection of a surface S on the xy-plane. Prove that the surface area of S is given by

$$\iint_R \sqrt{1 + \left(\frac{\partial z}{\partial x}\right)^2 + \left(\frac{\partial z}{\partial y}\right)^2}\, dx\, dy \text{ if the equation for } S \text{ is } z = f(x, y).$$

(b) What is the surface area if S has the equation $F(x, y, z) = 0$?

5.64. Find the surface area of the plane $x + 2y + 2z = 12$ cut off by: (a) $x = 0$, $y = 0$, $x = 1$, $y = 1$; (b) $x = 0$, $y = 0$, and $x^2 + y^2 = 16$.

5.65. Find the surface area of the region common to the intersecting cylinders $x^2 + y^2 = a^2$ and $x^2 + z^2 = a^2$.

5.66. Evaluate (a) $\iint_S (\nabla \times \mathbf{F}) \cdot \mathbf{n}\, dS$ and (b) $\iint_S \phi \mathbf{n}\, dS$ if $\mathbf{F} = (x + 2y)\mathbf{i} - 3z\mathbf{j} + x\mathbf{k}$, $\phi = 4x + 3y - 2z$, and S is the surface of $2x + y + 2z = 6$ bounded by $x = 0$, $x = 1$, $y = 0$ and $y = 2$.

5.67. Solve the preceding problem if S is the surface of $2x + y + 2z = 6$ bounded by $x = 0$, $y = 0$, and $z = 0$.

5.68. Evaluate $\iint_R \sqrt{x^2 + y^2}\, dx\, dy$ over the region R in the xy-plane bounded by $x^2 + y^2 = 36$.

5.69. Evaluate $\iiint_V (2x + y)\, dV$, where V is the closed region bounded by the cylinder $z = 4 - x^2$ and the planes $x = 0$, $y = 0$, $y = 2$, and $z = 0$.

5.70. Suppose $\mathbf{F} = (2x^2 - 3z)\mathbf{i} - 2xy\mathbf{j} - 4x\mathbf{k}$. Evaluate (a) $\iiint_V \nabla \cdot \mathbf{F}\, dV$ and (b) $\iiint_V \nabla \times \mathbf{F}\, dV$, where V is the closed region bounded by the planes $x = 0$, $y = 0$, $z = 0$, and $2x + 2y + z = 4$.

ANSWERS TO SUPPLEMENTARY PROBLEMS

5.28. (a) $(t^3 - t^2/2)\mathbf{i} + (2t - 3t^2)\mathbf{j} - 2t^2\mathbf{k} + \mathbf{c}$,
(b) $50\mathbf{i} - 32\mathbf{j} - 24\mathbf{k}$

5.29. $3\mathbf{i} + 2\mathbf{j}$

5.30. (a) 12, (b) $-24\mathbf{i} - \frac{40}{3}\mathbf{j} + \frac{65}{5}\mathbf{k}$

5.31. (a) 0, (b) $-\frac{87}{2}\mathbf{i} - \frac{44}{3}\mathbf{j} + \frac{15}{2}\mathbf{k}$

5.32. $\mathbf{v} = (1 - e^{-t})\mathbf{i} - (3t^2 + 6t)\mathbf{j} + (3 - 3\cos t)\mathbf{k}$,
$\mathbf{r} = (t - 1 + e^{-t})\mathbf{i} - (t^3 + 3t^2)\mathbf{j} + (3t - 3\sin t)\mathbf{k}$

5.33. $\mathbf{v} = v_0 \cos \theta_0 \mathbf{i} + (v_0 \sin \theta_0 - gt)\mathbf{j}$,
$\mathbf{r} = (v_0 \cos \theta_0)t\mathbf{i} + [(v_0 \sin \theta_0)t - \frac{1}{2}gt^2]\mathbf{j}$

5.34. 10

5.35. $\frac{1}{2}ab\omega\mathbf{k}$

5.37. (a) $288/35$, (b) 10, (c) 8

5.38. 35

5.39. 11

5.40. (a) 16, (b) 14.2, (c) 16

5.41. 6π, if C is traversed in the positive (counter-clockwise) direction

5.43. (a) $-14/3$, (b) $14/3$

5.44. $2/3$

5.45. 8π

5.46. (b) $\phi = 2x^2y - x^3z^2 + \text{constant}$

5.47. (b) $\phi = y^2 \sin x + xz^3 - 4y + 2z + \text{constant}$,
(c) $15 + 4\pi$

5.48. $\phi = \dfrac{r^4}{4} + \text{constant}$

5.49. non-conservative

5.51. 1

5.52. (a) $\phi = -\dfrac{r^3}{3} + \text{constant}$, (b) 0

5.53. $x^2 \cos y + xz \sin y = \text{constant}$

5.62. 320π

5.54. (a) $xe^{-y} + x^3y^2 = \text{constant}$,
(b) $xz + e^{-x} \sin y + y - 4z^2 = \text{constant}$

5.63. $\displaystyle\iint\limits_{R} \frac{\sqrt{\left(\dfrac{\partial \mathbf{F}}{\partial x}\right)^2 + \left(\dfrac{\partial \mathbf{F}}{\partial y}\right)^2 + \left(\dfrac{\partial \mathbf{F}}{\partial z}\right)^2}}{\left|\dfrac{\partial \mathbf{F}}{\partial z}\right|} \, dx \, dy$

5.55. (a) $\dfrac{19}{45}\mathbf{i} + \dfrac{11}{15}\mathbf{j} + \dfrac{75}{77}\mathbf{k}$, (b) $\dfrac{1}{2}\mathbf{j} + 2\mathbf{k}$

5.64. (a) 3/2, (b) 6π

5.56. $\left(2 - \dfrac{\pi}{4}\right)\mathbf{i} + \left(\pi - \dfrac{1}{2}\right)\mathbf{j}$

5.65. $16a^2$

5.57. $4\pi(7\mathbf{i} + 3\mathbf{j})$

5.66. (a) 1, (b) $2\mathbf{i} + \mathbf{j} + 2\mathbf{k}$

5.58. (a) 108, (b) 81

5.67. (a) 9/2, (b) $72\mathbf{i} + 36\mathbf{j} + 72\mathbf{k}$

5.59. 132

5.68. 144π

5.60. 18π

5.69. 80/3

5.61. (a) 3, (b) $4\pi a^3$

5.70. (a) $\dfrac{8}{3}$, (b) $\dfrac{8}{3}(\mathbf{j} - \mathbf{k})$

CHAPTER 6

Divergence Theorem, Stokes' Theorem, and Related Integral Theorems

6.1 Introduction

Elementary calculus tells us that the value of the definite integral of a continuous function $f(x)$ on a closed interval $[a, b]$ can be obtained from the anti-derivative of the function evaluated on the endpoints a and b (boundary) of the interval.

There is an analogous situation in the plane and space. That is, there is a relationship between a double integral over certain regions R in the plane, and a line integral over the boundary of the region R. Similarly, there is a relationship between the volume integral over certain volumes V in space and the double integral over the surface of the boundary of V.

We discuss these theorems and others in this chapter.

6.2 Main Theorems

The following theorems apply.

THEOREM 6.1 **(Divergence Theorem of Gauss)** Suppose V is the volume bounded by a closed surface S and $\mathbf{A}$ is a vector function of position with continuous derivatives. Then

$$\iiint_V \nabla \cdot \mathbf{A} \, dV = \iint_S \mathbf{A} \cdot \mathbf{n} \, dS = \oiint_S \mathbf{A} \cdot d\mathbf{S}$$

where $\mathbf{n}$ is the positive (outward drawn) normal to S.

THEOREM 6.2 **(Stokes' Theorem)** Suppose S is an open, two-sided surface bounded by a closed, nonintersecting curve C (simple closed curve), and suppose $\mathbf{A}$ is a vector function of position with continuous derivatives. Then

$$\oint_C \mathbf{A} \cdot d\mathbf{r} = \iint_S (\nabla \times \mathbf{A}) \cdot \mathbf{n} \, dS = \iint_S (\nabla \times \mathbf{A}) \cdot d\mathbf{S}$$

where C is traversed in the positive direction.

The *direction* of C is called *positive* if an observer, walking on the boundary of S in that direction, with his head pointing in the direction of the positive normal to S, has the surface on his left.

THEOREM 6.3   **(Green's Theorem in the Plane)**   Suppose R is a closed region in the xy-plane bounded by a simple closed curve C, and suppose M and N are continuous functions of x and y having continuous derivatives in R. Then

$$\oint_C M\,dx + N\,dy = \iint_R \left(\frac{\partial N}{\partial x} - \frac{\partial M}{\partial y} \right) dx\,dy$$

where C is traversed in the positive (counter-clockwise) direction.

Unless otherwise stated, we shall always assume $\oint$ to mean that the integral is described in the positive sense.

Green's theorem in the plane is a special case of Stokes' theorem (see Problem 6.4). Also, it is of interest to notice that Gauss' divergence theorem is a generalization of Green's theorem in the plane where the (plane) region R and its closed boundary (curve) C are replaced by a (space) region V and its closed boundary (surface) S. For this reason, the divergence theorem is often called *Green's theorem in space* (see Problem 6.4).

Green's theorem in the plane also holds for regions bounded by a finite number of simple closed curves that do not intersect (see Problems 6.10 and 6.11).

6.3   Related Integral Theorems

The following propositions apply.

PROPOSITION 6.4:   The following laws hold:

     (i) $\displaystyle\iiint_V [\phi\nabla^2\psi + (\nabla\phi)\cdot(\nabla\psi)]\,dV = \iint_S (\phi\nabla\psi)\cdot d\mathbf{S}$

         This is called *Green's first identity or theorem*.

     (ii) $\displaystyle\iiint_V (\phi\nabla^2\psi - \psi\nabla^2\phi)\,dV = \iint_S (\phi\nabla\psi - \psi\nabla\phi)\cdot d\mathbf{S}$

         This is called *Green's second identity or symmetrical theorem*. See Problem 6.21.

     (iii) $\displaystyle\iiint_V \nabla\times\mathbf{A}\,dV = \iint_S (\mathbf{n}\times\mathbf{A})\,dS = \iint_S d\mathbf{S}\times\mathbf{A}$

         Note that here the dot product of Gauss' divergence theorem is replaced by the cross product (see Problem 6.23).

     (iv) $\displaystyle\oint_C \phi\,d\mathbf{r} = \iint_S (\mathbf{n}\times\nabla\phi)\,dS = \iint_S d\mathbf{S}\times\nabla\phi$

PROPOSITION 6.5:   Let ψ represent either a vector or scalar function according as the symbol $\circ$ denotes a dot or cross product, or an ordinary multiplication. Then

     (i) $\displaystyle\iiint_V \nabla\circ\psi\,dV = \iint_S \mathbf{n}\circ\psi\,dS = \iint_S d\mathbf{S}\circ\psi$

(ii) $\displaystyle\oint_C d\mathbf{r} \circ \psi = \iint_S (\mathbf{n} \times \nabla) \circ \psi \, dS = \iint_S (d\mathbf{S} \times \nabla) \circ \psi$

Gauss' divergence theorem, Stokes' theorem and Proposition 6.4 (iii) and (iv) are special cases of these results (see Problems 6.22, 6.23, and 6.34).

Integral Operator Form for ∇

It is of interest that, using the terminology of Problem 6.19, the operator ∇ can be expressed symbolically in the form

$$\nabla \circ \equiv \lim_{\Delta V \to 0} \frac{1}{\Delta V} \oiint_{\Delta S} d\mathbf{S} \circ$$

where $\circ$ denotes a dot, cross, or an ordinary multiplication (see Problem 6.25). The result proves useful in extending the concepts of gradient, divergence and curl to coordinate systems other than rectangular (see Problems 6.19 and 6.24, and Chapter 7).

SOLVED PROBLEMS

Green's Theorem in the Plane

6.1. Prove Green's theorem in the plane where C is a closed curve which has the property that any straight line parallel to the coordinate axes cuts C in at most two points.

Solution

Let the equations of the curves AEB and AFB (see Fig. 6-1) be $y = Y_1(x)$ and $y = Y_2(x)$, respectively. If R is the region bounded by C, we have

$$\iint_R \frac{\partial M}{\partial y} \, dx \, dy = \int_{x=a}^{b} \left[\int_{y=Y_1(x)}^{Y_2(x)} \frac{\partial M}{\partial y} \, dy \right] dx = \int_{x=a}^{b} M(x,y) \Big|_{y=Y_1(x)}^{Y_2(x)} dx = \int_{a}^{b} [M(x,Y_2) - M(x,Y_1)] \, dx$$

$$= -\int_{a}^{b} M(x,Y_1) \, dx - \int_{b}^{a} M(x,Y_2) \, dx = -\oint_C M \, dx$$

Then

$$\oint_C M \, dx = -\iint_R \frac{\partial M}{\partial y} \, dx \, dy \qquad (1)$$

Similarly, let the equations of curves EAF and EBF be $x = X_1(y)$ and $x = X_2(y)$, respectively. Then

$$\iint_R \frac{\partial N}{\partial x} \, dx \, dy = \int_{y=e}^{f} \left[\int_{x=X_1(y)}^{X_2(y)} \frac{\partial N}{\partial x} \, dx \right] dy = \int_{e}^{f} [N(X_2,y) - N(X_1,y)] \, dy$$

$$= \int_{f}^{e} N(X_1,y) \, dy + \int_{e}^{f} N(X_2,y) \, dy = \oint_C N \, dy$$

Then

$$\oint_C N \, dy = \iint_R \frac{\partial N}{\partial x} \, dx \, dy \qquad (2)$$

Adding (1) and (2),

$$\oint_C M\,dx + N\,dy = \iint_R \left(\frac{\partial N}{\partial x} - \frac{\partial M}{\partial y}\right) dx\,dy.$$

Fig. 6-1   Fig. 6-2

6.2. Verify Green's theorem in the plane for $\oint_C (xy + y^2)\,dx + x^2\,dy$ where C is the closed curve of the region bounded by $y = x$ and $y = x^2$ (see Fig. 6-2).

Solution

In Fig. 6-2, $y = x$ and $y = x^2$ intersect at $(0, 0)$ and $(1, 1)$, and the positive direction in traversing C is also shown. Along $y = x^2$, the line integral equals

$$\int_0^1 [(x)(x^2) + x^4]\,dx + (x^2)(2x)\,dx = \int_0^1 (3x^3 + x^4)\,dx = \frac{19}{20}$$

Along $y = x$ from $(1, 1)$ to $(0, 0)$, the line integral equals

$$\int_1^0 [(x)(x) + x^2]\,dx + x^2\,dx = \int_1^0 3x^2\,dx = -1$$

Then the required line integral $= \dfrac{19}{20} - 1 = -\dfrac{1}{20}$.

$$\iint_R \left(\frac{\partial N}{\partial x} - \frac{\partial M}{\partial y}\right) dx\,dy = \iint_R \left[\frac{\partial}{\partial x}(x^2) - \frac{\partial}{\partial y}(xy + y^2)\right] dx\,dy$$

$$= \iint_R (x - 2y)\,dx\,dy = \int_{x=0}^1 \int_{y=x^2}^x (x - 2y)\,dy\,dx$$

$$= \int_0^1 \left[\int_{x^2}^x (x - 2y)\,dy\right] dx = \int_0^1 (xy - y^2)\Big|_{x^2}^x \,dx$$

$$= \int_0^1 (x^4 - x^3)\,dx = -\frac{1}{20}$$

so that the theorem is verified.

6.3. Extend the proof of Green's theorem in the plane given in Problem 6.1 to the curves C for which lines parallel to the coordinate axes may cut C in more than two points.

Solution

Consider a closed curve C such as shown in Fig. 6-3, in which lines parallel to the axes may meet C in more than two points. By constructing line ST, the region is divided into two regions (R_1 and R_2), which are of the type considered in Problem 6.1 and for which Green's theorem applies, that is,

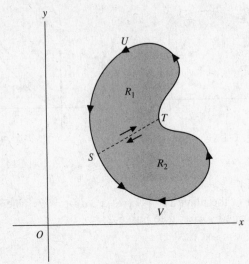

Fig. 6-3

$$\int_{STUS} M\,dx + N\,dy = \iint_{R_1} \left(\frac{\partial N}{\partial x} - \frac{\partial M}{\partial y} \right) dx\,dy \qquad (1)$$

$$\int_{SVTS} M\,dx + N\,dy = \iint_{R_2} \left(\frac{\partial N}{\partial x} - \frac{\partial M}{\partial y} \right) dx\,dy \qquad (2)$$

Adding the left-hand sides of (1) and (2), we have, omitting the integrand $M\,dx + N\,dy$ in each case,

$$\int_{STUS} + \int_{SVTS} = \int_{ST} + \int_{TUS} + \int_{SVT} + \int_{TS} = \int_{TUS} + \int_{SVT} = \int_{TUSVT}$$

using the fact that

$$\int_{ST} = -\int_{TS}$$

Adding the right-hand sides of (1) and (2), omitting the integrand,

$$\iint_{R_1} + \iint_{R_2} = \iint_{R}$$

where R consists of regions R_1 and R_2. Then

$$\int_{TUSVT} M\,dx + N\,dy = \iint_{R} \left(\frac{\partial N}{\partial x} - \frac{\partial M}{\partial y} \right) dx\,dy$$

and the theorem is proved.

A region R such as considered here and in Problem 6.1, for which any closed curve lying in R can be continuously shrunk to a point without leaving R, is called a *simply-connected region*. A region that is not simply-connected is called *multiply-connected*. We have shown here that Green's theorem in the plane applies to simply-connected regions bounded by closed curves. In Problem 6.10, the theorem is extended to multiply-connected regions.

For more complicated simply-connected regions, it may be necessary to construct more lines, such as *ST*, to establish the theorem.

6.4. Express Green's theorem in the plane in vector notation.

Solution

We have $M\,dx + N\,dy = (M\mathbf{i} + N\mathbf{j}) \cdot (dx\mathbf{i} + dy\mathbf{j}) = \mathbf{A} \cdot d\mathbf{r}$, where $\mathbf{A} = M\mathbf{i} + N\mathbf{j}$ and $\mathbf{r} = x\mathbf{i} + y\mathbf{j}$ so that $d\mathbf{r} = dx\mathbf{i} + dy\mathbf{j}$.

Also, if $\mathbf{A} = M\mathbf{i} + N\mathbf{j}$, then

$$\nabla \times \mathbf{A} = \begin{vmatrix} \mathbf{i} & \mathbf{j} & \mathbf{k} \\ \dfrac{\partial}{\partial x} & \dfrac{\partial}{\partial y} & \dfrac{\partial}{\partial z} \\ M & N & 0 \end{vmatrix} = -\frac{\partial N}{\partial z}\mathbf{i} + \frac{\partial M}{\partial z}\mathbf{j} + \left(\frac{\partial N}{\partial x} - \frac{\partial M}{\partial y}\right)\mathbf{k}$$

so that $(\nabla \times \mathbf{A}) \cdot \mathbf{k} = (\partial N/\partial x) - (\partial M/\partial y)$.

Then Green's theorem in the plane can be written

$$\oint_C \mathbf{A} \cdot d\mathbf{r} = \iint_R (\nabla \times \mathbf{A}) \cdot \mathbf{k}\, dR$$

where $dR = dx\,dy$.

A generalization of this to surfaces S in space having a curve C as a boundary leads quite naturally to *Stokes' theorem*, which is proved in Problem 6.31.

Another Method

As above, $M\,dx + N\,dy = \mathbf{A} \cdot d\mathbf{r} = \mathbf{A} \cdot (d\mathbf{r}/ds)\,ds = \mathbf{A} \cdot \mathbf{T}\,ds$, where $d\mathbf{r}/ds = \mathbf{T} = $ unit tangent vector to C (see Fig. 6-4). If $\mathbf{n}$ is the outward drawn unit normal to C, then $\mathbf{T} = \mathbf{k} \times \mathbf{n}$ so that

$$M\,dx + N\,dy = \mathbf{A} \cdot \mathbf{T}\,ds = \mathbf{A} \cdot (\mathbf{k} \times \mathbf{n})\,ds = (\mathbf{A} \times \mathbf{k}) \cdot \mathbf{n}\,ds$$

Since $\mathbf{A} = M\mathbf{i} + N\mathbf{j}$, $\mathbf{B} = \mathbf{A} \times \mathbf{k} = (M\mathbf{i} + N\mathbf{j}) \times \mathbf{k} = N\mathbf{i} - M\mathbf{j}$ and $(\partial N/\partial x) - (\partial M/\partial y) = \nabla \cdot \mathbf{B}$. Then Green's theorem in the plane becomes

$$\oint_C \mathbf{B} \cdot \mathbf{n}\,ds = \iint_R \nabla \cdot \mathbf{B}\, dR$$

where $dR = dx\,dy$.

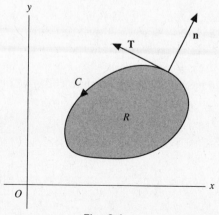

Fig. 6-4

Generalization of this to the case where the differential arc length ds of a closed curve C is replaced by the differential of surface area dS of a closed surface S, and the corresponding plane region R enclosed by C is replaced by the volume V enclosed by S, leads to *Gauss' divergence theorem* or *Green's theorem in space.*

$$\iint\limits_{S} \mathbf{B} \cdot \mathbf{n} \, dS = \iiint\limits_{V} \nabla \cdot \mathbf{B} \, dV$$

6.5. Interpret physically the first result of Problem 6.4.

Solution

If $\mathbf{A}$ denotes the force field acting on a particle, then $\oint_C \mathbf{A} \cdot d\mathbf{r}$ is the work done in moving the particle around a closed path C and is determined by the value of $\nabla \times \mathbf{A}$. It follows, in particular, that if $\nabla \times \mathbf{A} = \mathbf{0}$ or, equivalently, if $\mathbf{A} = \nabla \phi$, then the integral around a closed path is zero. This amounts to saying that the work done in moving the particle from one point in the plane to another is independent of the path in the plane joining the points or that the force field is conservative. These results have already been demonstrated for force fields and curves in space (see Chapter 5).

Conversely, if the integral is independent of the path joining any two points of a region, that is, if the integral around any closed path is zero, then $\nabla \times \mathbf{A} = \mathbf{0}$. In the plane, the condition $\nabla \times \mathbf{A} = \mathbf{0}$ is equivalent to the condition $\partial M / \partial y = \partial N / \partial x$ where $\mathbf{A} = M\mathbf{i} + N\mathbf{j}$.

6.6. Evaluate $\int_{(0,0)}^{(2,1)} (10x^4 - 2xy^3) \, dx - 3x^2 y^2 \, dy$ along the path $x^4 - 6xy^3 = 4y^2$.

Solution

A direct evaluation is difficult. However, noting that $M = 10x^4 - 2xy^3$, $N = -3x^2 y^2$ and $\partial M / \partial y = -6xy^2 = \partial N / \partial x$, it follows that the integral is independent of the path. Then we can use any path, for example the path consisting of straight line segments from (0, 0) to (2, 0) and then from (2, 0) to (2, 1).

Along the straight line path from (0, 0) to (2, 0), $y = 0$, $dy = 0$ and the integral equals $\int_{x=0}^{2} 10x^4 \, dx = 64$.

Along the straight line path from (2, 0) to (2, 1), $x = 2$, $dx = 0$ and the integral equals $\int_{y=0}^{1} -12y^2 \, dy = -4$.

Then the required value of the line integral $= 64 - 4 = 60$.

Another Method

Since $\partial M / \partial y = \partial N / \partial x, (10x^4 - 2xy^3) \, dx - 3x^2 y^2 \, dy$ is an exact differential (of $2x^5 - x^2 y^3$). Then

$$\int\limits_{(0,0)}^{(2,1)} (10x^4 - 2xy^3) \, dx - 3x^2 y^2 \, dy = \int\limits_{(0,0)}^{(2,1)} d(2x^5 - x^2 y^3) = 2x^5 - x^2 y^3 \Big|_{(0,0)}^{(2,1)} = 60$$

6.7. Show that the area bounded by a simple closed curve C is given by $\frac{1}{2} \oint_C x \, dy - y \, dx$.

Solution

In Green's theorem, put $M = -y, N = x$. Then

$$\oint\limits_{C} x \, dy - y \, dx = \iint\limits_{R} \left(\frac{\partial}{\partial x}(x) - \frac{\partial}{\partial y}(-y) \right) dx \, dy = 2 \iint\limits_{R} dx \, dy = 2A$$

where A is the required area. Thus $A = \frac{1}{2} \oint_C x \, dy - y \, dx$.

6.8. Find the area of the ellipse $x = a\cos\theta, y = b\sin\theta$.

Solution

$$\text{Area} = \frac{1}{2}\oint_C x\,dy - y\,dx = \frac{1}{2}\int_0^{2\pi}(a\cos\theta)(b\cos\theta)\,d\theta - (b\sin\theta)(-a\sin\theta)\,d\theta$$

$$= \frac{1}{2}\int_0^{2\pi}ab(\cos^2\theta + \sin^2\theta)\,d\theta = \frac{1}{2}\int_0^{2\pi}ab\,d\theta = \pi ab$$

6.9. Evaluate $\oint_C(y - \sin x)\,dx + \cos x\,dy$, where C is the triangle shown in Fig. 6-5, (a) directly, and (b) by using Green's theorem in the plane.

Solution

(a) Along OA, $y = 0$, $dy = 0$ and the integral equals

$$\int_0^{\pi/2}(0 - \sin x)\,dx + (\cos x)(0) = \int_0^{\pi/2} -\sin x\,dx = \cos x\Big|_0^{\pi/2} = -1$$

Along AB, $x = \pi/2$, $dx = 0$, and the integral equals

$$\int_0^1 (y - 1)0 + 0\,dy = 0$$

Along $BO, y = 2x/\pi, dy = (2/\pi)\,dx$, and the integral equals

$$\int_{\pi/2}^0\left(\frac{2x}{\pi} - \sin x\right)dx + \frac{2}{\pi}\cos x\,dx = \left(\frac{x^2}{\pi} + \cos x + \frac{2}{\pi}\sin x\right)\Big|_{\pi/2}^0 = 1 - \frac{\pi}{4} - \frac{2}{\pi}$$

Then the integral along $C = -1 + 0 + 1 - \frac{\pi}{4} - \frac{2}{\pi} = -\frac{\pi}{4} - \frac{2}{\pi}$.

(b) $M = y - \sin x, N = \cos x, \partial N/\partial x = -\sin x, \partial M/\partial y = 1$ and

$$\oint_C M\,dx + N\,dy = \iint_R\left(\frac{\partial N}{\partial x} - \frac{\partial M}{\partial y}\right)dx\,dy = \iint_R(-\sin x - 1)\,dy\,dx$$

$$= \int_{x=0}^{\pi/2}\left[\int_{y=0}^{2x/\pi}(-\sin x - 1)\,dy\right]dx = \int_{x=0}^{\pi/2}(-y\sin x - y)\Big|_0^{2x/\pi}\,dx$$

$$= \int_0^{\pi/2}\left(-\frac{2x}{\pi}\sin x - \frac{2x}{\pi}\right)dx = -\frac{2}{\pi}(-x\cos x + \sin x) - \frac{x^2}{\pi}\Big|_0^{\pi/2} = -\frac{2}{\pi} - \frac{\pi}{4}$$

in agreement with part (a).

Note that although there exist lines parallel to the coordinate axes (coincident with the coordinate axes in this case), which meet C in an *infinite* number of points, Green's theorem in the plane still holds. In general, the theorem is valid when C is composed of a finite number of straight line segments.

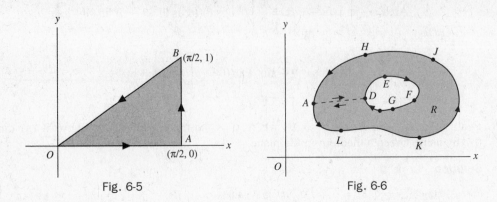

Fig. 6-5   Fig. 6-6

6.10. Show that Green's theorem in the plane is also valid for a multiply-connected region R such as shown in Fig. 6-6.

Solution

The shaded region R, shown in Fig. 6-6, is multiply-connected since not every closed curve lying in R can be shrunk to a point without leaving R, as is observed by considering a curve surrounding $DEFGD$ for example. The boundary of R, which consists of the exterior boundary $AHJKLA$ and the interior boundary $DEFGD$, is to be traversed in the positive direction, so that a person traveling in this direction always has the region on his left. Positive directions are those indicated in Fig. 6-6.

In order to establish the theorem, construct a line, such as AD, called a *cross-cut*, connecting the exterior and interior boundaries. The region bounded by $ADEFGDALKJHA$ is simply-connected, and so Green's theorem is valid. Then

$$\oint_{ADEFGDALKJHA} M\,dx + N\,dy = \iint_R \left(\frac{\partial N}{\partial x} - \frac{\partial M}{\partial y} \right) dx\,dy$$

But the integral on the left, leaving out the integrand, is equal to

$$\int_{AD} + \int_{DEFGD} + \int_{DA} + \int_{ALKJHA} = \int_{DEFGD} + \int_{ALKJHA}$$

since $\int_{AD} = -\int_{DA}$. Thus, if C_1 is the curve $ALKJHA$, C_2 is the curve $DEFGD$, and C is the boundary of R consisting of C_1 and C_2 (traversed in the positive directions), then $\int_{C_1} + \int_{C_2} = \int_C$ and so

$$\oint_C M\,dx + N\,dy = \iint_R \left(\frac{\partial N}{\partial x} - \frac{\partial M}{\partial y} \right) dx\,dy$$

6.11. Show that Green's theorem in the plane holds for the region R, of Fig. 6-7, bounded by the simple closed curves $C_1(ABDEFGA)$, $C_2(HKLPH)$, $C_3(QSTUQ)$, and $C_4(VWXYV)$.

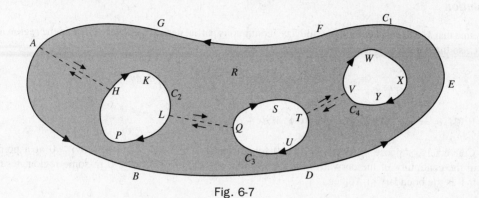

Fig. 6-7

Solution

Construct the cross-cuts AH, LQ, and TV. Then the region bounded by $AHKLQSTVWXYVTUQLPHABDEFGA$ is simply-connected and Green's theorem applies. The integral over this boundary is equal to

$$\int_{AH} + \int_{HKL} + \int_{LQ} + \int_{QST} + \int_{TV} + \int_{VWXYV} + \int_{VT} + \int_{TUQ} + \int_{QL} + \int_{LPH} + \int_{HA} + \int_{ABDEFGA}$$

Since the integrals along AH and HA, LQ and QL, and TV and VT cancel out in pairs, this becomes

$$\int_{HKL} + \int_{QST} + \int_{VWXYV} + \int_{TUQ} + \int_{LPH} + \int_{ABDEFGA}$$

$$= \left(\int_{HKL} + \int_{LPH} \right) + \left(\int_{QST} + \int_{TUQ} \right) + \int_{VWXYV} + \int_{ABDEFGA}$$

$$= \int_{HKLPH} + \int_{QSTUQ} + \int_{VWXYV} + \int_{ABDEFGA}$$

$$= \int_{C_2} + \int_{C_3} + \int_{C_4} + \int_{C_1} = \int_{C}$$

where C is the boundary consisting of C_1, C_2, C_3, and C_4. Then

$$\oint_C M\,dx + N\,dy = \iint_R \left(\frac{\partial N}{\partial x} - \frac{\partial M}{\partial y} \right) dx\,dy$$

as required.

6.12. Consider a closed curve C in a simply-connected region. Prove that $\oint_C M\,dx + N\,dy = 0$ if and only if $\partial M/\partial y = \partial N/\partial x$ everywhere in the region.

Solution

Assume that M and N are continuous and have continuous partial derivatives everywhere in the region R bounded by C, so that Green's theorem is applicable. Then

$$\oint_C M\,dx + N\,dy = \iint_R \left(\frac{\partial N}{\partial x} - \frac{\partial M}{\partial y}\right) dx\,dy$$

If $\partial M/\partial y = \partial N/\partial x$ in R, then clearly $\oint_C M\,dx + N\,dy = 0$.

Conversely, suppose $\oint_C M\,dx + N\,dy = 0$ for all curves C. If $(\partial N/\partial x) - (\partial M/\partial y) > 0$ at a point P, then from the continuity of the derivatives it follows that $(\partial N/\partial x) - (\partial M/\partial y) > 0$ in some region A surrounding P. If Γ is the boundary of A, then

$$\oint_\Gamma M\,dx + N\,dy = \iint_A \left(\frac{\partial N}{\partial x} - \frac{\partial M}{\partial y}\right) dx\,dy > 0$$

which contradicts the assumption that the line integral is zero around every closed curve. Similarly, the assumption $(\partial N/\partial x) - (\partial M/\partial y) < 0$ leads to a contradiction. Thus, $(\partial N/\partial x) - (\partial M/\partial y) = 0$ at all points.

Note that the condition $(\partial M/\partial y) = (\partial N/\partial x)$ is equivalent to the condition $\nabla \times \mathbf{A} = \mathbf{0}$ where $\mathbf{A} = M\mathbf{i} + N\mathbf{j}$ (see Problems 5.10 and 5.11). For a generalization to space curves, see Problem 6.31.

6.13. Let $\mathbf{F} = -y\mathbf{i} + x\mathbf{j}/(x^2 + y^2)$. (a) Calculate $\nabla \times \mathbf{F}$. (b) Evaluate $\oint \mathbf{F} \cdot d\mathbf{r}$ around any closed path and explain the results.

Solution

(a) $\nabla \times \mathbf{F} = \begin{vmatrix} \mathbf{i} & \mathbf{j} & \mathbf{k} \\ \dfrac{\partial}{\partial x} & \dfrac{\partial}{\partial y} & \dfrac{\partial}{\partial z} \\ \dfrac{-y}{x^2 + y^2} & \dfrac{x}{x^2 + y^2} & 0 \end{vmatrix} = \mathbf{0}$ in any region excluding $(0, 0)$.

(b) $\oint \mathbf{F} \cdot d\mathbf{r} = \oint \dfrac{-y\,dx + x\,dy}{x^2 + y^2}$. Let $x = \rho\cos\phi, y = \rho\sin\phi$, where (ρ, ϕ) are polar coordinates. Then

$$dx = -\rho\sin\phi\,d\phi + d\rho\,\cos\phi, \quad dy = \rho\cos\phi\,d\phi + d\rho\,\sin\phi$$

and so

$$\frac{-y\,dx + x\,dy}{x^2 + y^2} = d\phi = d\left(\arctan\frac{y}{x}\right)$$

For a closed curve $ABCDA$ (see Fig. 6-8a) surrounding the origin, $\phi = 0$ at A and $\phi = 2\pi$ after a complete circuit back to A. In this case, the line integral equals $\int_0^{2\pi} d\phi = 2\pi$.

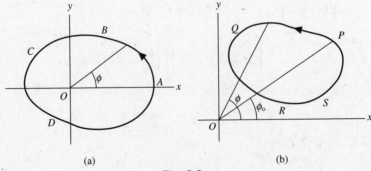

(a)　　　　　　　　(b)

Fig. 6-8

For a closed curve $PQRSP$ (see Fig. 6-8b) not surrounding the origin, $\phi = \phi_0$ at P and $\phi = \phi_0$ after a complete circuit back to P. In this case, the line integral equals $\int_{\phi_0}^{\phi_0} d\phi = 0$.

Since $\mathbf{F} = M\mathbf{i} + N\mathbf{j}$, $\nabla \times \mathbf{F} = \mathbf{0}$ is equivalent to $\partial M/\partial y = \partial N/\partial x$ and the results would seem to contradict those of Problem 6.12. However, no contradiction exists since $M = -y/(x^2 + y^2)$ and $N = x/(x^2 + y^2)$ do not have continuous derivatives throughout any region including $(0, 0)$, and this was assumed in Problem 6.12.

The Divergence Theorem

6.14. (a) Express the divergence theorem in words and (b) write it in rectangular form.

Solution

(a) The surface integral of the normal component of a vector $\mathbf{A}$ taken over a closed surface is equal to the integral of the divergence of $\mathbf{A}$ taken over the volume enclosed by the surface.

(b) Let $\mathbf{A} = A_1\mathbf{i} + A_2\mathbf{j} + A_3\mathbf{k}$. Then div $\mathbf{A} = \nabla \cdot \mathbf{A} = \dfrac{\partial A_1}{\partial x} + \dfrac{\partial A_2}{\partial y} + \dfrac{\partial A_3}{\partial z}$.

The unit normal to S is $\mathbf{n} = n_1\mathbf{i} + n_2\mathbf{j} + n_3\mathbf{k}$. Then $n_1 = \mathbf{n} \cdot \mathbf{i} = \cos \alpha$, $n_2 = \mathbf{n} \cdot \mathbf{j} = \cos \beta$, and $n_3 = \mathbf{n} \cdot \mathbf{k} = \cos \gamma$, where $\alpha, \beta,$ and γ are the angles that $\mathbf{n}$ makes with the positive x, y, z axes or $\mathbf{i}, \mathbf{j}, \mathbf{k}$ directions, respectively. The quantities $\cos \alpha$, $\cos \beta$, and $\cos \gamma$ are the direction cosines of $\mathbf{n}$. Then

$$\mathbf{A} \cdot \mathbf{n} = (A_1\mathbf{i} + A_2\mathbf{j} + A_3\mathbf{k}) \cdot (\cos \alpha\,\mathbf{i} + \cos \beta\,\mathbf{j} + \cos \gamma\,\mathbf{k})$$
$$= A_1 \cos \alpha + A_2 \cos \beta + A_3 \cos \gamma$$

and the divergence theorem can be written

$$\iiint\limits_{V} \left(\frac{\partial A_1}{\partial x} + \frac{\partial A_2}{\partial y} + \frac{\partial A_3}{\partial z} \right) dx\,dy\,dz = \iint\limits_{S} (A_1 \cos \alpha + A_2 \cos \beta + A_3 \cos \gamma)\,dS$$

6.15. Demonstrate the divergence theorem physically.

Solution

Let $\mathbf{A}$ = velocity $\mathbf{v}$ at any point of a moving fluid. From Figure 6.9(a), we have:

$$\text{Volume of fluid crossing } dS \text{ in } \Delta t \text{ seconds}$$
$$= \text{volume contained in cylinder of base } dS \text{ and slant height } \mathbf{v}\Delta t$$
$$= (\mathbf{v}\Delta t) \cdot \mathbf{n}\,dS = \mathbf{v} \cdot \mathbf{n}\,dS\,\Delta t$$

Then, volume per second of fluid crossing $dS = \mathbf{v} \cdot \mathbf{n}\,dS$

(a) (b)

Fig. 6-9

From Figure 6-9(b), we have

$$\text{Total volume per second of fluid emerging from closed surface } S = \iint_S \mathbf{v} \cdot \mathbf{n} \, dS$$

From Problem 4.21 of Chapter 4, $\nabla \cdot \mathbf{v} \, dV$ is the volume per second of fluid emerging from a volume element dV. Then

$$\text{Total volume per second of fluid emerging from all volume elements in } S = \iiint_V \nabla \cdot \mathbf{v} \, dV$$

Thus

$$\iint_S \mathbf{v} \cdot \mathbf{n} \, dS = \iiint_V \nabla \cdot \mathbf{v} \, dV$$

6.16. Prove the divergence theorem.

Fig. 6-10

Solution

Let S be a closed surface such that any line parallel to the coordinate axes cuts S in, at most, two points. Assume the equations of the lower and upper portions, S_1 and S_2, to be $z = f_1(x, y)$ and $z = f_2(x, y)$, respectively. Denote the projection of the surface on the xy-plane by R (see Fig. 6-10). Consider:

$$\iiint_V \frac{\partial A_3}{\partial z} \, dV = \iiint_V \frac{\partial A_3}{\partial z} \, dz \, dy \, dx = \iint_R \left[\int_{z=f_1(x,y)}^{f_2(x,y)} \frac{\partial A_3}{\partial z} \, dz \right] dy \, dx$$

$$= \iint_R A_3(x, y, z) \Big|_{z=f_1}^{f_2} dy \, dx = \iint_R [A_3(x, y, f_2) - A_3(x, y, f_1)] \, dy \, dx$$

For the upper portion S_2, $dy \, dx = \cos \gamma_2 \, dS_2 = \mathbf{k} \cdot \mathbf{n}_2 \, dS_2$ since the normal $\mathbf{n}_2$ to S_2 makes an acute angle γ_2 with $\mathbf{k}$.

For the lower portion S_1, $dy \, dx = -\cos \gamma_1 \, dS_1 = -\mathbf{k} \cdot \mathbf{n}_1 \, dS_1$ since the normal $\mathbf{n}_1$ to S_1 makes an obtuse angle γ_1 with $\mathbf{k}$.

Then

$$\iint\limits_{R} A_3(x, y, f_2) \, dy \, dx = \iint\limits_{S_2} A_3 \mathbf{k} \cdot \mathbf{n}_2 \, dS_2$$

$$\iint\limits_{R} A_3(x, y, f_1) \, dy \, dx = -\iint\limits_{S_1} A_3 \mathbf{k} \cdot \mathbf{n}_1 \, dS_1$$

and

$$\iint\limits_{R} A_3(x, y, f_2) \, dy \, dx - \iint\limits_{R} A_3(x, y, f_1) \, dy \, dx = \iint\limits_{S_2} A_3 \mathbf{k} \cdot \mathbf{n}_2 \, dS_2 + \iint\limits_{S_1} A_3 \mathbf{k} \cdot \mathbf{n}_1 \, dS_1$$

$$= \iint\limits_{S} A_3 \mathbf{k} \cdot \mathbf{n} \, dS$$

so that

$$\iiint\limits_{V} \frac{\partial A_3}{\partial z} \, dV = \iint\limits_{S} A_3 \mathbf{k} \cdot \mathbf{n} \, dS \tag{1}$$

Similarly, by projecting S on the other coordinate planes,

$$\iiint\limits_{V} \frac{\partial A_1}{\partial x} \, dV = \iint\limits_{S} A_1 \mathbf{i} \cdot \mathbf{n} \, dS \tag{2}$$

$$\iiint\limits_{V} \frac{\partial A_2}{\partial y} \, dV = \iint\limits_{S} A_2 \mathbf{j} \cdot \mathbf{n} \, dS \tag{3}$$

Adding (1), (2), and (3),

$$\iiint\limits_{V} \left(\frac{\partial A_1}{\partial x} + \frac{\partial A_2}{\partial y} + \frac{\partial A_3}{\partial z} \right) dV = \iint\limits_{S} (A_1 \mathbf{i} + A_2 \mathbf{j} + A_3 \mathbf{k}) \cdot \mathbf{n} \, dS$$

or

$$\iiint\limits_{V} \nabla \cdot \mathbf{A} \, dV = \iint\limits_{S} \mathbf{A} \cdot \mathbf{n} \, dS$$

The theorem can be extended to surfaces where lines parallel to the coordinate axes meet them in more than two points. To establish this extension, subdivide the region bounded by S into subregions whose surfaces do satisfy this condition. The procedure is analogous to that used in Green's theorem for the plane.

6.17. Evaluate $\iint_S \mathbf{F} \cdot \mathbf{n}\, dS$, where $\mathbf{F} = 4xz\mathbf{i} - y^2\mathbf{j} + yz\mathbf{k}$ and S is the surface of the cube bounded by $x = 0, x = 1, y = 0, y = 1, z = 0, z = 1$.

Solution

By the divergence theorem, the required integral is equal to

$$\iiint_V \nabla \cdot \mathbf{F}\, dV = \iiint_V \left[\frac{\partial}{\partial x}(4xz) + \frac{\partial}{\partial y}(-y^2) + \frac{\partial}{\partial z}(yz) \right] dV$$

$$= \iiint_V (4z - y)\, dV = \int_{x=0}^{1} \int_{y=0}^{1} \int_{z=0}^{1} (4z - y)\, dz\, dy\, dx$$

$$= \int_{x=0}^{1} \int_{y=0}^{1} \left. 2z^2 - yz \right|_{z=0}^{1} dy\, dx = \int_{x=0}^{1} \int_{y=0}^{1} (2 - y)\, dy\, dx = \frac{3}{2}$$

The surface integral may also be evaluated directly as in Problem 5.23.

6.18. Verify the divergence theorem for $\mathbf{A} = 4x\mathbf{i} - 2y^2\mathbf{j} + z^2\mathbf{k}$ taken over the region bounded by $x^2 + y^2 = 4$, $z = 0$, and $z = 3$.

Solution

$$\text{Volume integral} = \iiint_V \nabla \cdot \mathbf{A}\, dV = \iiint_V \left[\frac{\partial}{\partial x}(4x) + \frac{\partial}{\partial y}(-2y^2) + \frac{\partial}{\partial z}(z^2) \right] dV$$

$$= \iiint_V (4 - 4y + 2z)\, dV = \int_{x=-2}^{2} \int_{y=-\sqrt{4-x^2}}^{\sqrt{4-x^2}} \int_{z=0}^{3} (4 - 4y + 2z)\, dz\, dy\, dx = 84\pi$$

The surface S of the cylinder consists of a base S_1 ($z = 0$), the top S_2 ($z = 3$) and the convex portion S_3 ($x^2 + y^2 = 4$). Then

$$\text{Surface integral} = \iint_S \mathbf{A} \cdot \mathbf{n}\, dS = \iint_{S_1} \mathbf{A} \cdot \mathbf{n}\, dS_1 + \iint_{S_2} \mathbf{A} \cdot \mathbf{n}\, dS_2 + \iint_{S_3} \mathbf{A} \cdot \mathbf{n}\, dS_3$$

On S_1 ($z = 0$), $\mathbf{n} = -\mathbf{k}, \mathbf{A} = 4x\mathbf{i} - 2y^2\mathbf{j}$ and $\mathbf{A} \cdot \mathbf{n} = 0$, so that $\iint_{S_1} \mathbf{A} \cdot \mathbf{n}\, dS_1 = 0$.

On S_2 ($z = 3$), $\mathbf{n} = \mathbf{k}, \mathbf{A} = 4x\mathbf{i} - 2y^2\mathbf{j} + 9\mathbf{k}$ and $\mathbf{A} \cdot \mathbf{n} = 9$, so that

$$\iint_{S_2} \mathbf{A} \cdot \mathbf{n}\, dS_2 = 9 \iint_{S_2} dS_2 = 36\pi, \quad \text{since area of } S_2 = 4\pi$$

On S_3 ($x^2 + y^2 = 4$). A perpendicular to $x^2 + y^2 = 4$ has the direction $\nabla(x^2 + y^2) = 2x\mathbf{i} + 2y\mathbf{j}$.

Then a unit normal is $\mathbf{n} = \dfrac{2x\mathbf{i} + 2y\mathbf{j}}{\sqrt{4x^2 + 4y^2}} = \dfrac{x\mathbf{i} + y\mathbf{j}}{2}$ since $x^2 + y^2 = 4$.

$$\mathbf{A} \cdot \mathbf{n} = (4x\mathbf{i} - 2y^2\mathbf{j} + z^2\mathbf{k}) \cdot \left(\frac{x\mathbf{i} + y\mathbf{j}}{2} \right) = 2x^2 - y^3$$

Fig. 6-11

From Fig. 6-11, $x = 2 \cos \theta, y = 2 \sin \theta, dS_3 = 2 \, d\theta \, dz$ and so

$$\iint_{S_3} \mathbf{A} \cdot \mathbf{n} \, dS_3 = \int_{\theta=0}^{2\pi} \int_{z=0}^{3} [2(2 \cos \theta)^2 - (2 \sin \theta)^3] 2 \, dz \, d\theta$$

$$= \int_{\theta=0}^{2\pi} (48 \cos^2 \theta - 48 \sin^3 \theta) \, d\theta = \int_{\theta=0}^{2\pi} 48 \cos^2 \theta \, d\theta = 48\pi$$

Then the surface integral $= 0 + 36\pi + 48\pi = 84\pi$, agreeing with the volume integral and verifying the divergence theorem.

Note that evaluation of the surface integral over S_3 could also have been done by projection of S_3 on the xz- or yz-coordinate planes.

6.19. Suppose div $\mathbf{A}$ denotes the divergence of a vector field $\mathbf{A}$ at a point P. Show that

$$\text{div } \mathbf{A} = \lim_{\Delta V \to 0} \frac{\iint_{\Delta S} \mathbf{A} \cdot \mathbf{n} \, dS}{\Delta V}$$

where ΔV is the volume enclosed by the surface ΔS and the limit is obtained by shrinking ΔV to the point P.

Solution

By the divergence theorem,

$$\iiint_{\Delta V} \text{div } \mathbf{A} \, dV = \iint_{\Delta S} \mathbf{A} \cdot \mathbf{n} \, dS$$

By the mean-value theorem for integrals, the left side can be written

$$\overline{\text{div } \mathbf{A}} \iiint_{\Delta V} dV = \overline{\text{div } \mathbf{A}} \, \Delta V$$

where $\overline{\text{div } \mathbf{A}}$ is some value intermediate between the maximum and minimum of div $\mathbf{A}$ throughout ΔV. Then

$$\overline{\text{div } \mathbf{A}} = \frac{\iint_{\Delta S} \mathbf{A} \cdot \mathbf{n} \, dS}{\Delta V}$$

Taking the limit as $\Delta V \to 0$ such that P is always interior to ΔV, $\overline{\text{div } \mathbf{A}}$ approaches the value div $\mathbf{A}$ at point P; hence

$$\text{div } \mathbf{A} = \lim_{\Delta V \to 0} \frac{\iint_{\Delta S} \mathbf{A} \cdot \mathbf{n} \, dS}{\Delta V}$$

This result can be taken as a starting point for defining the divergence of $\mathbf{A}$, and from it all the properties may be derived including proof of the divergence theorem. In Chapter 7, we use this definition to extend the concept of divergence of a vector to coordinate systems other than rectangular. Physically,

$$\frac{\iint_{\Delta S} \mathbf{A} \cdot \mathbf{n} \, dS}{\Delta V}$$

represents the flux or net outflow per unit volume of the vector $\mathbf{A}$ from the surface ΔS. If div $\mathbf{A}$ is positive in the neighborhood of a point P, it means that the outflow from P is positive and we call P a *source*. Similarly, if div $\mathbf{A}$ is negative in the neighborhood of P, the outflow is really an inflow and P is called a *sink*. If in a region there are no sources or sinks, then div $\mathbf{A} = 0$ and we call $\mathbf{A}$ a *solenoidal* vector field.

6.20. Evaluate $\iint_S \mathbf{r} \cdot \mathbf{n} \, dS$, where S is a closed surface.

Solution

By the divergence theorem,

$$\iint_S \mathbf{r} \cdot \mathbf{n} \, dS = \iiint_V \nabla \cdot \mathbf{r} \, dV = \iiint_V \left(\frac{\partial}{\partial x}\mathbf{i} + \frac{\partial}{\partial y}\mathbf{j} + \frac{\partial}{\partial z}\mathbf{k} \right) \cdot (x\mathbf{i} + y\mathbf{j} + z\mathbf{k}) \, dV$$

$$= \iiint_V \left(\frac{\partial x}{\partial x} + \frac{\partial y}{\partial y} + \frac{\partial z}{\partial z} \right) dV = 3 \iiint_V dV = 3V$$

where V is the volume enclosed by S.

6.21. Prove $\iiint_V (\phi \nabla^2 \psi - \psi \nabla^2 \phi) \, dV = \iint_S (\phi \nabla \psi - \psi \nabla \phi) \cdot d\mathbf{S}$.

Solution

Let $\mathbf{A} = \phi \nabla \psi$ in the divergence theorem. Then

$$\iiint_V \nabla \cdot (\phi \nabla \psi) \, dV = \iint_S (\phi \nabla \psi) \cdot \mathbf{n} \, dS = \iint_S (\phi \nabla \psi) \cdot d\mathbf{S}$$

But

$$\nabla \cdot (\phi \nabla \psi) = \phi (\nabla \cdot \nabla \psi) + (\nabla \phi) \cdot (\nabla \psi) = \phi \nabla^2 \psi + (\nabla \phi) \cdot (\nabla \psi)$$

Thus

$$\iiint_V \nabla \cdot (\phi \nabla \psi) \, dV = \iiint_V [\phi \nabla^2 \psi + (\nabla \phi) \cdot (\nabla \psi)] \, dV$$

or

$$\iiint_V [\phi \nabla^2 \psi + (\nabla \phi) \cdot (\nabla \psi)] \, dV = \iint_S (\phi \nabla \psi) \cdot d\mathbf{S} \tag{1}$$

which proves *Green's first identity*. Interchanging ϕ and ψ in (1),

$$\iiint_V [\psi \nabla^2 \phi + (\nabla \psi) \cdot (\nabla \phi)] \, dV = \iint_S (\psi \nabla \phi) \cdot d\mathbf{S} \tag{2}$$

Subtracting (2) from (1), we have

$$\iiint_V (\phi \nabla^2 \psi - \psi \nabla^2 \phi)\, dV = \iint_S (\phi \nabla \psi - \psi \nabla \phi) \cdot d\mathbf{S} \tag{3}$$

which is *Green's second identity* or *symmetrical theorem*. In the proof, we have assumed that ϕ and ψ are scalar functions of position with continuous derivatives of the second order at least.

6.22. Prove $\displaystyle\iiint_V \nabla \phi\, dV = \iint_S \phi \mathbf{n}\, dS$.

Solution

In the divergence theorem, let $\mathbf{A} = \phi \mathbf{C}$ where $\mathbf{C}$ is a constant vector. Then

$$\iiint_V \nabla \cdot (\phi \mathbf{C})\, dV = \iint_S \phi \mathbf{C} \cdot \mathbf{n}\, dS$$

Since $\nabla \cdot (\phi \mathbf{C}) = (\nabla \phi) \cdot \mathbf{C} = \mathbf{C} \cdot \nabla \phi$ and $\phi \mathbf{C} \cdot \mathbf{n} = \mathbf{C} \cdot (\phi \mathbf{n})$,

$$\iiint_V \mathbf{C} \cdot \nabla \phi\, dV = \iint_S \mathbf{C} \cdot (\phi \mathbf{n})\, dS$$

Taking $\mathbf{C}$ outside the integrals,

$$\mathbf{C} \cdot \iiint_V \nabla \phi\, dV = \mathbf{C} \cdot \iint_S \phi \mathbf{n}\, dS$$

and since $\mathbf{C}$ is an arbitrary constant vector,

$$\iiint_V \nabla \phi\, dV = \iint_S \phi \mathbf{n}\, dS$$

6.23. Prove $\displaystyle\iiint_V \nabla \times \mathbf{B}\, dV = \iint_S \mathbf{n} \times \mathbf{B}\, dS$.

Solution

In the divergence theorem, let $\mathbf{A} = \mathbf{B} \times \mathbf{C}$ where $\mathbf{C}$ is a constant vector. Then

$$\iiint_V \nabla \cdot (\mathbf{B} \times \mathbf{C})\, dV = \iint_S (\mathbf{B} \times \mathbf{C}) \cdot \mathbf{n}\, dS$$

Since $\nabla \cdot (\mathbf{B} \times \mathbf{C}) = \mathbf{C} \cdot (\nabla \times \mathbf{B})$ and $(\mathbf{B} \times \mathbf{C}) \cdot \mathbf{n} = \mathbf{B} \cdot (\mathbf{C} \times \mathbf{n}) = (\mathbf{C} \times \mathbf{n}) \cdot \mathbf{B} = \mathbf{C} \cdot (\mathbf{n} \times \mathbf{B})$,

$$\iiint_V \mathbf{C} \cdot (\nabla \times \mathbf{B})\, dV = \iint_S \mathbf{C} \cdot (\mathbf{n} \times \mathbf{B})\, dS$$

Taking $\mathbf{C}$ outside the integrals,

$$\mathbf{C} \cdot \iiint_V \nabla \times \mathbf{B}\, dV = \mathbf{C} \cdot \iint_S \mathbf{n} \times \mathbf{B}\, dS$$

and since $\mathbf{C}$ is an arbitrary constant vector,

$$\iiint_V \nabla \times \mathbf{B}\, dV = \iint_S \mathbf{n} \times \mathbf{B}\, dS$$

6.24. Show that at any point P

$$\text{(a)} \quad \nabla\phi = \lim_{\Delta V \to 0} \frac{\iint_{\Delta S} \phi\mathbf{n}\, dS}{\Delta V} \quad \text{and} \quad \text{(b)} \quad \nabla \times \mathbf{A} = \lim_{\Delta V \to 0} \frac{\iint_{\Delta S} \mathbf{n} \times \mathbf{A}\, dS}{\Delta V}$$

where ΔV is the volume enclosed by the surface ΔS, and the limit is obtained by shrinking ΔV to the point P.

Solution

(a) From Problem 6.22, $\iiint_{\Delta V} \nabla\phi\, dV = \iint_{\Delta S} \phi\mathbf{n}\, dS$. Then $\iiint_{\Delta V} \nabla\phi \cdot \mathbf{i}\, dV = \iint_{\Delta S} \phi\mathbf{n} \cdot \mathbf{i}\, dS$. Using the same principle employed in Problem 6.19, we have

$$\overline{\nabla\phi \cdot \mathbf{i}} = \frac{\iint_{\Delta S} \phi\mathbf{n} \cdot \mathbf{i}\, dS}{\Delta V}$$

where $\overline{\nabla\phi \cdot \mathbf{i}}$ is some value intermediate between the maximum and minimum of $\nabla\phi \cdot \mathbf{i}$ throughout ΔV. Taking the limit as $\Delta V \to 0$ in such a way that P is always interior to ΔV, $\nabla\phi \cdot \mathbf{i}$ approaches the value

$$\nabla\phi \cdot \mathbf{i} = \lim_{\Delta V \to 0} \frac{\iint_S \phi\mathbf{n} \cdot \mathbf{i}\, dS}{\Delta V} \tag{1}$$

Similarly, we find

$$\nabla\phi \cdot \mathbf{j} = \lim_{\Delta V \to 0} \frac{\iint_S \phi\mathbf{n} \cdot \mathbf{j}\, dS}{\Delta V} \tag{2}$$

$$\nabla\phi \cdot \mathbf{k} = \lim_{\Delta V \to 0} \frac{\iint_S \phi\mathbf{n} \cdot \mathbf{k}\, dS}{\Delta V} \tag{3}$$

Multiplying (1), (2), and (3) by $\mathbf{i}$, $\mathbf{j}$, and $\mathbf{k}$, respectively, and adding, using

$$\nabla\phi = (\nabla\phi \cdot \mathbf{i})\mathbf{i} + (\nabla\phi \cdot \mathbf{j})\mathbf{j} + (\nabla\phi \cdot \mathbf{k})\mathbf{k}, \quad \mathbf{n} = (\mathbf{n} \cdot \mathbf{i})\mathbf{i} + (\mathbf{n} \cdot \mathbf{j})\mathbf{j} + (\mathbf{n} \cdot \mathbf{k})\mathbf{k}$$

(see Problem 2.17) the result follows.

(b) From Problem 6.23, replacing $\mathbf{B}$ by $\mathbf{A}$, $\iiint_{\Delta V} \nabla \times \mathbf{A}\, dV = \iint_{\Delta S} \mathbf{n} \times \mathbf{A}\, dS$. Then, as in part (a), we can show that

$$(\nabla \times \mathbf{A}) \cdot \mathbf{i} = \lim_{\Delta V \to 0} \frac{\iint_{\Delta S} (\mathbf{n} \times \mathbf{A}) \cdot \mathbf{i}\, dS}{\Delta V}$$

and similar results with $\mathbf{j}$ and $\mathbf{k}$ replacing $\mathbf{i}$. Multiplying by $\mathbf{i}$, $\mathbf{j}$, and $\mathbf{k}$ adding, the result follows.

The results obtained can be taken as starting points for definition of gradient and curl. Using these definitions, extensions can be made to coordinate systems other than rectangular.

6.25. Establish the operator equivalence

$$\nabla \circ \equiv \lim_{\Delta V \to 0} \frac{1}{\Delta V} \oiint_{\Delta S} d\mathbf{S} \circ$$

where $\circ$ indicates a dot product, cross product, or ordinary product.

Solution

To establish the equivalence, the results of the operation on a vector or scalar field must be consistent with already established results.

If $\circ$ is the dot product, then for a vector $\mathbf{A}$,

$$\nabla \circ \mathbf{A} = \lim_{\Delta V \to 0} \frac{1}{\Delta V} \iint_{\Delta S} d\mathbf{S} \circ \mathbf{A}$$

or

$$\operatorname{div}\mathbf{A} = \lim_{\Delta V \to 0}\frac{1}{\Delta V}\iint_{\Delta S}d\mathbf{S}\cdot\mathbf{A}$$

$$= \lim_{\Delta V \to 0}\frac{1}{\Delta V}\iint_{\Delta S}\mathbf{A}\cdot\mathbf{n}\,dS$$

established in Problem 6.19.

Similarly, if ∘ is the cross product,

$$\operatorname{curl}\mathbf{A} = \nabla\times\mathbf{A} = \lim_{\Delta V \to 0}\frac{1}{\Delta V}\iint_{\Delta S}d\mathbf{S}\times\mathbf{A}$$

$$= \lim_{\Delta V \to 0}\frac{1}{\Delta V}\iint_{\Delta S}\mathbf{n}\times\mathbf{A}\,dS$$

established in Problem 6.24(b).

Also, if ∘ is ordinary multiplication, then for a scalar ϕ,

$$\nabla\circ\phi = \lim_{\Delta V \to 0}\frac{1}{\Delta V}\iint_{\Delta S}d\mathbf{S}\circ\phi \quad \text{or} \quad \nabla\phi = \lim_{\Delta V \to 0}\frac{1}{\Delta V}\iint_{\Delta S}\phi\,d\mathbf{S}$$

established in Problem 6.24(a).

6.26. Let S be a closed surface and let $\mathbf{r}$ denote the position vector of any point (x, y, z) measured from an origin O. Prove that

$$\iint_S \frac{\mathbf{n}\cdot\mathbf{r}}{r^3}dS$$

is equal to (a) zero if O lies outside S; (b) 4π if O lies inside S. This result is known as *Gauss' theorem*.

Solution

(a) By the divergence theorem,

$$\iint_S \frac{\mathbf{n}\cdot\mathbf{r}}{r^3}dS = \iiint_V \nabla\cdot\frac{\mathbf{r}}{r^3}\,dV$$

But $\nabla\cdot(\mathbf{r}/r^3) = 0$ (Problem 4.19) everywhere within V provided $r \neq 0$ in V, that is, provided O is outside of V and thus outside of S. Then

$$\iint_S \frac{\mathbf{n}\cdot\mathbf{r}}{r^3}dS = 0$$

(b) If O is inside S, surround O by a small sphere s of radius a. Let τ denote the region bounded by S and s. Then, by the divergence theorem

$$\iint_{S+s}\frac{\mathbf{n}\cdot\mathbf{r}}{r^3}dS = \iint_S\frac{\mathbf{n}\cdot\mathbf{r}}{r^3}dS + \iint_s\frac{\mathbf{n}\cdot\mathbf{r}}{r^3}dS = \iiint_\tau \nabla\cdot\frac{\mathbf{r}}{r^3}dV = 0$$

since $r \neq 0$ in τ. Thus

$$\iint_S \frac{\mathbf{n}\cdot\mathbf{r}}{r^3}dS = -\iint_s\frac{\mathbf{n}\cdot\mathbf{r}}{r^3}dS$$

Now on $s, r = a, \mathbf{n} = -\dfrac{\mathbf{r}}{a}$ so that $\dfrac{\mathbf{n} \cdot \mathbf{r}}{r^3} = \dfrac{(-\mathbf{r}/a) \cdot \mathbf{r}}{a^3} = -\dfrac{\mathbf{r} \cdot \mathbf{r}}{a^4} = -\dfrac{a^2}{a^4} = -\dfrac{1}{a^2}$ and

$$\iint\limits_S \frac{\mathbf{n} \cdot \mathbf{r}}{r^3} dS = -\iint\limits_s \frac{\mathbf{n} \cdot \mathbf{r}}{r^3} dS = \iint\limits_s \frac{1}{a^2} dS = \frac{1}{a^2} \iint\limits_s dS = \frac{4\pi a^2}{a^2} = 4\pi$$

6.27. Interpret Gauss' theorem (Problem 6.26) geometrically.

Solution

Let dS denote an element of surface area and connect all points on the boundary of dS to O (see Fig. 6-12), thereby forming a cone. Let $d\Omega$ be the area of that portion of a sphere with O as center and radius r which is cut out by this cone; then the *solid angle* subtended by dS at O is defined as $d\omega = d\Omega/r^2$ and is numerically equal to the area of that portion of a sphere with center O and unit radius cut out by the cone. Let $\mathbf{n}$ be the positive unit normal to dS and call θ the angle between $\mathbf{n}$ and $\mathbf{r}$; then $\cos\theta = \mathbf{n} \cdot \mathbf{r}/r$. Also,

$$d\Omega = \pm dS \cos\theta = \pm(\mathbf{n} \cdot \mathbf{r}/r) dS \text{ so that } d\omega = \pm(\mathbf{n} \cdot \mathbf{r}/r^3) dS,$$

the $+$ or $-$ being chosen according as $\mathbf{n}$ and $\mathbf{r}$ form an acute or an obtuse angle θ with each other.

Fig. 6-12

Let S be a surface, as in Fig. 6-13(a), such that any line meets S in not more than two points. If O lies outside S, then at a position such as 1, $(\mathbf{n} \cdot \mathbf{r}/r^3) dS = d\omega$; whereas at the corresponding position 2, $(\mathbf{n} \cdot \mathbf{r}/r^3) dS = -d\omega$. An integration over these two regions gives zero, since the contributions to the solid angle cancel out. When the integration is performed over S, it thus follows that $\iint_S (\mathbf{n} \cdot \mathbf{r}/r^3) dS = 0$, since for every positive contribution, there is a negative one.

In case O is inside S, however, then at a position such as 3, $(\mathbf{n} \cdot \mathbf{r}/r^3) dS = d\omega$ and at 4, $(\mathbf{n} \cdot \mathbf{r}/r^3) dS = d\omega$ so that the contributions add instead of cancel. The total solid angle in this case is equal to the area of a unit sphere, which is 4π, so that $\iint_S (\mathbf{n} \cdot \mathbf{r}/r^3) dS = 4\pi$.

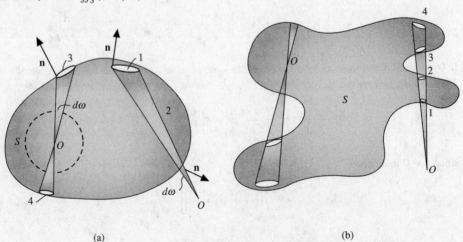

(a)   (b)

Fig. 6-13

For surfaces S, such that a line may meet S in more than two points, an exactly similar situation holds as is seen by reference to Fig. 6-13. If O is outside S, for example, then a cone with vertex at O intersects S at an even number of places and the contribution to the surface integral is zero since the solid angles subtended at O cancel out in pairs. If O is inside S, however, a cone having vertex at O intersects S at an old number of places and since cancellation occurs only for an even number of these, there will always be a contribution of 4π for the entire surface S.

6.28. A fluid of density $\rho(x, y, z, t)$ moves with velocity $\mathbf{v}(x, y, z, t)$. If there are no sources or sinks, prove that

$$\nabla \cdot \mathbf{J} + \frac{\partial \rho}{\partial t} = 0$$

where $\mathbf{J} = \rho \mathbf{v}$.

Solution

Consider an arbitrary surface enclosing a volume V of the fluid. At any time, the mass of fluid within V is

$$M = \iiint_V \rho \, dV$$

The time rate of increase of this mass is

$$\frac{\partial M}{\partial t} = \frac{\partial}{\partial t} \iiint_V \rho \, dV = \iiint_V \frac{\partial \rho}{\partial t} \, dV$$

The mass of fluid per unit time leaving V is

$$\iint_S \rho \mathbf{v} \cdot \mathbf{n} \, dS$$

(see Problem 6.15) and the time rate of increase in mass is therefore

$$-\iint_S \rho \mathbf{v} \cdot \mathbf{n} \, dS = -\iiint_V \nabla \cdot (\rho \mathbf{v}) \, dV$$

by the divergence theorem. Then

$$\iiint_V \frac{\partial \rho}{\partial t} \, dV = -\iiint_V \nabla \cdot (\rho \mathbf{v}) \, dV$$

or

$$\iiint_V \left(\nabla \cdot (\rho \mathbf{v}) + \frac{\partial \rho}{\partial t} \right) dV = 0$$

Since V is arbitrary, the integrand, assumed continuous, must be identically zero, by reasoning similar to that used in Problem 6.12. Then

$$\nabla \cdot \mathbf{J} + \frac{\partial \rho}{\partial t} = 0$$

where $\mathbf{J} = \rho \mathbf{v}$. The equation is called the *continuity equation*. If ρ is a constant, the fluid is incompressible and $\nabla \cdot \mathbf{v} = 0$, that is, $\mathbf{v}$ is solenoidal.

The continuity equation also arises in electromagnetic theory, where ρ is the *charge density* and $\mathbf{J} = \rho \mathbf{v}$ is the *current density*.

6.29. If the temperature at any point (x, y, z) of a solid at time t is $U(x, y, z, t)$ and if κ, ρ, and c are, respectively, the thermal conductivity, density, and specific heat of the solid, assumed constant, show that

$$\frac{\partial U}{\partial t} = k\nabla^2 U$$

where $k = \kappa/\rho c$.

Solution

Let V be an arbitrary volume lying within the solid and let S denote its surface. The total flux of heat across S, or the quantity of heat leaving S per unit time, is

$$\iint_S (-\kappa\nabla U) \cdot \mathbf{n}\, dS$$

Thus, the quantity of heat entering S per unit time is

$$\iint_S (\kappa\nabla U) \cdot \mathbf{n}\, dS = \iiint_V \nabla \cdot (\kappa\nabla U)\, dV \tag{1}$$

by the divergence theorem. The heat contained in a volume V is given by

$$\iiint_V c\rho U\, dV$$

Then the time rate of increase of heat is

$$\frac{\partial}{\partial t} \iiint_V c\rho U\, dV = \iiint_V c\rho \frac{\partial U}{\partial t}\, dV \tag{2}$$

Equating the right hand sides of (1) and (2),

$$\iiint_V \left[c\rho \frac{\partial U}{\partial t} - \nabla \cdot (\kappa\nabla U) \right] dV = 0$$

and since V is arbitrary, the integrand, assumed continuous, must be identically zero so that

$$c\rho \frac{\partial U}{\partial t} = \nabla \cdot (\kappa\nabla U)$$

or if κ, c, ρ are constants,

$$\frac{\partial U}{\partial t} = \frac{\kappa}{c\rho} \nabla \cdot \nabla U = k\nabla^2 U$$

The quantity k is called the *diffusivity*. For steady-state heat flow (i.e. $\partial U/\partial t = 0$ or U is independent of time), the equation reduces to Laplace's equation $\nabla^2 U = 0$.

Stokes' Theorem

6.30. (a) Express Stokes' theorem in words and (b) write it in rectangular form.

Solution

(a) The line integral of the tangential component of a vector $\mathbf{A}$ taken around a simple closed curve C is equal to the surface integral of the normal component of the curl of $\mathbf{A}$ taken over any surface S having C as its boundary.

(b) As in Problem 6.14(b),

$$\mathbf{A} = A_1\mathbf{i} + A_2\mathbf{j} + A_3\mathbf{k}, \quad \mathbf{n} = \cos\alpha\mathbf{i} + \cos\beta\mathbf{j} + \cos\gamma\mathbf{k}$$

Then

$$\nabla \times \mathbf{A} = \begin{vmatrix} \mathbf{i} & \mathbf{j} & \mathbf{k} \\ \dfrac{\partial}{\partial x} & \dfrac{\partial}{\partial y} & \dfrac{\partial}{\partial z} \\ A_1 & A_2 & A_3 \end{vmatrix} = \left(\frac{\partial A_3}{\partial y} - \frac{\partial A_2}{\partial z} \right)\mathbf{i} + \left(\frac{\partial A_1}{\partial z} - \frac{\partial A_3}{\partial x} \right)\mathbf{j} + \left(\frac{\partial A_2}{\partial x} - \frac{\partial A_1}{\partial y} \right)\mathbf{k}$$

$$(\nabla \times \mathbf{A}) \cdot \mathbf{n} = \left(\frac{\partial A_3}{\partial y} - \frac{\partial A_2}{\partial z} \right) \cos \alpha + \left(\frac{\partial A_1}{\partial z} - \frac{\partial A_3}{\partial x} \right) \cos \beta + \left(\frac{\partial A_2}{\partial x} - \frac{\partial A_1}{\partial y} \right) \cos \gamma$$

$$\mathbf{A} \cdot d\mathbf{r} = (A_1\mathbf{i} + A_2\mathbf{j} + A_3\mathbf{k}) \cdot (dx\,\mathbf{i} + dy\,\mathbf{j} + dz\,\mathbf{k}) = A_1\,dx + A_2\,dy + A_3\,dz$$

and Stokes' theorem becomes

$$\iint_S \left[\left(\frac{\partial A_3}{\partial y} - \frac{\partial A_2}{\partial z} \right) \cos \alpha + \left(\frac{\partial A_1}{\partial z} - \frac{\partial A_3}{\partial x} \right) \cos \beta + \left(\frac{\partial A_2}{\partial x} - \frac{\partial A_1}{\partial y} \right) \cos \gamma \right] dS = \oint_C A_1\,dx + A_2\,dy + A_3\,dz$$

6.31. Prove Stokes' theorem.

Solution

Let S be a surface such that its projections on the xy-, yz-, and xz-planes are regions bounded by simple closed curves, as indicated in Fig. 6-14. Assume S to have representation $z = f(x, y)$ or $x = g(y, z)$ or $y = h(x, z)$, where f, g, and h are single-valued, continuous, and differentiable functions, respectively. We must show that

$$\iint_S (\nabla \times \mathbf{A}) \cdot \mathbf{n}\, dS = \iint_S [\nabla \times (A_1\mathbf{i} + A_2\mathbf{j} + A_3\mathbf{k})] \cdot \mathbf{n}\, dS$$

$$= \oint_C \mathbf{A} \cdot d\mathbf{r}$$

where C is the boundary of S.

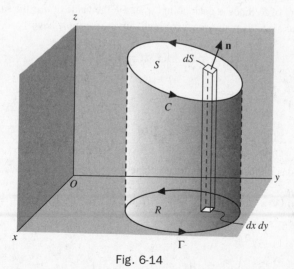

Fig. 6-14

Consider first $\iint_S [\nabla \times (A_1\mathbf{i})] \cdot \mathbf{n}\, dS$. Since

$$\nabla \times (A_1\mathbf{i}) = \begin{vmatrix} \mathbf{i} & \mathbf{j} & \mathbf{k} \\ \dfrac{\partial}{\partial x} & \dfrac{\partial}{\partial y} & \dfrac{\partial}{\partial z} \\ A_1 & 0 & 0 \end{vmatrix} = \frac{\partial A_1}{\partial z}\mathbf{j} - \frac{\partial A_1}{\partial y}\mathbf{k}$$

then

$$[\nabla \times (A_1 \mathbf{i})] \cdot \mathbf{n} \, dS = \left(\frac{\partial A_1}{\partial z} \mathbf{n} \cdot \mathbf{j} - \frac{\partial A_1}{\partial y} \mathbf{n} \cdot \mathbf{k} \right) dS \tag{1}$$

If $z = f(x, y)$ is taken as the equation of S, then the position vector to any point of S is $\mathbf{r} = x\mathbf{i} + y\mathbf{j} + z\mathbf{k} = x\mathbf{i} + y\mathbf{j} + f(x, y)\mathbf{k}$ so that $\partial \mathbf{r}/\partial y = \mathbf{j} + (\partial z/\partial y)\mathbf{k} = \mathbf{j} + (\partial f/\partial y)\mathbf{k}$. But $\partial \mathbf{r}/\partial y$ is a vector tangent to S (see Problem 3.25) and is thus perpendicular to $\mathbf{n}$, so that

$$\mathbf{n} \cdot \frac{\partial \mathbf{r}}{\partial y} = \mathbf{n} \cdot \mathbf{j} + \frac{\partial z}{\partial y} \mathbf{n} \cdot \mathbf{k} = 0 \quad \text{or} \quad \mathbf{n} \cdot \mathbf{j} = -\frac{\partial z}{\partial y} \mathbf{n} \cdot \mathbf{k}$$

Substitute in (1) to obtain

$$\left(\frac{\partial A_1}{\partial z} \mathbf{n} \cdot \mathbf{j} - \frac{\partial A_1}{\partial y} \mathbf{n} \cdot \mathbf{k} \right) dS = \left(-\frac{\partial A_1}{\partial z} \frac{\partial z}{\partial y} \mathbf{n} \cdot \mathbf{k} - \frac{\partial A_1}{\partial y} \mathbf{n} \cdot \mathbf{k} \right) dS$$

or

$$[\nabla \times (A_1 \mathbf{i})] \cdot \mathbf{n} \, dS = -\left(\frac{\partial A_1}{\partial y} + \frac{\partial A_1}{\partial z} \frac{\partial z}{\partial y} \right) \mathbf{n} \cdot \mathbf{k} \, dS \tag{2}$$

Now on S, $A_1(x, y, z) = A_1(x, y, f(x, y)) = F(x, y)$; hence $\dfrac{\partial A_1}{\partial y} + \dfrac{\partial A_1}{\partial z} \dfrac{\partial z}{\partial y} = \dfrac{\partial F}{\partial y}$ and (2) becomes

$$[\nabla \times (A_1 \mathbf{i})] \cdot \mathbf{n} \, dS = -\frac{\partial F}{\partial y} \mathbf{n} \cdot \mathbf{k} \, dS = -\frac{\partial F}{\partial y} \, dx \, dy$$

Then

$$\iint_S [\nabla \times (A_1 \mathbf{i})] \cdot \mathbf{n} \, dS = \iint_R -\frac{\partial F}{\partial y} \, dx \, dy$$

where R is the projection of S on the xy-plane. By Green's theorem for the plane, the last integral equals $\oint_\Gamma F \, dx$ where Γ is the boundary of R. Since at each point (x, y) of Γ the value of F is the same as the value of A_1 at each point (x, y, z) of C, and since dx is the same for both curves, we must have

$$\oint_\Gamma F \, dx = \oint_C A_1 \, dx$$

or

$$\iint_S [\nabla \times (A_1 \mathbf{i})] \cdot \mathbf{n} \, dS = \oint_C A_1 \, dx$$

Similarly, by projections on the other coordinate planes,

$$\iint_S [\nabla \times (A_2 \mathbf{j})] \cdot \mathbf{n} \, dS = \oint_C A_2 \, dy \quad \text{and} \quad \iint_S [\nabla \times (A_3 \mathbf{k})] \cdot \mathbf{n} \, dS = \oint_C A_3 \, dz$$

Thus, by addition,

$$\iint_S (\nabla \times \mathbf{A}) \cdot \mathbf{n} \, dS = \oint_C \mathbf{A} \cdot d\mathbf{r}$$

The theorem is also valid for surfaces S which may not satisfy the restrictions imposed above. Specifically suppose S can be subdivided into surfaces $S_1, S_2, \ldots, S_k$ with boundaries $C_1, C_2, \ldots, C_k$ which do satisfy the restrictions. Then Stokes' theorem holds for each such surface. Adding these surface integrals, the total surface integral over S is obtained. Adding the corresponding line integrals over $C_1, C_2, \ldots, C_k$, the line integral over C is obtained.

6.32. Verify Stokes' theorem for $\mathbf{A} = (2x - y)\mathbf{i} - yz^2\mathbf{j} - y^2z\mathbf{k}$, where S is the upper half surface of the sphere $x^2 + y^2 + z^2 = 1$ and C is its boundary. Let R be the projection of S on the xy-plane.

Solution

The boundary C of S is a circle in the xy-plane of radius one and center at the origin. Let $x = \cos t, y = \sin t, z = 0, 0 \le t < 2\pi$ be parametric equations of C. Then

$$\oint_C \mathbf{A} \cdot d\mathbf{r} = \oint_C (2x - y)\,dx - yz^2\,dy - y^2z\,dz$$

$$= \int_0^{2\pi} (2\cos t - \sin t)(-\sin t)\,dt = \pi$$

Also,

$$\nabla \times \mathbf{A} = \begin{vmatrix} \mathbf{i} & \mathbf{j} & \mathbf{k} \\ \dfrac{\partial}{\partial x} & \dfrac{\partial}{\partial y} & \dfrac{\partial}{\partial z} \\ 2x - y & -yz^2 & -y^2z \end{vmatrix} = \mathbf{k}$$

Then

$$\iint_S (\nabla \times \mathbf{A}) \cdot \mathbf{n}\,dS = \iint_S \mathbf{k} \cdot \mathbf{n}\,dS = \iint_R dx\,dy$$

since $\mathbf{n} \cdot \mathbf{k}\,dS = dx\,dy$ and R is the projection of S on the xy-plane. This last integral equals

$$\int_{x=-1}^{1} \int_{y=-\sqrt{1-x^2}}^{\sqrt{1-x^2}} dy\,dx = 4 \int_0^1 \int_0^{\sqrt{1-x^2}} dy\,dx = 4 \int_0^1 \sqrt{1 - x^2}\,dx = \pi$$

and Stokes' theorem is verified.

6.33. Prove that a necessary and sufficient condition that $\oint_C \mathbf{A} \cdot d\mathbf{r} = 0$ for every closed curve C is that $\nabla \times \mathbf{A} = \mathbf{0}$ identically.

Solution

Sufficiency. Suppose $\nabla \times \mathbf{A} = \mathbf{0}$. Then, by Stokes' theorem

$$\oint_C \mathbf{A} \cdot d\mathbf{r} = \iint_S (\nabla \times \mathbf{A}) \cdot \mathbf{n}\,dS = 0$$

Necessity. Suppose $\oint_C \mathbf{A} \cdot d\mathbf{r} = 0$ around every closed path C, and assume $\nabla \times \mathbf{A} \ne \mathbf{0}$ at some point P. Then, assuming $\nabla \times \mathbf{A}$ is continuous, there will be a region with P as an interior point, where $\nabla \times \mathbf{A} \ne \mathbf{0}$. Let S be a surface contained in this region whose normal $\mathbf{n}$ at each point has the same direction as $\nabla \times \mathbf{A}$, that is, where $\nabla \times \mathbf{A} = \alpha\mathbf{n}$ where α is a positive constant. Let C be the boundary of S. Then, by Stokes' theorem

$$\oint_C \mathbf{A} \cdot d\mathbf{r} = \iint_S (\nabla \times \mathbf{A}) \cdot \mathbf{n}\,dS = \alpha \iint_S \mathbf{n} \cdot \mathbf{n}\,dS > 0$$

which contradicts the hypothesis that $\oint_C \mathbf{A} \cdot d\mathbf{r} = 0$ and shows that $\nabla \times \mathbf{A} = \mathbf{0}$.

It follows that $\nabla \times \mathbf{A} = \mathbf{0}$ is also a necessary and sufficient condition for a line integral $\int_{P_1}^{P_2} \mathbf{A} \cdot d\mathbf{r}$ to be independent of the path joining points P_1 and P_2 (see Problems 5.10 and 5.11).

6.34. Prove $\oint d\mathbf{r} \times \mathbf{B} = \iint_S (\mathbf{n} \times \nabla) \times \mathbf{B} \, dS$.

Solution

In Stokes' theorem, let $\mathbf{A} = \mathbf{B} \times \mathbf{C}$ where $\mathbf{C}$ is a constant vector. Then

$$\oint d\mathbf{r} \cdot (\mathbf{B} \times \mathbf{C}) = \iint_S [\nabla \times (\mathbf{B} \times \mathbf{C})] \cdot \mathbf{n} \, dS$$

$$\oint \mathbf{C} \cdot (d\mathbf{r} \times \mathbf{B}) = \iint_S [(\mathbf{C} \cdot \nabla)\mathbf{B} - \mathbf{C}(\nabla \cdot \mathbf{B})] \cdot \mathbf{n} \, dS$$

$$\mathbf{C} \cdot \oint d\mathbf{r} \times \mathbf{B} = \iint_S [(\mathbf{C} \cdot \nabla)\mathbf{B}] \cdot \mathbf{n} \, dS - \iint_S [\mathbf{C}(\nabla \cdot \mathbf{B})] \cdot \mathbf{n} \, dS$$

$$= \iint_S \mathbf{C} \cdot [\nabla(\mathbf{B} \cdot \mathbf{n})] \, dS - \iint_S \mathbf{C} \cdot [\mathbf{n}(\nabla \cdot \mathbf{B})] \, dS$$

$$= \mathbf{C} \cdot \iint_S [\nabla(\mathbf{B} \cdot \mathbf{n}) - \mathbf{n}(\nabla \cdot \mathbf{B})] \, dS = \mathbf{C} \cdot \iint_S (\mathbf{n} \times \nabla) \times \mathbf{B} \, dS$$

Since $\mathbf{C}$ is an arbitrary constant vector $\oint d\mathbf{r} \times \mathbf{B} = \iint_S (\mathbf{n} \times \nabla) \times \mathbf{B} \, dS$.

6.35. Suppose ΔS is a surface bounded by a simple closed curve C, P is any point of ΔS not on C, and $\mathbf{n}$ is a unit normal to ΔS at P. Show that at P

$$(\text{curl}\, \mathbf{A}) \cdot \mathbf{n} = \lim_{\Delta S \to 0} \frac{\oint_C \mathbf{A} \cdot d\mathbf{r}}{\Delta S}$$

where the limit is taken in such a way that ΔS shrinks to P.

Solution

By Stokes' theorem, $\iint_{\Delta S} (\text{curl}\, \mathbf{A}) \cdot \mathbf{n} \, dS = \oint_C \mathbf{A} \cdot d\mathbf{r}$.

Using the mean value theorem for integrals as in Problems 6.19 and 6.24, this can be written

$$\overline{(\text{curl}\, \mathbf{A}) \cdot \mathbf{n}} = \frac{\oint_C \mathbf{A} \cdot d\mathbf{r}}{\Delta S}$$

and the required result follows upon taking the limit as $\Delta S \to 0$.

This can be used as a starting point for defining curl $\mathbf{A}$ (see Problem 6.36) and is useful in obtaining curl $\mathbf{A}$ in coordinate systems other than rectangular. Since $\oint_C \mathbf{A} \cdot d\mathbf{r}$ is called the circulation of $\mathbf{A}$ about C, the normal component of the curl can be interpreted physically as the limit of the circulation per unit area, thus accounting for the synonym rotation of $\mathbf{A}$ (rot $\mathbf{A}$) instead of curl of $\mathbf{A}$.

6.36. Suppose curl $\mathbf{A}$ is defined according to the limiting process of Problem 6.35. Find the z component of curl $\mathbf{A}$.

Fig. 6-15

Let *EFGH* be a rectangle parallel to the *xy*-plane with interior point $P(x, y, z)$ taken as midpoint, as shown in Fig. 6-15. Let A_1 and A_2 be the components of **A** at *P* in the positive *x* and *y* directions, respectively.

If *C* is the boundary of the rectangle, then

$$\oint_C \mathbf{A} \cdot d\mathbf{r} = \int_{EF} \mathbf{A} \cdot d\mathbf{r} + \int_{FG} \mathbf{A} \cdot d\mathbf{r} + \int_{GH} \mathbf{A} \cdot d\mathbf{r} + \int_{HE} \mathbf{A} \cdot d\mathbf{r}$$

But

$$\int_{EF} \mathbf{A} \cdot d\mathbf{r} = \left(A_1 - \frac{1}{2}\frac{\partial A_1}{\partial y}\Delta y\right)\Delta x \qquad \int_{GH} \mathbf{A} \cdot d\mathbf{r} = -\left(A_1 + \frac{1}{2}\frac{\partial A_1}{\partial y}\Delta y\right)\Delta x$$

$$\int_{FG} \mathbf{A} \cdot d\mathbf{r} = \left(A_2 + \frac{1}{2}\frac{\partial A_2}{\partial x}\Delta x\right)\Delta y \qquad \int_{HE} \mathbf{A} \cdot d\mathbf{r} = -\left(A_2 - \frac{1}{2}\frac{\partial A_2}{\partial x}\Delta x\right)\Delta y$$

except for infinitesimals of higher order than $\Delta x\, \Delta y$.

Adding, we have approximately

$$\oint_C \mathbf{A} \cdot d\mathbf{r} = \left(\frac{\partial A_2}{\partial x} - \frac{\partial A_1}{\partial y}\right)\Delta x\, \Delta y$$

Then, since $\Delta S = \Delta x\, \Delta y$,

$$z \text{ component of curl } \mathbf{A} = (\text{curl } \mathbf{A}) \cdot \mathbf{k} = \lim_{\Delta S \to 0} \frac{\oint_C \mathbf{A} \cdot d\mathbf{r}}{\Delta S}$$

$$= \lim_{\substack{\Delta x \to 0 \\ \Delta y \to 0}} \frac{\left(\dfrac{\partial A_2}{\partial x} - \dfrac{\partial A_1}{\partial y}\right)\Delta x\, \Delta y}{\Delta x\, \Delta y}$$

$$= \frac{\partial A_2}{\partial x} - \frac{\partial A_1}{\partial y}$$

SUPPLEMENTARY PROBLEMS

6.37. Verify Green's theorem in the plane for $\oint_C (3x^2 - 8y^2)\, dx + (4y - 6xy)\, dy$, where *C* is the boundary of the region defined by: (a) $y = \sqrt{x}$, $y = x^2$; (b) $x = 0$, $y = 0$, $x + y = 1$.

6.38. Evaluate $\oint_C (3x + 4y)\, dx + (2x - 3y)\, dy$ where *C*, a circle of radius two with center at the origin of the *xy*-plane, is traversed in the positive sense.

6.39. Work the previous problem for the line integral $\oint_C (x^2 + y^2)\, dx + 3xy^2\, dy$.

6.40. Evaluate $\oint (x^2 - 2xy)\, dx + (x^2 y + 3)\, dy$ around the boundary of the region defined by $y^2 = 8x$ and $x = 2$ (a) directly, and (b) by using Green's theorem.

6.41. Evaluate $\int_{(0,0)}^{(\pi, 2)} (6xy - y^2)\, dx + (3x^2 - 2xy)\, dy$ along the cycloid $x = \theta - \sin\theta$, $y = 1 - \cos\theta$.

6.42. Evaluate $\oint (3x^2 + 2y)\, dx - (x + 3\cos y)\, dy$ around the parallelogram having vertices at (0, 0), (2, 0), (3, 1), and (1, 1).

6.43. Find the area bounded by one arch of the cycloid $x = a(\theta - \sin \theta)$, $y = a(1 - \cos \theta)$, $a > 0$, and the x axis.

6.44. Find the area bounded by the hypocycloid $x^{2/3} + y^{2/3} = a^{2/3}$, $a > 0$.
Hint: Parametric equations are $x = a \cos^3 \theta$, $y = a \sin^3 \theta$.

6.45. Show that in polar coordinates (ρ, ϕ), the expression $x \, dy - y \, dx = \rho^2 \, d\phi$. Interpret $\frac{1}{2} \int x \, dy - y \, dx$.

6.46. Find the area of a loop of the four-leafed rose $\rho = 3 \sin 2\phi$.

6.47. Find the area of both loops of the lemniscate $\rho^2 = a^2 \cos 2\phi$.

6.48. Find the area of the loop of the folium of Descartes $x^3 + y^3 = 3axy$, $a > 0$ (see Fig. 6-16).

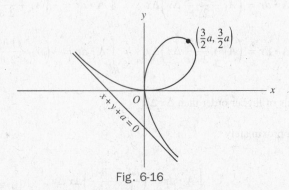

Fig. 6-16

Hint: Let $y = tx$ and obtain the parametric equations of the curve. Then use the fact that

$$\text{Area} = \frac{1}{2} \oint x \, dy - y \, dx = \frac{1}{2} \oint x^2 d\left(\frac{y}{x}\right) = \frac{1}{2} \oint x^2 \, dt$$

6.49. Verify Green's theorem in the plane for $\oint_C (2x - y^3) \, dx - xy \, dy$, where C is the boundary of the region enclosed by the circles $x^2 + y^2 = 1$ and $x^2 + y^2 = 9$.

6.50. Evaluate $\displaystyle\int_{(1,0)}^{(-1,0)} \frac{-y \, dx + x \, dy}{x^2 + y^2}$ along the following paths:

(a) Straight line segments from $(1, 0)$ to $(1, 1)$, then to $(-1, 1)$, then to $(-1, 0)$.

(b) Straight line segments from $(1, 0)$ to $(1, -1)$, then to $(-1, -1)$, then to $(-1, 0)$.
Show that although $\partial M / \partial y = \partial N / \partial x$, the line integral *is dependent* on the path joining $(1, 0)$ to $(-1, 0)$ and explain.

6.51. By changing variables from (x, y) to (u, v) according to the transformation $x = x(u, v)$, $y = y(u, v)$, show that the area A of a region R bounded by a simple closed curve C is given by

$$A = \iint_R \left| J\left(\frac{x, y}{u, v}\right) \right| du \, dv \quad \text{where} \quad J\left(\frac{x, y}{u, v}\right) \equiv \begin{vmatrix} \dfrac{\partial x}{\partial u} & \dfrac{\partial y}{\partial u} \\ \dfrac{\partial x}{\partial v} & \dfrac{\partial y}{\partial v} \end{vmatrix}$$

is the Jacobian of x and y with respect to u and v. What restrictions should you make? Illustrate the result where u and v are polar coordinates.
Hint: Use the result $A = \frac{1}{2} \int x \, dy - y \, dx$, transform to u, v coordinates and then use Green's theorem.

6.52. Evaluate $\iint_S \mathbf{F} \cdot \mathbf{n} \, dS$, where $\mathbf{F} = 2xy\mathbf{i} + yz^2\mathbf{j} + xz\mathbf{k}$ and S is:

(a) The surface of the parallelepiped bounded by $x = 0, y = 0, z = 0, x = 2, y = 1$, and $z = 3$.

(b) The surface of the region bounded by $x = 0, y = 0, y = 3, z = 0$, and $x + 2z = 6$.

6.53. Verify the divergence theorem for $\mathbf{A} = 2x^2 y \mathbf{i} - y^2 \mathbf{j} + 4xz^2 \mathbf{k}$ taken over the region in the first octant bounded by $y^2 + z^2 = 9$ and $x = 2$.

6.54. Evaluate $\iint_S \mathbf{r} \cdot \mathbf{n}\, dS$ where (a) S is the sphere of radius 2 with center at $(0, 0, 0)$, (b) S is the surface of the cube bounded by $x = -1, y = -1, z = -1, x = 1, y = 1, z = 1$, (c) S is the surface bounded by the paraboloid $z = 4 - (x^2 + y^2)$ and the xy-plane.

6.55. Suppose S is any closed surface enclosing a volume V and $\mathbf{A} = ax\mathbf{i} + by\mathbf{j} + cz\mathbf{k}$. Prove that
$$\iint_S \mathbf{A} \cdot \mathbf{n}\, dS = (a + b + c)V.$$

6.56. Suppose $\mathbf{H} = \text{curl } \mathbf{A}$. Prove that $\iint_S \mathbf{H} \cdot \mathbf{n}\, dS = 0$ for any closed surface S.

6.57. Suppose $\mathbf{n}$ is the unit outward drawn normal to any closed surface of area S. Show that $\iiint_V \text{div } \mathbf{n}\, dV = S$.

6.58. Prove $\iiint_V \dfrac{dV}{r^2} = \iint_S \dfrac{\mathbf{r} \cdot \mathbf{n}}{r^2}\, dS.$

6.59. Prove $\iint_S r^5 \mathbf{n}\, dS = \iiint_V 5r^3 \mathbf{r}\, dV.$

6.60. Prove $\iint_S \mathbf{n}\, dS = \mathbf{0}$ for any closed surface S.

6.61. Show that Green's second identity can be written $\iiint_V (\phi \nabla^2 \psi - \psi \nabla^2 \phi)\, dV = \iint_S \left(\phi \dfrac{d\psi}{dn} - \psi \dfrac{d\phi}{dn} \right) dS.$

6.62. Prove $\iint_S \mathbf{r} \times d\mathbf{S} = \mathbf{0}$ for any closed surface S.

6.63. Verify Stokes' theorem for $\mathbf{A} = (y - z + 2)\mathbf{i} + (yz + 4)\mathbf{j} - xz\mathbf{k}$, where S is the surface of the cube $x = 0$, $y = 0, z = 0, x = 2, y = 2, z = 2$ above the xy-plane.

6.64. Verify Stokes' theorem for $\mathbf{F} = xz\mathbf{i} - y\mathbf{j} + x^2 y\mathbf{k}$, where S is the surface of the region bounded by $x = 0$, $y = 0, z = 0, 2x + y + 2z = 8$, which is not included in the xz-plane.

6.65. Evaluate $\iint_S (\nabla \times \mathbf{A}) \cdot \mathbf{n}\, dS$, where $\mathbf{A} = (x^2 + y - 4)\mathbf{i} + 3xy\mathbf{j} + (2xz + z^2)\mathbf{k}$ and S is the surface of (a) the hemisphere $x^2 + y^2 + z^2 = 16$ above the xy-plane, (b) the paraboloid $z = 4 - (x^2 + y^2)$ above the xy-plane.

6.66. Let $\mathbf{A} = 2yz\mathbf{i} - (x + 3y - 2)\mathbf{j} + (x^2 + z)\mathbf{k}$. Evaluate $\iint_S (\nabla \times \mathbf{A}) \cdot \mathbf{n}\, dS$ over the surface of intersection of the cylinders $x^2 + y^2 = a^2, x^2 + z^2 = a^2$, which is included in the first octant.

6.67. A vector $\mathbf{B}$ is always normal to a given closed surface S. Show that $\iiint_V \text{curl } \mathbf{B}\, dV = \mathbf{0}$, where V is the region bounded by S.

6.68. Let $\oint_C \mathbf{E} \cdot d\mathbf{r} = -\dfrac{1}{c} \dfrac{\partial}{\partial t} \iint_S \mathbf{H} \cdot d\mathbf{S}$, where S is any surface bounded by the curve C. Show that $\nabla \times \mathbf{E} = -\dfrac{1}{c} \dfrac{\partial \mathbf{H}}{\partial t}$.

6.69. Prove $\oint_C \phi\, d\mathbf{r} = \iint_S d\mathbf{S} \times \nabla \phi.$

6.70. Use the operator equivalence of Solved Problem 6.25 to arrive at (a) $\nabla \phi$, (b) $\nabla \cdot \mathbf{A}$, (c) $\nabla \times \mathbf{A}$ in rectangular coordinates.

6.71. Prove $\iiint_V \nabla \phi \cdot \mathbf{A}\, dV = \iint_S \phi \mathbf{A} \cdot \mathbf{n}\, dS - \iiint_V \phi \nabla \cdot \mathbf{A}\, dV.$

6.72. Let $\mathbf{r}$ be the position vector of any point relative to an origin O. Suppose ϕ has continuous derivatives of order two, at least, and let S be a closed surface bounding a volume V. Denote ϕ at O by ϕ_0. Show that

$$\iint_S \left[\frac{1}{r} \nabla \phi - \phi \nabla \left(\frac{1}{r} \right) \right] \cdot d\mathbf{S} = \iiint_V \frac{\nabla^2 \phi}{r}\, dV + \alpha$$

where $\alpha = 0$ or $4\pi \phi_0$ according as O is outside or inside S.

6.73. The potential $\phi(P)$ at a point $P(x, y, z)$ due to a system of charges (or masses) $q_1, q_2, \ldots, q_n$ having position vectors $\mathbf{r}_1, \mathbf{r}_2, \ldots, \mathbf{r}_n$ with respect to P is given by

$$\phi = \sum_{m=1}^{n} \frac{q_m}{r_m}$$

Prove *Gauss' law*

$$\iint_S \mathbf{E} \cdot d\mathbf{S} = 4\pi Q$$

where $\mathbf{E} = -\nabla\phi$ is the electric field intensity, S is a surface enclosing all the charges and $Q = \sum_{m=1}^{n} q_m$ is the total charge within S.

6.74. If a region V bounded by a surface S has a continuous charge (or mass) distribution of density ρ, then the potential $\phi(P)$ at a point P is defined by

$$\phi = \iiint_V \frac{\rho \, dV}{r}.$$

Deduce the following under suitable assumptions:

(a) $\iint_S \mathbf{E} \cdot d\mathbf{S} = 4\pi \iiint_V \rho \, dV$, where $\mathbf{E} = -\nabla\phi$.

(b) $\nabla^2\phi = -4\pi\rho$ (Poisson's equation) at all points P where charges exist, and $\nabla^2\phi = 0$ (Laplace's equation) where no charges exist.

ANSWERS TO SUPPLEMENTARY PROBLEMS

6.37. (a) common value $= 3/2$,
  (b) common value $= 5/3$

6.38. -8π

6.39. 12π

6.40. $128/5$

6.41. $6\pi^2 - 4\pi$

6.42. -6

6.43. $3\pi a^2$

6.44. $3\pi a^2/8$

6.46. $9\pi/8$

6.47. a^2

6.48. $3a^2/2$

6.49. common value $= 60\pi$

6.50. (a) π, (b) $-\pi$

6.52. (a) 30, (b) 351/2

6.53. 180

6.54. (a) 32π, (b) 24, (c) 24π

6.63. common value $= -4$

6.64. common value $= 32/3$

6.65. (a) -16π, (b) -4π

6.66. $-a^2(3\pi + 8a)/12$

CHAPTER 7

Curvilinear Coordinates

7.1 Introduction

The reader is familiar with the rectangular coordinate system, (x, y), and the polar coordinate system, (r, θ), in the plane. The two systems are related by the equations

$$x = r \cos \theta, \quad y = r \sin \theta \quad \text{and} \quad r = \sqrt{x^2 + y^2}, \quad \theta = \text{arc } \tan(y/x)$$

This chapter treats general coordinate systems in space.

7.2 Transformation of Coordinates

Suppose the rectangular coordinates (x, y, z) of any point in space are each expressed as functions of (u_1, u_2, u_3). Say,

$$x = x(u_1, u_2, u_3), \quad y = y(u_1, u_2, u_3), \quad z = z(u_1, u_2, u_3) \tag{1}$$

Suppose that (1) can be solved for u_1, u_2, u_3 in terms of x, y, z, that is,

$$u_1 = u_1(x, y, z), \quad u_2 = u_2(x, y, z), \quad u_3 = u_3(x, y, z) \tag{2}$$

The functions in (1) and (2) are assumed to be single-valued and to have continuous derivatives so that the correspondence between (x, y, z) and (u_1, u_2, u_3) is unique. In practice, this assumption may not apply at certain points and special consideration is required.

Given a point P with rectangular coordinates (x, y, z), we can, from (2), associate a unique set of coordinates (u_1, u_2, u_3) called the *curvilinear coordinates* of P. The sets of equations (1) or (2) define a *transformation of coordinates*.

7.3 Orthogonal Curvilinear Coordinates

The surfaces $u_1 = c_1, u_2 = c_2, u_3 = c_3$, where c_1, c_2, c_3 are constants, are called *coordinate surfaces* and each pair of these surfaces intersect in curves called coordinate curves or lines (see Fig. 7-1). If the coordinate surfaces intersect at right angles, the curvilinear coordinate system is called *orthogonal*. The $u_1, u_2,$

and u_3 coordinate curves of a curvilinear system are analogous to the x, y, and z coordinate axes of a rectangular system.

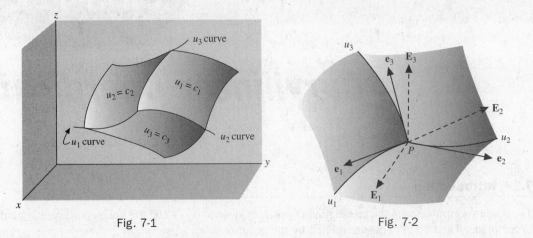

Fig. 7-1   Fig. 7-2

7.4 Unit Vectors in Curvilinear Systems

Let $\mathbf{r} = x\mathbf{i} + y\mathbf{j} + z\mathbf{k}$ be the position vector of a point P in space. Then (1) can be written $\mathbf{r} = \mathbf{r}(u_1, u_2, u_3)$. A tangent vector to the u_1 curve at P (for which u_2 and u_3 are constants) is $\partial \mathbf{r}/\partial u_1$. Then a unit tangent vector in this direction is $\mathbf{e}_1 = (\partial \mathbf{r}/\partial u_1)/|\partial \mathbf{r}/\partial u_1|$ so that $\partial \mathbf{r}/\partial u_1 = h_1 \mathbf{e}_1$ where $h_1 = |\partial \mathbf{r}/\partial u_1|$. Similarly, if $\mathbf{e}_2$ and $\mathbf{e}_3$ are unit tangent vectors to the u_2 and u_3 curves at P, respectively, then $\partial \mathbf{r}/\partial u_2 = h_2 \mathbf{e}_2$ and $\partial \mathbf{r}/\partial u_3 = h_3 \mathbf{e}_3$ where $h_2 = |\partial \mathbf{r}/\partial u_2|$ and $h_3 = |\partial \mathbf{r}/\partial u_3|$. The quantities h_1, h_2, h_3 are called *scale factors*. The unit vectors $\mathbf{e}_1, \mathbf{e}_2, \mathbf{e}_3$ are in the directions of increasing u_1, u_2, u_3, respectively.

Since ∇u_1 is a vector at P normal to the surface $u_1 = c_1$, a unit vector in this direction is given by $\mathbf{E}_1 = \nabla u_1/|\nabla u_1|$. Similarly, the unit vectors $\mathbf{E}_2 = \nabla u_2/|\nabla u_2|$ and $\mathbf{E}_3 = \nabla u_3/|\nabla u_3|$ at P are normal to the surfaces $u_2 = c_2$ and $u_3 = c_3$, respectively.

Thus, at each point P of a curvilinear system, there exist, in general, two sets of unit vectors, $\mathbf{e}_1, \mathbf{e}_2, \mathbf{e}_3$ tangent to the coordinate curves and $\mathbf{E}_1, \mathbf{E}_2, \mathbf{E}_3$ normal to the coordinate surfaces (see Fig. 7-2). The sets become identical if and only if the curvilinear coordinate system is orthogonal (see Problem 7.19). Both sets are analogous to the $\mathbf{i}, \mathbf{j}, \mathbf{k}$ unit vectors in rectangular coordinates but are unlike them in that they may change directions from point to point. It can be shown (see Problem 7.15) that the sets $\partial \mathbf{r}/\partial u_1, \partial \mathbf{r}/\partial u_2, \partial \mathbf{r}/\partial u_3$ and $\nabla u_1, \nabla u_2, \nabla u_3$ constitute reciprocal systems of vectors.

A vector $\mathbf{A}$ can be represented in terms of the unit base vectors $\mathbf{e}_1, \mathbf{e}_2, \mathbf{e}_3$ or $\mathbf{E}_1, \mathbf{E}_2, \mathbf{E}_3$ in the form

$$\mathbf{A} = A_1 \mathbf{e}_1 + A_2 \mathbf{e}_2 + A_3 \mathbf{e}_3 = a_1 \mathbf{E}_1 + a_2 \mathbf{E}_2 + a_3 \mathbf{E}_3$$

where A_1, A_2, A_3 and a_1, a_2, a_3 are the respective *components* of $\mathbf{A}$ in each system.

We can also represent $\mathbf{A}$ in terms of the base vectors $\partial \mathbf{r}/\partial u_1, \partial \mathbf{r}/\partial u_2, \partial \mathbf{r}/\partial u_3$ or $\nabla u_1, \nabla u_2, \nabla u_3$, which are called *unitary base vectors* but *are not* unit vectors in general. In this case

$$\mathbf{A} = C_1 \frac{\partial \mathbf{r}}{\partial u_1} + C_2 \frac{\partial \mathbf{r}}{\partial u_2} + C_3 \frac{\partial \mathbf{r}}{\partial u_3} = C_1 \boldsymbol{\alpha}_1 + C_2 \boldsymbol{\alpha}_2 + C_3 \boldsymbol{\alpha}_3$$

and

$$\mathbf{A} = c_1 \nabla u_1 + c_2 \nabla u_2 + c_3 \nabla u_3 = c_1 \boldsymbol{\beta}_1 + c_2 \boldsymbol{\beta}_2 + c_3 \boldsymbol{\beta}_3$$

where C_1, C_2, C_3 are called the *contravariant components* of $\mathbf{A}$ and c_1, c_2, c_3 are called the *covariant components* of $\mathbf{A}$ (see Problems 7.33 and 7.34). Note that $\boldsymbol{\alpha}_p = \partial \mathbf{r}/\partial u_p, \boldsymbol{\beta}_p = \nabla u_p, p = 1, 2, 3$.

7.5 Arc Length and Volume Elements

Recall first that the position vector of a point P can be written in the form $\mathbf{r} = \mathbf{r}(u_1, u_2, u_3)$. Then

$$d\mathbf{r} = \frac{\partial \mathbf{r}}{\partial u_1} du_1 + \frac{\partial \mathbf{r}}{\partial u_2} du_2 + \frac{\partial \mathbf{r}}{\partial u_3} du_3 = h_1 \, du_1 \mathbf{e}_1 + h_2 \, du_2 \mathbf{e}_2 + h_3 \, du_3 \mathbf{e}_3$$

Then the differential of arc length ds is determined from $ds^2 = d\mathbf{r} \cdot d\mathbf{r}$. For orthogonal systems, $\mathbf{e}_1 \cdot \mathbf{e}_2 = \mathbf{e}_2 \cdot \mathbf{e}_3 = \mathbf{e}_3 \cdot \mathbf{e}_1 = 0$ and

$$ds^2 = h_1^2 \, du_1^2 + h_2^2 \, du_2^2 + h_3^2 \, du_3^2$$

For non-orthogonal or general curvilinear systems, see Problem 7.17.

Along a u_1 curve, u_2 and u_3 are constants so that $d\mathbf{r} = h_1 \, du_1 \mathbf{e}_1$. Then the differential of arc length ds_1 along u_1 at P is $h_1 \, du_1$. Similarly, the differential arc lengths along u_2 and u_3 at P are $ds_2 = h_2 \, du_2$, $ds_3 = h_3 \, du_3$.

Referring to Fig. 7-3, the volume element for an orthogonal curvilinear coordinate system is given by

$$dV = |(h_1 \, du_1 \mathbf{e}_1) \cdot (h_2 \, du_2 \mathbf{e}_2) \times (h_3 \, du_3 \mathbf{e}_3)| = h_1 h_2 h_3 \, du_1 \, du_2 \, du_3$$

since $|\mathbf{e}_1 \cdot \mathbf{e}_2 \times \mathbf{e}_3| = 1$.

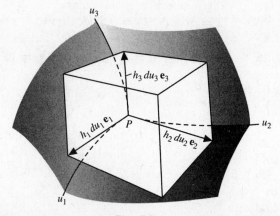

Fig. 7-3

7.6 Gradient, Divergence, Curl

The operations of gradient, divergence, and curl can be expressed in terms of curvilinear coordinates. Specifically, the following proposition applies.

PROPOSITION 7.1: Suppose Φ is a scalar function and $\mathbf{A} = A_1 \mathbf{e}_1 + A_2 \mathbf{e}_2 + A_3 \mathbf{e}_3$ is a vector function of orthogonal curvilinear coordinates u_1, u_2, u_3. Then the following laws hold.

(i) $\nabla \Phi = \operatorname{grad} \Phi = \dfrac{1}{h_1} \dfrac{\partial \Phi}{\partial u_1} \mathbf{e}_1 + \dfrac{1}{h_2} \dfrac{\partial \Phi}{\partial u_2} \mathbf{e}_2 + \dfrac{1}{h_3} \dfrac{\partial \Phi}{\partial u_3} \mathbf{e}_3$

(ii) $\nabla \cdot \mathbf{A} = \operatorname{div} \mathbf{A} = \dfrac{1}{h_1 h_2 h_3} \left[\dfrac{\partial}{\partial u_1}(h_2 h_3 A_1) + \dfrac{\partial}{\partial u_2}(h_3 h_1 A_2) + \dfrac{\partial}{\partial u_3}(h_1 h_2 A_3) \right]$

(iii) $\nabla \times \mathbf{A} = \operatorname{curl} \mathbf{A} = \dfrac{1}{h_1 h_2 h_3} \begin{vmatrix} h_1 \mathbf{e}_1 & h_2 \mathbf{e}_2 & h_3 \mathbf{e}_3 \\ \dfrac{\partial}{\partial u_1} & \dfrac{\partial}{\partial u_2} & \dfrac{\partial}{\partial u_3} \\ h_1 A_1 & h_2 A_2 & h_3 A_3 \end{vmatrix}$

(iv) $\nabla^2\Phi = $ Laplacian of Φ

$$= \frac{1}{h_1 h_2 h_3}\left[\frac{\partial}{\partial u_1}\left(\frac{h_2 h_3}{h_1}\frac{\partial\Phi}{\partial u_1}\right) + \frac{\partial}{\partial u_2}\left(\frac{h_3 h_1}{h_2}\frac{\partial\Phi}{\partial u_2}\right) + \frac{\partial}{\partial u_3}\left(\frac{h_1 h_2}{h_3}\frac{\partial\Phi}{\partial u_3}\right)\right]$$

Observe that if $h_1 = h_2 = h_3 = 1$, and $\mathbf{e}_1, \mathbf{e}_2, \mathbf{e}_3$ are replaced by $\mathbf{i}, \mathbf{j}, \mathbf{k}$, then the above laws reduce to the usual expressions in rectangular coordinates where (u_1, u_2, u_3) is replaced by (x, y, z).

Extensions of the above results are achieved by a more general theory of curvilinear systems using the methods of *tensor analysis*, which is considered in Chapter 8.

7.7   Special Orthogonal Coordinate Systems

The following is a list of nine special orthogonal coordinate systems beside the usual rectangular coordinates (x, y, z).

1. **Cylindrical Coordinates (ρ, ϕ, z).**
   See Fig. 7-4. Here

$$x = \rho\cos\phi, \quad y = \rho\sin\phi, \quad z = z$$

where $\rho \geq 0$, $0 \leq \phi < 2\pi$, $-\infty < z < \infty$, $h_\rho = 1$, $h_\phi = \rho$, and $h_z = 1$.

2. **Spherical Coordinates (r, θ, ϕ).**
   See Fig. 7-5. Here

$$x = r\sin\theta\cos\phi, \quad y = r\sin\theta\sin\phi, \quad z = r\cos\theta$$

where $r \geq 0$, $0 \leq \phi < 2\pi$, $0 \leq \theta \leq \pi$, $h_r = 1$, $h_\theta = r$, and $h_\phi = r\sin\theta$.

 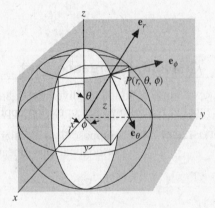

Fig. 7-4   Fig. 7-5

3. **Parabolic Cylindrical Coordinates (u, v, z).**
   See Fig. 7-6. Here

$$x = \tfrac{1}{2}(u^2 - v^2), \quad y = uv, \quad z = z$$

where $-\infty < u < \infty$, $v \geq 0$, $-\infty < z < \infty$, $h_u = h_v = \sqrt{u^2 + v^2}$, and $h_z = 1$.

In cylindrical coordinates, $u = \sqrt{2\rho}\,\cos{(\phi/2)}$, $v = \sqrt{2\rho}\,\sin{(\phi/2)}$, $z = z$.

The traces of the coordinate surfaces on the *xy*-plane are shown in Fig. 7-6. They are confocal parabolas with a common axis.

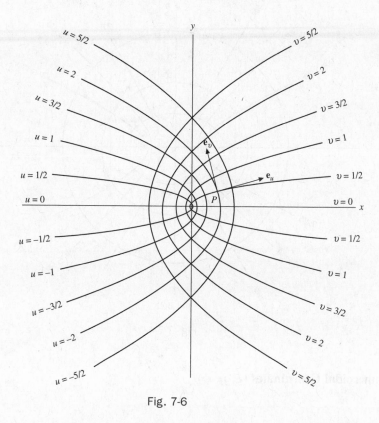

Fig. 7-6

4. **Paraboloidal Coordinates (u, v, ϕ).**
   Here

$$x = uv\cos\phi, \quad y = uv\sin\phi, \quad z = \tfrac{1}{2}(u^2 - v^2)$$

where $u \geq 0$, $v \geq 0$, $0 \leq \phi < 2\pi$, $h_u = h_v = \sqrt{u^2 + v^2}$, and $h_\phi = uv$.

   Two sets of coordinate surfaces are obtained by revolving the parabolas of Fig. 7-6 above about the *x*-axis which is relabeled the *z*-axis. The third set of coordinate surfaces are planes passing through this axis.

5. **Elliptic Cylindrical Coordinates (u, v, z).**
   See Fig. 7-7. Here

$$x = a\cosh u\,\cos v, \quad y = a\sinh u\,\sin v, \quad z = z$$

where $u \geq 0$, $0 \leq v < 2\pi$, $-\infty < z < \infty$, $h_u = h_v = a\sqrt{\sinh^2 u + \sin^2 v}$, and $h_z = 1$.

The traces of the coordinate surfaces on the *xy*-plane are shown in Fig. 7-7. They are confocal ellipses and hyperbolas.

Fig. 7-7

6. **Prolate Spheroidal Coordinates (ξ, η, ϕ).**
   Here:

$$x = a \sinh \xi \sin \eta \cos \phi, \quad y = a \sinh \xi \sin \eta \sin \phi, \quad z = a \cosh \xi \cos \eta$$

where $\xi \geq 0$, $0 \leq \eta \leq \pi$, $0 \leq \phi < 2\pi$, $h_\xi = h_\eta = a\sqrt{\sinh^2\xi + \sin^2\eta}$, and $h_\phi = a \sinh \xi \sin \eta$.

Two sets of coordinate surfaces are obtained by revolving the curves of Fig. 7-7 about the *x*-axis which is relabeled the *z*-axis. The third set of coordinate surfaces are planes passing through this axis.

7. **Oblate Spheroidal Coordinates (ξ, η, ϕ).**
   Here:

$$x = a \cosh \xi \cos \eta \cos \phi, \quad y = a \cosh \xi \cos \eta \sin \phi, \quad z = a \sinh \xi \sin \eta$$

where $\xi \geq 0$, $-\pi/2 \leq \eta \leq \pi/2$, $0 \leq \phi < 2\pi$, $h_\xi = h_\eta = a\sqrt{\sinh^2\xi + \sin^2\eta}$, and $h_\phi = a \cosh \xi \cos \eta$.

Two sets of coordinate surfaces are obtained by revolving the curves of Fig. 7-7 about the *y*-axis which is relabeled the *z*-axis. The third set of coordinate surfaces are planes passing through this axis.

8. **Ellipsoidal Coordinates (λ, μ, ν).**
   Here

$$\frac{x^2}{a^2 - \lambda} + \frac{y^2}{b^2 - \lambda} + \frac{z^2}{c^2 - \lambda} = 1, \quad \lambda < c^2 < b^2 < a^2$$

$$\frac{x^2}{a^2 - \mu} + \frac{y^2}{b^2 - \mu} + \frac{z^2}{c^2 - \mu} = 1, \quad c^2 < \mu < b^2 < a^2$$

$$\frac{x^2}{a^2 - \nu} + \frac{y^2}{b^2 - \nu} + \frac{z^2}{c^2 - \nu} = 1, \quad c^2 < b^2 < \nu < a^2$$

$$h_\lambda = \frac{1}{2}\sqrt{\frac{(\mu - \lambda)(\nu - \lambda)}{(a^2 - \lambda)(b^2 - \lambda)(c^2 - \lambda)}},$$

$$h_\mu = \frac{1}{2}\sqrt{\frac{(\nu - \mu)(\lambda - \mu)}{(a^2 - \mu)(b^2 - \mu)(c^2 - \mu)}}$$

$$h_\nu = \frac{1}{2}\sqrt{\frac{(\lambda - \nu)(\mu - \nu)}{(a^2 - \nu)(b^2 - \nu)(c^2 - \nu)}}$$

9. **Bipolar Coordinates (u, v, z).**
   See Fig. 7-8. Here

$$x^2 + (y - a\cot u)^2 = a^2\csc^2 u, \quad (x - a\coth v)^2 + y^2 = a^2\operatorname{csch}^2 v, \quad z = z$$

or

$$x = \frac{a\sinh v}{\cosh v - \cos u}, \quad y = \frac{a\sin u}{\cosh v - \cos u}, \quad z = z$$

where $0 \le u < 2\pi$, $-\infty < v < \infty$, $-\infty < z < \infty$, $h_u = h_v = a/(\cosh v - \cos u)$, and $h_z = 1$.

The traces of the coordinate surfaces on the xy-plane are shown in Fig. 7-8. By revolving the curves of Fig. 7-8 about the y-axis and relabeling this the z-axis, a *toroidal coordinate system* is obtained.

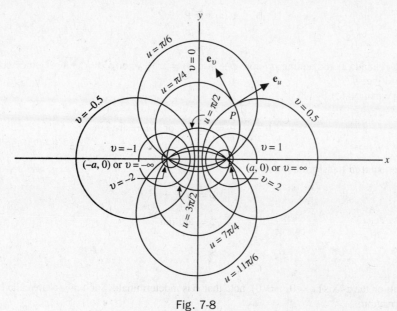

Fig. 7-8

SOLVED PROBLEMS

7.1. Describe the coordinate surfaces and coordinate curves for (a) cylindrical and (b) spherical coordinates.

Solution

(a)　The coordinate surfaces (or level surfaces) are:

$\rho = c_1$　cylinders coaxial with the z-axis (or z-axis if $c_1 = 0$).
$\phi = c_2$　planes through the z-axis.
$z = c_3$　planes perpendicular to the z-axis.

The coordinate curves are:

Intersection of $\rho = c_1$　and　$\phi = c_2$ (z curve) is a straight line.
Intersection of $\rho = c_1$　and　$z = c_3$ (ϕ curve) is a circle (or point).
Intersection of $\phi = c_2$　and　$z = c_3$ (ρ curve) is a straight line.

(b)　The coordinate surfaces are:

$r = c_1$　spheres having center at the origin (or origin if $c_1 = 0$).
$\theta = c_2$　cones having vertex at the origin (lines if $c_2 = 0$ or π, and the xy-plane if
　　$c_2 = \pi/2$).
$\phi = c_3$　planes through the z-axis.

The coordinate curves are:

Intersection of $r = c_1$　and　$\theta = c_2$ (ϕ curve) is a circle (or point).
Intersection of $r = c_1$　and　$\phi = c_3$ (θ curve) is a semi-circle ($c_1 \neq 0$).
Intersection of $\theta = c_2$　and　$\phi = c_3$ (r curve) is a line.

7.2. Determine the transformation from cylindrical to rectangular coordinates.

Solution

The equations defining the transformation from rectangular to cylindrical coordinates are

$$x = \rho \cos \phi \tag{1}$$
$$y = \rho \sin \phi \tag{2}$$
$$z = z \tag{3}$$

Squaring (1) and (2) and adding $\rho^2(\cos^2 \phi + \sin^2 \phi) = x^2 + y^2$ or $\rho = \sqrt{x^2 + y^2}$, since $\cos^2 \phi + \sin^2 \phi = 1$ and ρ is positive.

Dividing equation (2) by (1),

$$\frac{y}{x} = \frac{\rho \sin \phi}{\rho \cos \phi} = \tan \phi \quad \text{or} \quad \phi = \text{arc tan} \frac{y}{x}$$

Then the required transformation is

$$\rho = \sqrt{x^2 + y^2} \tag{4}$$
$$\phi = \text{arc tan} \frac{y}{x} \tag{5}$$
$$z = z \tag{6}$$

For points on the z-axis ($x = 0$, $y = 0$), note that ϕ is indeterminate. Such points are called *singular points* of the transformation.

7.3. Prove that a cylindrical coordinate system is orthogonal.

Solution

The position vector of any point in cylindrical coordinates is

$$\mathbf{r} = x\mathbf{i} + y\mathbf{j} + z\mathbf{k} = \rho\cos\phi\mathbf{i} + \rho\sin\phi\mathbf{j} + z\mathbf{k}$$

The tangent vectors to the ρ, ϕ, and z curves are given respectively by $\partial\mathbf{r}/\partial\rho$, $\partial\mathbf{r}/\partial\phi$, and $\partial\mathbf{r}/\partial z$ where

$$\frac{\partial\mathbf{r}}{\partial\rho} = \cos\phi\mathbf{i} + \sin\phi\mathbf{j}, \qquad \frac{\partial\mathbf{r}}{\partial\phi} = -\rho\sin\phi\mathbf{i} + \rho\cos\phi\mathbf{j}, \qquad \frac{\partial\mathbf{r}}{\partial z} = \mathbf{k}$$

The unit vectors in these directions are

$$\mathbf{e}_1 = \mathbf{e}_\rho = \frac{\partial\mathbf{r}/\partial\rho}{|\partial\mathbf{r}/\partial\rho|} = \frac{\cos\phi\mathbf{i} + \sin\phi\mathbf{j}}{\sqrt{\cos^2\phi + \sin^2\phi}} = \cos\phi\mathbf{i} + \sin\phi\mathbf{j}$$

$$\mathbf{e}_2 = \mathbf{e}_\phi = \frac{\partial\mathbf{r}/\partial\phi}{|\partial\mathbf{r}/\partial\phi|} = \frac{-\rho\sin\phi\mathbf{i} + \rho\cos\phi\mathbf{j}}{\sqrt{\rho^2\sin^2\phi + \rho^2\cos^2\phi}} = -\sin\phi\mathbf{i} + \cos\phi\mathbf{j}$$

$$\mathbf{e}_3 = \mathbf{e}_z = \frac{\partial\mathbf{r}/\partial z}{|\partial\mathbf{r}/\partial z|} = \mathbf{k}$$

Then

$$\mathbf{e}_1 \cdot \mathbf{e}_2 = (\cos\phi\mathbf{i} + \sin\phi\mathbf{j}) \cdot (-\sin\phi\mathbf{i} + \cos\phi\mathbf{j}) = 0$$
$$\mathbf{e}_1 \cdot \mathbf{e}_3 = (\cos\phi\mathbf{i} + \sin\phi\mathbf{j}) \cdot (\mathbf{k}) = 0$$
$$\mathbf{e}_2 \cdot \mathbf{e}_3 = (-\sin\phi\mathbf{i} + \cos\phi\mathbf{j}) \cdot (\mathbf{k}) = 0$$

and so $\mathbf{e}_1$, $\mathbf{e}_2$, and $\mathbf{e}_3$ are mutually perpendicular and the coordinate system is orthogonal.

7.4. Represent the vector $\mathbf{A} = z\mathbf{i} - 2x\mathbf{j} + y\mathbf{k}$ in cylindrical coordinates. Thus determine A_ρ, A_ϕ, and A_z.

Solution

From Problem 7.3,

$$\mathbf{e}_\rho = \cos\phi\mathbf{i} + \sin\phi\mathbf{j} \tag{1}$$
$$\mathbf{e}_\phi = -\sin\phi\mathbf{i} + \cos\phi\mathbf{j} \tag{2}$$
$$\mathbf{e}_z = \mathbf{k} \tag{3}$$

Solving (1) and (2) simultaneously,

$$\mathbf{i} = \cos\phi\mathbf{e}_\rho - \sin\phi\mathbf{e}_\phi, \quad \mathbf{j} = \sin\phi\mathbf{e}_\rho + \cos\phi\mathbf{e}_\phi.$$

Then

$$\mathbf{A} = z\mathbf{i} - 2x\mathbf{j} + y\mathbf{k}$$
$$= z(\cos\phi\mathbf{e}_\rho - \sin\phi\mathbf{e}_\phi) - 2\rho\cos\phi(\sin\phi\mathbf{e}_\rho + \cos\phi\mathbf{e}_\phi) + \rho\sin\phi\mathbf{e}_z$$
$$= (z\cos\phi - 2\rho\cos\phi\sin\phi)\mathbf{e}_\rho - (z\sin\phi + 2\rho\cos^2\phi)\mathbf{e}_\phi + \rho\sin\phi\mathbf{e}_z$$

and

$$A_\rho = z\cos\phi - 2\rho\cos\phi\sin\phi, \qquad A_\phi = -z\sin\phi - 2\rho\cos^2\phi, \qquad A_z = \rho\sin\phi.$$

7.5. Prove $\dfrac{d}{dt}\mathbf{e}_\rho = \dot\phi\mathbf{e}_\phi$, $\dfrac{d}{dt}\mathbf{e}_\phi = -\dot\phi\mathbf{e}_\rho$ where dots denote differentiation with respect to time t.

Solution

From Problem 7.3

$$\mathbf{e}_\rho = \cos\phi\mathbf{i} + \sin\phi\mathbf{j}, \quad \mathbf{e}_\phi = -\sin\phi\mathbf{i} + \cos\phi\mathbf{j}$$

Then

$$\frac{d}{dt}\mathbf{e}_\rho = -(\sin\phi)\dot\phi\mathbf{i} + (\cos\phi)\dot\phi\mathbf{j} = (-\sin\phi\mathbf{i} + \cos\phi\mathbf{j})\dot\phi = \dot\phi\mathbf{e}_\phi$$

$$\frac{d}{dt}\mathbf{e}_\phi = -(\cos\phi)\dot\phi\mathbf{i} - (\sin\phi)\dot\phi\mathbf{j} = -(\cos\phi\mathbf{i} + \sin\phi\mathbf{j})\dot\phi = -\dot\phi\mathbf{e}_\rho$$

7.6. Express the velocity $\mathbf{v}$ and acceleration $\mathbf{a}$ of a particle in cylindrical coordinates.

Solution

In rectangular coordinates, the position vector is $\mathbf{r} = x\mathbf{i} + y\mathbf{j} + z\mathbf{k}$ and the velocity and acceleration vectors are

$$\mathbf{v} = \frac{d\mathbf{r}}{dt} = \dot{x}\mathbf{i} + \dot{y}\mathbf{j} + \dot{z}\mathbf{k} \quad \text{and} \quad \mathbf{a} = \frac{d^2\mathbf{r}}{dt^2} = \ddot{x}\mathbf{i} + \ddot{y}\mathbf{j} + \ddot{z}\mathbf{k}$$

In cylindrical coordinates, using Problem 7.4.

$$\mathbf{r} = x\mathbf{i} + y\mathbf{j} + z\mathbf{k} = (\rho\cos\phi)(\cos\phi\mathbf{e}_\rho - \sin\phi\mathbf{e}_\phi) + (\rho\sin\phi)(\sin\phi\mathbf{e}_\rho + \cos\phi\mathbf{e}_\phi) + z\mathbf{e}_z$$
$$= \rho\mathbf{e}_\rho + z\mathbf{e}_z$$

Then

$$\mathbf{v} = \frac{d\mathbf{r}}{dt} = \frac{d\rho}{dt}\mathbf{e}_\rho + \rho\frac{d\mathbf{e}_\rho}{dt} + \frac{dz}{dt}\mathbf{e}_z = \dot\rho\mathbf{e}_\rho + \rho\dot\phi\mathbf{e}_\phi + \dot{z}\mathbf{e}_z$$

using Problem 7.5. Differentiating again,

$$\mathbf{a} = \frac{d^2\mathbf{r}}{dt^2} = \frac{d}{dt}(\dot\rho\mathbf{e}_\rho + \rho\dot\phi\mathbf{e}_\phi + \dot{z}\mathbf{e}_z)$$
$$= \dot\rho\frac{d\mathbf{e}_\rho}{dt} + \ddot\rho\mathbf{e}_\rho + \rho\dot\phi\frac{d\mathbf{e}_\phi}{dt} + \rho\ddot\phi\mathbf{e}_\phi + \dot\rho\dot\phi\mathbf{e}_\phi + \ddot{z}\mathbf{e}_z$$
$$= \dot\rho\dot\phi\mathbf{e}_\phi + \ddot\rho\mathbf{e}_\rho + \rho\dot\phi(-\dot\phi\mathbf{e}_\rho) + \rho\ddot\phi\mathbf{e}_\phi + \dot\rho\dot\phi\mathbf{e}_\phi + \ddot{z}\mathbf{e}_z$$
$$= (\ddot\rho - \rho\dot\phi^2)\mathbf{e}_\rho + (\rho\ddot\phi + 2\dot\rho\dot\phi)\mathbf{e}_\phi + \ddot{z}\mathbf{e}_z$$

using Problem 7.5.

7.7. Find the square of the element of arc length in cylindrical coordinates and determine the corresponding scale factors.

Solution

First Method.

$$x = \rho\cos\phi, \quad y = \rho\sin\phi, \quad z = z$$
$$dx = -\rho\sin\phi\,d\phi + \cos\phi\,d\rho, \quad dy = \rho\cos\phi\,d\phi + \sin\phi\,d\rho, \quad dz = dz$$

Then

$$ds^2 = dx^2 + dy^2 + dz^2 = (-\rho\sin\phi\,d\phi + \cos\phi\,d\rho)^2 + (\rho\cos\phi\,d\phi + \sin\phi\,d\rho)^2 + (dz)^2$$
$$= (d\rho)^2 + \rho^2(d\phi)^2 + (dz)^2 = h_1^2(d\rho)^2 + h_2^2(d\phi)^2 + h_3^2(dz)^2$$

and $h_1 = h_\rho = 1$, $h_2 = h_\phi = \rho$, $h_3 = h_z = 1$ are the scale factors.

Second Method. The position vector is $\mathbf{r} = \rho\cos\phi\mathbf{i} + \rho\sin\phi\mathbf{j} + z\mathbf{k}$. Then

$$d\mathbf{r} = \frac{\partial\mathbf{r}}{\partial\rho}d\rho + \frac{\partial\mathbf{r}}{\partial\phi}d\phi + \frac{\partial\mathbf{r}}{\partial z}dz$$
$$= (\cos\phi\mathbf{i} + \sin\phi\mathbf{j})\,d\rho + (-\rho\sin\phi\mathbf{i} + \rho\cos\phi\mathbf{j})\,d\phi + \mathbf{k}\,dz$$
$$= (\cos\phi\,d\rho - \rho\sin\phi\,d\phi)\mathbf{i} + (\sin\phi\,d\rho + \rho\cos\phi\,d\phi)\mathbf{j} + \mathbf{k}\,dz$$

Thus

$$ds^2 = d\mathbf{r}\cdot d\mathbf{r} = (\cos\phi\,d\rho - \rho\sin\phi\,d\phi)^2 + (\sin\phi\,d\rho + \rho\cos\phi\,d\phi)^2 + (dz)^2$$
$$= (d\rho)^2 + \rho^2(d\phi)^2 + (dz)^2$$

7.8. Work Problem 7.7 for (a) spherical and (b) parabolic cylindrical coordinates.

Solution

(a)
$$x = r \sin \theta \cos \phi, \quad y = r \sin \theta \sin \phi, \quad z = r \cos \theta$$

Then

$$dx = -r \sin \theta \sin \phi \, d\phi + r \cos \theta \cos \phi \, d\theta + \sin \theta \cos \phi \, dr$$
$$dy = r \sin \theta \cos \phi \, d\phi + r \cos \theta \sin \phi \, d\theta + \sin \theta \sin \phi \, dr$$
$$dz = -r \sin \theta \, d\theta + \cos \theta \, dr$$

and

$$(ds)^2 = (dx)^2 + (dy)^2 + (dz)^2 = (dr)^2 + r^2(d\theta)^2 + r^2 \sin^2 \theta (d\phi)^2$$

The scale factors are $h_1 = h_r = 1$, $h_2 = h_\theta = r$, $h_3 = h_\phi = r \sin \theta$.

(b)
$$x = \tfrac{1}{2}(u^2 - v^2), \quad y = uv, \quad z = z$$

Then

$$dx = u \, du - v \, dv, \quad dy = u \, dv + v \, du, \quad dz = dz$$

and

$$(ds)^2 = (dx)^2 + (dy)^2 + (dz)^2 = (u^2 + v^2)(du)^2 + (u^2 + v^2)(dv)^2 + (dz)^2$$

The scale factors are $h_1 = h_u = \sqrt{u^2 + v^2}$, $h_2 = h_v = \sqrt{u^2 + v^2}$, $h_3 = h_z = 1$.

7.9. Sketch a volume element in (a) cylindrical and (b) spherical coordinates giving the magnitudes of its edges.

Solution

(a) The edges of the volume element in cylindrical coordinates (Fig. 7-9(a)) have magnitudes $\rho \, d\phi$, $d\rho$, and dz. This could also be seen from the fact that the edges are given by

$$ds_1 = h_1 du_1 = (1)(d\rho) = d\rho, \quad ds_2 = h_2 du_2 = \rho \, d\phi, \quad ds_3 = (1)(dz) = dz$$

using the scale factors obtained from Problem 7.7.

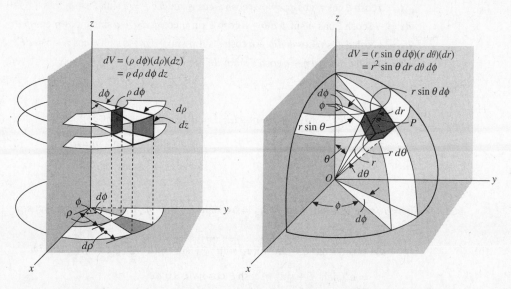

(a) Volume element in cylindrical coordinates.

(b) Volume element in spherical coordinates.

Fig. 7-9

(b) The edges of the volume element in spherical coordinates (Fig. 7-9(b)) have magnitudes dr, $r\,d\theta$, and $r\sin\theta\,d\phi$. This could also be seen from the fact that the edges are given by

$$ds_1 = h_1\,du_1 = (1)(dr) = dr, \quad ds_2 = h_2\,du_2 = r\,d\theta, \quad ds_3 = h_3\,du_3 = r\sin\theta\,d\phi$$

using the scale factors obtained from Problem 7.8(a).

7.10. Find the volume element dV in (a) cylindrical, (b) spherical, and (c) parabolic cylindrical coordinates.

Solution

The volume element in orthogonal curvilinear coordinates u_1, u_2, u_3 is

$$dV = h_1 h_2 h_3\,du_1\,du_2\,du_3$$

(a) In cylindrical coordinates, $u_1 = \rho$, $u_2 = \phi$, $u_3 = z$, $h_1 = 1$, $h_2 = \rho$, $h_3 = 1$ (see Problem 7.7). Then

$$dV = (1)(\rho)(1)\,d\rho\,d\phi\,dz = \rho\,d\rho\,d\phi\,dz$$

This can also be observed directly from Fig. 7-9(a) of Problem 7.9.

(b) In spherical coordinates, $u_1 = r$, $u_2 = \theta$, $u_3 = \phi$, $h_1 = 1$, $h_2 = r$, $h_3 = r\sin\theta$ (see Problem 7.8(a)). Then

$$dV = (1)(r)(r\sin\theta)\,dr\,d\theta\,d\phi = r^2\sin\theta\,dr\,d\theta\,d\phi$$

This can also be observed directly from Fig. 7-9(b).

(c) In parabolic cylindrical coordinates, $u_1 = u$, $u_2 = v$, $u_3 = z$, $h_1 = \sqrt{u^2 + v^2}$, $h_2 = \sqrt{u^2 + v^2}$, $h_3 = 1$ (see Problem 7.8(b)). Then

$$dV = (\sqrt{u^2 + v^2})(\sqrt{u^2 + v^2})(1)\,du\,dv\,dz = (u^2 + v^2)\,du\,dv\,dz$$

7.11. Find (a) the scale factors and (b) the volume element dV in oblate spheroidal coordinates.

Solution

(a) $\qquad x = a\cosh\xi\cos\eta\cos\phi, \;\; y = a\cosh\xi\cos\eta\sin\phi, \;\; z = a\sinh\xi\sin\eta$

$dx = -a\cosh\xi\cos\eta\sin\phi\,d\phi - a\cosh\xi\sin\eta\cos\phi\,d\eta + a\sinh\xi\cos\eta\cos\phi\,d\xi$

$dy = a\cosh\xi\cos\eta\cos\phi\,d\phi - a\cosh\xi\sin\eta\sin\phi\,d\eta + a\sinh\xi\cos\eta\sin\phi\,d\xi$

$dz = a\sinh\xi\cos\eta\,d\eta + a\cosh\xi\sin\eta\,d\xi$

Then

$$(ds)^2 = (dx)^2 + (dy)^2 + (dz)^2 = a^2(\sinh^2\xi + \sin^2\eta)(d\xi)^2$$
$$+ a^2(\sinh^2\xi + \sin^2\eta)(d\eta)^2$$
$$+ a^2\cosh^2\xi\cos^2\eta(d\phi)^2$$

and $h_1 = h_\xi = a\sqrt{\sinh^2\xi + \sin^2\eta}$, $h_2 = h_\eta = a\sqrt{\sinh^2\xi + \sin^2\eta}$, $h_3 = h_\phi = a\cosh\xi\cos\eta$.

(b) $\qquad dV = (a\sqrt{\sinh^2\xi + \sin^2\eta})(a\sqrt{\sinh^2\xi + \sin^2\eta})(a\cosh\xi\cos\eta)\,d\xi\,d\eta\,d\phi$

$$= a^3(\sinh^2\xi + \sin^2\eta)\cosh\xi\cos\eta\,d\xi\,d\eta\,d\phi$$

7.12. Find expressions for the elements of area in orthogonal curvilinear coordinates.

Solution

Referring to Fig. 7-3, the area elements are given by

$$dA_1 = |(h_2 \, du_2 \mathbf{e}_2) \times (h_3 \, du_3 \mathbf{e}_3)| = h_2 h_3 |\mathbf{e}_2 \times \mathbf{e}_3| du_2 \, du_3 = h_2 h_3 \, du_2 \, du_3$$

since $|\mathbf{e}_2 \times \mathbf{e}_3| = |\mathbf{e}_1| = 1$. Similarly

$$dA_2 = |(h_1 \, du_1 \mathbf{e}_1) \times (h_3 \, du_3 \mathbf{e}_3)| = h_1 h_3 \, du_1 \, du_3$$
$$dA_3 = |(h_1 \, du_1 \mathbf{e}_1) \times (h_2 \, du_2 \mathbf{e}_2)| = h_1 h_2 \, du_1 \, du_2$$

7.13. Suppose u_1, u_2, u_3 are orthogonal curvilinear coordinates. Show that the Jacobian of x, y, z with respect to u_1, u_2, u_3 is

$$J\left(\frac{x, \, y, \, z}{u_1, \, u_2, \, u_3}\right) = \frac{\partial(x, \, y, \, z)}{\partial(u_1, \, u_2, \, u_3)} = \begin{vmatrix} \dfrac{\partial x}{\partial u_1} & \dfrac{\partial y}{\partial u_1} & \dfrac{\partial z}{\partial u_1} \\[2mm] \dfrac{\partial x}{\partial u_2} & \dfrac{\partial y}{\partial u_2} & \dfrac{\partial z}{\partial u_2} \\[2mm] \dfrac{\partial x}{\partial u_3} & \dfrac{\partial y}{\partial u_3} & \dfrac{\partial z}{\partial u_3} \end{vmatrix} = h_1 h_2 h_3$$

Solution

By Problem 2.38, the given determinant equals

$$\left(\frac{\partial x}{\partial u_1}\mathbf{i} + \frac{\partial y}{\partial u_1}\mathbf{j} + \frac{\partial z}{\partial u_1}\mathbf{k}\right) \cdot \left(\frac{\partial x}{\partial u_2}\mathbf{i} + \frac{\partial y}{\partial u_2}\mathbf{j} + \frac{\partial z}{\partial u_2}\mathbf{k}\right) \times \left(\frac{\partial x}{\partial u_3}\mathbf{i} + \frac{\partial y}{\partial u_3}\mathbf{j} + \frac{\partial z}{\partial u_3}\mathbf{k}\right)$$

$$= \frac{\partial \mathbf{r}}{\partial u_1} \cdot \frac{\partial \mathbf{r}}{\partial u_2} \times \frac{\partial \mathbf{r}}{\partial u_3} = h_1 \mathbf{e}_1 \cdot h_2 \mathbf{e}_2 \times h_3 \mathbf{e}_3$$

$$= h_1 h_2 h_3 \mathbf{e}_1 \cdot \mathbf{e}_2 \times \mathbf{e}_3 = h_1 h_2 h_3$$

If the Jacobian equals zero identically, then $\partial \mathbf{r}/\partial u_1$, $\partial \mathbf{r}/\partial u_2$, $\partial \mathbf{r}/\partial u_3$ are coplanar vectors and the curvilinear coordinate transformation breaks down, that is, there is a relation between x, y, z having the form $F(x, y, z) = 0$. We shall therefore require the Jacobian to be different from zero.

7.14. Evaluate $\iiint_V (x^2 + y^2 + z^2) \, dx \, dy \, dz$ where V is a sphere having center at the origin and radius equal to a.

Solution

The required integral is equal to eight times the integral evaluated over that part of the sphere contained in the first octant (see Fig. 7-10(a)).

Fig. 7-10

Then, in rectangular coordinates, the integral equals

$$8 \int\limits_{x=0}^{a} \int\limits_{y=0}^{\sqrt{a^2-x^2}} \int\limits_{z=0}^{\sqrt{a^2-x^2-y^2}} (x^2 + y^2 + z^2)\, dz\, dy\, dx$$

but the evaluation, although possible, is tedious. It is easier to use spherical coordinates for the evaluation. In changing to spherical coordinates, the integrand $x^2 + y^2 + z^2$ is replaced by its equivalent r^2 while the volume element $dx\, dy\, dz$ is replaced by the volume element $r^2 \sin\theta\, dr\, d\theta\, d\phi$ (see Problem 7.10(b)). To cover the required region in the first octant, fix θ and ϕ (see Fig. 7-10(b)) and integrate from $r = 0$ to $r = a$; then keep ϕ constant and integrate from $\theta = 0$ to $\pi/2$; finally, integrate with respect to ϕ from $\phi = 0$ to $\phi = \pi/2$. Here, we have performed the integration in the order r, θ, ϕ although any order can be used. The result is

$$8 \int\limits_{\phi=0}^{\pi/2} \int\limits_{\theta=0}^{\pi/2} \int\limits_{r=0}^{a} (r^2)(r^2 \sin\theta\, dr\, d\theta\, d\phi) = 8 \int\limits_{\phi=0}^{\pi/2} \int\limits_{\theta=0}^{\pi/2} \int\limits_{r=0}^{a} r^4 \sin\theta\, dr\, d\theta\, d\phi$$

$$= 8 \int\limits_{\phi=0}^{\pi/2} \int\limits_{\theta=0}^{\pi/2} \frac{r^5}{5} \sin\theta \Big|_{r=0}^{a} d\theta\, d\phi = \frac{8a^5}{5} \int\limits_{\phi=0}^{\pi/2} \int\limits_{\theta=0}^{\pi/2} \sin\theta\, d\theta\, d\phi$$

$$= \frac{8a^5}{5} \int\limits_{\phi=0}^{\pi/2} -\cos\theta \Big|_{\theta=0}^{\pi/2} d\phi = \frac{8a^5}{5} \int\limits_{\phi=0}^{\pi/2} d\phi = \frac{4\pi a^5}{5}$$

Physically, the integral represents the moment of inertia of the sphere with respect to the origin, that is, the polar moment of inertia, if the sphere has unit density.

In general, when transforming multiple integrals from rectangular to orthogonal curvilinear coordinates, the volume element $dx\, dy\, dz$ is replaced by $h_1 h_2 h_3\, du_1\, du_2\, du_3$ or the equivalent

$$J\left(\frac{x, y, z}{u_1, u_2, u_3}\right) du_1\, du_2\, du_3$$

where J is the Jacobian of the transformation from x, y, z to u_1, u_2, u_3 (see Problem 7.13).

7.15. Let u_1, u_2, u_3 be general coordinates. Show that $\partial\mathbf{r}/\partial u_1$, $\partial\mathbf{r}/\partial u_2$, $\partial\mathbf{r}/\partial u_3$ and ∇u_1, ∇u_2, ∇u_3 are reciprocal systems of vectors.

Solution

We must show that

$$\frac{\partial\mathbf{r}}{\partial u_p} \cdot \nabla u_q = \begin{cases} 1 & \text{if } p = q \\ 0 & \text{if } p \neq q \end{cases}$$

where p and q can have any of the values 1, 2, 3. We have

$$d\mathbf{r} = \frac{\partial\mathbf{r}}{\partial u_1} du_1 + \frac{\partial\mathbf{r}}{\partial u_2} du_2 + \frac{\partial\mathbf{r}}{\partial u_3} du_3$$

Multiply by $\nabla u_1 \cdot$. Then

$$\nabla u_1 \cdot d\mathbf{r} = du_1 = \left(\nabla u_1 \cdot \frac{\partial\mathbf{r}}{\partial u_1}\right) du_1 + \left(\nabla u_1 \cdot \frac{\partial\mathbf{r}}{\partial u_2}\right) du_2 + \left(\nabla u_1 \cdot \frac{\partial\mathbf{r}}{\partial u_3}\right) du_3$$

or

$$\nabla u_1 \cdot \frac{\partial\mathbf{r}}{\partial u_1} = 1, \quad \nabla u_1 \cdot \frac{\partial\mathbf{r}}{\partial u_2} = 0, \quad \nabla u_1 \cdot \frac{\partial\mathbf{r}}{\partial u_3} = 0$$

Similarly, upon multiplying by $\nabla u_2 \cdot$ and $\nabla u_3 \cdot$ the remaining relations are proved.

7.16. Prove $\left\{\dfrac{\partial \mathbf{r}}{\partial u_1} \cdot \dfrac{\partial \mathbf{r}}{\partial u_2} \times \dfrac{\partial \mathbf{r}}{\partial u_3}\right\}\{\nabla u_1 \cdot \nabla u_2 \times \nabla u_3\} = 1.$

Solution

From Problem 7.15, $\partial \mathbf{r}/\partial u_1$, $\partial \mathbf{r}/\partial u_2$, $\partial \mathbf{r}/\partial u_3$ and ∇u_1, ∇u_2, ∇u_3 are reciprocal systems of vectors. Then, the required result follows from Problem 2.53(c).

The result is equivalent to a theorem on Jacobians for

$$\nabla u_1 \cdot \nabla u_2 \times \nabla u_3 = \begin{vmatrix} \dfrac{\partial u_1}{\partial x} & \dfrac{\partial u_1}{\partial y} & \dfrac{\partial u_1}{\partial z} \\[2mm] \dfrac{\partial u_2}{\partial x} & \dfrac{\partial u_2}{\partial y} & \dfrac{\partial u_2}{\partial z} \\[2mm] \dfrac{\partial u_3}{\partial x} & \dfrac{\partial u_3}{\partial y} & \dfrac{\partial u_3}{\partial z} \end{vmatrix} = J\left(\dfrac{u_1, u_2, u_3}{x, y, z}\right)$$

and so $J\left(\dfrac{x, y, z}{u_1, u_2, u_3}\right) J\left(\dfrac{u_1, u_2, u_3}{x, y, z}\right) = 1$ using Problem 7.13.

7.17. Show that the square of the element of arc length in general curvilinear coordinates can be expressed by

$$ds^2 = \sum_{p=1}^{3} \sum_{q=1}^{3} g_{pq}\, du_p\, du_q$$

Solution

We have

$$d\mathbf{r} = \frac{\partial \mathbf{r}}{\partial u_1} du_1 + \frac{\partial \mathbf{r}}{\partial u_2} du_2 + \frac{\partial \mathbf{r}}{\partial u_3} du_3 = \boldsymbol{\alpha}_1\, du_1 + \boldsymbol{\alpha}_2\, du_2 + \boldsymbol{\alpha}_3\, du_3$$

Then

$$ds^2 = d\mathbf{r} \cdot d\mathbf{r} = \boldsymbol{\alpha}_1 \cdot \boldsymbol{\alpha}_1\, du_1^2 + \boldsymbol{\alpha}_1 \cdot \boldsymbol{\alpha}_2\, du_1\, du_2 + \boldsymbol{\alpha}_1 \cdot \boldsymbol{\alpha}_3\, du_1\, du_3$$

$$+ \boldsymbol{\alpha}_2 \cdot \boldsymbol{\alpha}_1\, du_2\, du_1 + \boldsymbol{\alpha}_2 \cdot \boldsymbol{\alpha}_2\, du_2^2 + \boldsymbol{\alpha}_2 \cdot \boldsymbol{\alpha}_3\, du_2\, du_3$$

$$+ \boldsymbol{\alpha}_3 \cdot \boldsymbol{\alpha}_1\, du_3\, du_1 + \boldsymbol{\alpha}_3 \cdot \boldsymbol{\alpha}_2\, du_3\, du_2 + \boldsymbol{\alpha}_3 \cdot \boldsymbol{\alpha}_3\, du_3^2$$

$$= \sum_{p=1}^{3} \sum_{q=1}^{3} g_{pq}\, du_p\, du_q$$

where $g_{pq} = \boldsymbol{\alpha}_p \cdot \boldsymbol{\alpha}_q$.

This is called the *fundamental quadratic form* or *metric form*. The quantities g_{pq} are called *metric coefficients* and are symmetric, that is, $g_{pq} = g_{qp}$. If $g_{pq} = 0$, $p \neq q$, then the coordinate system is orthogonal. In this case, $g_{11} = h_1^2$, $g_{22} = h_2^2$, $g_{33} = h_3^2$. The metric form extended to higher dimensional space is of fundamental importance in the theory of relativity (see Chapter 8).

Gradient, Divergence, and Curl in Orthogonal Coordinates

7.18. Derive an expression for $\nabla \Phi$ in orthogonal curvilinear coordinates.

Solution

Let $\nabla \Phi = f_1 \mathbf{e}_1 + f_2 \mathbf{e}_2 + f_3 \mathbf{e}_3$ where f_1, f_2, f_3 are to be determined. Since

$$d\mathbf{r} = \frac{\partial \mathbf{r}}{\partial u_1}\, du_1 + \frac{\partial \mathbf{r}}{\partial u_2}\, du_2 + \frac{\partial \mathbf{r}}{\partial u_3}\, du_3$$

$$= h_1 \mathbf{e}_1\, du_1 + h_2 \mathbf{e}_2\, du_2 + h_3 \mathbf{e}_3\, du_3$$

we have

$$d\Phi = \nabla\Phi \cdot d\mathbf{r} = h_1 f_1 \, du_1 + h_2 f_2 \, du_2 + h_3 f_3 \, du_3 \tag{1}$$

But

$$d\Phi = \frac{\partial\Phi}{\partial u_1} du_1 + \frac{\partial\Phi}{\partial u_2} du_2 + \frac{\partial\Phi}{\partial u_3} du_3 \tag{2}$$

Equating (1) and (2),

$$f_1 = \frac{1}{h_1}\frac{\partial\Phi}{\partial u_1}, \quad f_2 = \frac{1}{h_2}\frac{\partial\Phi}{\partial u_2}, \quad f_3 = \frac{1}{h_3}\frac{\partial\Phi}{\partial u_3}.$$

Then

$$\nabla\Phi = \frac{\mathbf{e}_1}{h_1}\frac{\partial\Phi}{\partial u_1} + \frac{\mathbf{e}_2}{h_2}\frac{\partial\Phi}{\partial u_2} + \frac{\mathbf{e}_3}{h_3}\frac{\partial\Phi}{\partial u_3}$$

This indicates the operator equivalence

$$\nabla \equiv \frac{\mathbf{e}_1}{h_1}\frac{\partial}{\partial u_1} + \frac{\mathbf{e}_2}{h_2}\frac{\partial}{\partial u_2} + \frac{\mathbf{e}_3}{h_3}\frac{\partial}{\partial u_3}$$

which reduces to the usual expression for the operator ∇ in rectangular coordinates.

7.19. Let u_1, u_2, u_3 be orthogonal coordinates. (a) Prove that $|\nabla u_p| = h_p^{-1}$, $p = 1, 2, 3$. (b) Show that $\mathbf{e}_p = \mathbf{E}_p$.

Solution

(a) Let $\Phi = u_1$ in Problem 7.18. Then $\nabla u_1 = \mathbf{e}_1/h_1$ and so $|\nabla u_1| = |\mathbf{e}_1|/h_1 = h_1^{-1}$, since $|\mathbf{e}_1| = 1$. Similarly, by letting $\Phi = u_2$ and u_3, $|\nabla u_2| = h_2^{-1}$ and $|\nabla u_3| = h_3^{-1}$.

(b) By definition, $\mathbf{E}_p = \dfrac{\nabla u_p}{|\nabla u_p|}$. From part (a), this can be written $\mathbf{E}_p = h_p \nabla u_p = \mathbf{e}_p$ and the result is proved.

7.20. Prove $\mathbf{e}_1 = h_2 h_3 \nabla u_2 \times \nabla u_3$ with similar equations for $\mathbf{e}_2$ and $\mathbf{e}_3$, where u_1, u_2, u_3 are orthogonal coordinates.

Solution

From Problem 7.19,

$$\nabla u_1 = \frac{\mathbf{e}_1}{h_1}, \qquad \nabla u_2 = \frac{\mathbf{e}_2}{h_2}, \qquad \nabla u_3 = \frac{\mathbf{e}_3}{h_3}.$$

Then

$$\nabla u_2 \times \nabla u_3 = \frac{\mathbf{e}_2 \times \mathbf{e}_3}{h_2 h_3} = \frac{\mathbf{e}_1}{h_2 h_3} \qquad \text{and} \qquad \mathbf{e}_1 = h_2 h_3 \nabla u_2 \times \nabla u_3.$$

Similarly

$$\mathbf{e}_2 = h_3 h_1 \nabla u_3 \times \nabla u_1 \qquad \text{and} \qquad \mathbf{e}_3 = h_1 h_2 \nabla u_1 \times \nabla u_2.$$

7.21. Show that in orthogonal coordinates

(a) $\nabla \cdot (A_1 \mathbf{e}_1) = \dfrac{1}{h_1 h_2 h_3}\dfrac{\partial}{\partial u_1}(A_1 h_2 h_3)$

(b) $\nabla \times (A_1 \mathbf{e}_1) = \dfrac{\mathbf{e}_2}{h_3 h_1}\dfrac{\partial}{\partial u_3}(A_1 h_1) - \dfrac{\mathbf{e}_3}{h_1 h_2}\dfrac{\partial}{\partial u_2}(A_1 h_1)$

with similar results for vectors $A_2 \mathbf{e}_2$ and $A_3 \mathbf{e}_3$.

Solution

(a) From Problem 7.20,

$$\nabla \cdot (A_1 \mathbf{e}_1) = \nabla \cdot (A_1 h_2 h_3 \nabla u_2 \times \nabla u_3)$$

$$= \nabla(A_1 h_2 h_3) \cdot \nabla u_2 \times \nabla u_3 + A_1 h_2 h_3 \nabla \cdot (\nabla u_2 \times \nabla u_3)$$

$$= \nabla(A_1 h_2 h_3) \cdot \frac{\mathbf{e}_2}{h_2} \times \frac{\mathbf{e}_3}{h_3} + 0 = \nabla(A_1 h_2 h_3) \cdot \frac{\mathbf{e}_1}{h_2 h_3}$$

$$= \left[\frac{\mathbf{e}_1}{h_1} \frac{\partial}{\partial u_1}(A_1 h_2 h_3) + \frac{\mathbf{e}_2}{h_2} \frac{\partial}{\partial u_2}(A_1 h_2 h_3) + \frac{\mathbf{e}_3}{h_3} \frac{\partial}{\partial u_3}(A_1 h_2 h_3) \right] \cdot \frac{\mathbf{e}_1}{h_2 h_3}$$

$$= \frac{1}{h_1 h_2 h_3} \frac{\partial}{\partial u_1}(A_1 h_2 h_3)$$

(b) $\nabla \times (A_1 \mathbf{e}_1) = \nabla \times (A_1 h_1 \nabla u_1)$

$$= \nabla(A_1 h_1) \times \nabla u_1 + A_1 h_1 \nabla \times \nabla u_1$$

$$= \nabla(A_1 h_1) \times \frac{\mathbf{e}_1}{h_1} + \mathbf{0}$$

$$= \left[\frac{\mathbf{e}_1}{h_1} \frac{\partial}{\partial u_1}(A_1 h_1) + \frac{\mathbf{e}_2}{h_2} \frac{\partial}{\partial u_2}(A_1 h_1) + \frac{\mathbf{e}_3}{h_3} \frac{\partial}{\partial u_3}(A_1 h_1) \right] \times \frac{\mathbf{e}_1}{h_1}$$

$$= \frac{\mathbf{e}_2}{h_3 h_1} \frac{\partial}{\partial u_3}(A_1 h_1) - \frac{\mathbf{e}_3}{h_1 h_2} \frac{\partial}{\partial u_2}(A_1 h_1)$$

7.22. Express div $\mathbf{A} = \nabla \cdot \mathbf{A}$ in orthogonal coordinates.

Solution

$$\nabla \cdot \mathbf{A} = \nabla \cdot (A_1 \mathbf{e}_1 + A_2 \mathbf{e}_2 + A_3 \mathbf{e}_3) = \nabla \cdot (A_1 \mathbf{e}_1) + \nabla \cdot (A_2 \mathbf{e}_2) + \nabla \cdot (A_3 \mathbf{e}_3)$$

$$= \frac{1}{h_1 h_2 h_3} \left[\frac{\partial}{\partial u_1}(A_1 h_2 h_3) + \frac{\partial}{\partial u_2}(A_2 h_3 h_1) + \frac{\partial}{\partial u_3}(A_3 h_1 h_2) \right]$$

using Problem 7.21(a).

7.23. Express curl $\mathbf{A} = \nabla \times \mathbf{A}$ in orthogonal coordinates.

Solution

$$\nabla \times \mathbf{A} = \nabla \times (A_1 \mathbf{e}_1 + A_2 \mathbf{e}_2 + A_3 \mathbf{e}_3) = \nabla \times (A_1 \mathbf{e}_1) + \nabla \times (A_2 \mathbf{e}_2) + \nabla \times (A_3 \mathbf{e}_3)$$

$$= \frac{\mathbf{e}_2}{h_3 h_1} \frac{\partial}{\partial u_3}(A_1 h_1) - \frac{\mathbf{e}_3}{h_1 h_2} \frac{\partial}{\partial u_2}(A_1 h_1) + \frac{\mathbf{e}_3}{h_1 h_2} \frac{\partial}{\partial u_1}(A_2 h_2) - \frac{\mathbf{e}_1}{h_2 h_3} \frac{\partial}{\partial u_3}(A_2 h_2)$$

$$+ \frac{\mathbf{e}_1}{h_2 h_3} \frac{\partial}{\partial u_2}(A_3 h_3) - \frac{\mathbf{e}_2}{h_3 h_1} \frac{\partial}{\partial u_1}(A_3 h_3)$$

$$= \frac{\mathbf{e}_1}{h_2 h_3} \left[\frac{\partial}{\partial u_2}(A_3 h_3) - \frac{\partial}{\partial u_3}(A_2 h_2) \right] + \frac{\mathbf{e}_2}{h_3 h_1} \left[\frac{\partial}{\partial u_3}(A_1 h_1) - \frac{\partial}{\partial u_1}(A_3 h_3) \right]$$

$$+ \frac{\mathbf{e}_3}{h_1 h_2} \left[\frac{\partial}{\partial u_1}(A_2 h_2) - \frac{\partial}{\partial u_2}(A_1 h_1) \right]$$

using Problem 7.21(b). This can be written

$$\nabla \times \mathbf{A} = \frac{1}{h_1 h_2 h_3} \begin{vmatrix} h_1 \mathbf{e}_1 & h_2 \mathbf{e}_2 & h_3 \mathbf{e}_3 \\ \dfrac{\partial}{\partial u_1} & \dfrac{\partial}{\partial u_2} & \dfrac{\partial}{\partial u_3} \\ A_1 h_1 & A_2 h_2 & A_3 h_3 \end{vmatrix}$$

7.24. Express $\nabla^2 \psi$ in orthogonal curvilinear coordinates.

Solution

From Problem 7.18,

$$\nabla\psi = \frac{\mathbf{e}_1}{h_1}\frac{\partial\psi}{\partial u_1} + \frac{\mathbf{e}_2}{h_2}\frac{\partial\psi}{\partial u_2} + \frac{\mathbf{e}_3}{h_3}\frac{\partial\psi}{\partial u_3}.$$

If $\mathbf{A} = \nabla\psi$, then $A_1 = \dfrac{1}{h_1}\dfrac{\partial\psi}{\partial u_1}$, $A_2 = \dfrac{1}{h_2}\dfrac{\partial\psi}{\partial u_2}$, $A_3 = \dfrac{1}{h_3}\dfrac{\partial\psi}{\partial u_3}$ and by Problem 7.22,

$$\nabla\cdot\mathbf{A} = \nabla\cdot\nabla\psi = \nabla^2\psi$$

$$= \frac{1}{h_1 h_2 h_3}\left[\frac{\partial}{\partial u_1}\left(\frac{h_2 h_3}{h_1}\frac{\partial\psi}{\partial u_1}\right) + \frac{\partial}{\partial u_2}\left(\frac{h_3 h_1}{h_2}\frac{\partial\psi}{\partial u_2}\right) + \frac{\partial}{\partial u_3}\left(\frac{h_1 h_2}{h_3}\frac{\partial\psi}{\partial u_3}\right)\right]$$

7.25. Use the integral definition

$$\operatorname{div}\mathbf{A} = \nabla\cdot\mathbf{A} = \lim_{\Delta V\to 0}\frac{\iint_{\Delta S}\mathbf{A}\cdot\mathbf{n}\,dS}{\Delta V}$$

(see Problem 6.19) to express $\nabla\cdot\mathbf{A}$ in orthogonal curvilinear coordinates.

Solution

Consider the volume element ΔV (see Fig. 7-11) having edges $h_1\,\Delta u_1$, $h_2\,\Delta u_2$, $h_3\,\Delta u_3$.

Let $\mathbf{A} = A_1\mathbf{e}_1 + A_2\mathbf{e}_2 + A_3\mathbf{e}_3$ and let $\mathbf{n}$ be the outward drawn unit normal to the surface ΔS of ΔV. On face $JKLP$, $\mathbf{n} = -\mathbf{e}_1$. Then, we have approximately,

$$\iint_{JKLP}\mathbf{A}\cdot\mathbf{n}\,dS = (\mathbf{A}\cdot\mathbf{n}\ \text{at point } P)(\text{Area of } JKLP)$$

$$= [(A_1\mathbf{e}_1 + A_2\mathbf{e}_2 + A_3\mathbf{e}_3)\cdot(-\mathbf{e}_1)](h_2 h_3\,\Delta u_2\,\Delta u_3)$$

$$= -A_1 h_2 h_3\,\Delta u_2\,\Delta u_3$$

On face $EFGH$, the surface integral is

$$A_1 h_2 h_3\,\Delta u_2\,\Delta u_3 + \frac{\partial}{\partial u_1}(A_1 h_2 h_3\,\Delta u_2\,\Delta u_3)\,\Delta u_1$$

apart from infinitesimals of order higher than $\Delta u_1\,\Delta u_2\,\Delta u_3$. Then the net contribution to the surface integral from these two faces is

$$\frac{\partial}{\partial u_1}(A_1 h_2 h_3\,\Delta u_2\,\Delta u_3)\,\Delta u_1 = \frac{\partial}{\partial u_1}(A_1 h_2 h_3)\,\Delta u_1\,\Delta u_2\,\Delta u_3$$

The contribution from all six faces of ΔV is

$$\left[\frac{\partial}{\partial u_1}(A_1 h_2 h_3) + \frac{\partial}{\partial u_2}(A_2 h_1 h_3) + \frac{\partial}{\partial u_3}(A_3 h_1 h_2)\right]\Delta u_1\,\Delta u_2\,\Delta u_3$$

Dividing this by the volume $h_1 h_2 h_3\,\Delta u_1\,\Delta u_2\,\Delta u_3$ and taking the limit as Δu_1, Δu_2, Δu_3 approach zero, we find

$$\operatorname{div}\mathbf{A} = \nabla\cdot\mathbf{A} = \frac{1}{h_1 h_2 h_3}\left[\frac{\partial}{\partial u_1}(A_1 h_2 h_3) + \frac{\partial}{\partial u_2}(A_2 h_1 h_3) + \frac{\partial}{\partial u_3}(A_3 h_1 h_2)\right]$$

Note that the same result would be obtained had we chosen the volume element ΔV such that P is at its center. In this case, the calculation would proceed in a manner analogous to that of Problem 4.21.

Fig. 7-11

Fig. 7-12

7.26. Use the integral definition

$$(\text{curl } \mathbf{A}) \cdot \mathbf{n} = (\nabla \times \mathbf{A}) \cdot \mathbf{n} = \lim_{\Delta S \to 0} \frac{\oint_C \mathbf{A} \cdot d\mathbf{r}}{\Delta S}$$

(see Problem 6.35) to express $\nabla \times \mathbf{A}$ in orthogonal curvilinear coordinates.

Solution

Let us first calculate $(\text{curl } \mathbf{A}) \cdot \mathbf{e}_1$. To do this, consider the surface S_1 normal to $\mathbf{e}_1$ at P, as shown in Fig. 7-12. Denote the boundary of S_1 by C_1. Let $\mathbf{A} = A_1 \mathbf{e}_1 + A_2 \mathbf{e}_2 + A_3 \mathbf{e}_3$. We have

$$\oint_{C_1} \mathbf{A} \cdot d\mathbf{r} = \int_{PQ} \mathbf{A} \cdot d\mathbf{r} + \int_{QL} \mathbf{A} \cdot d\mathbf{r} + \int_{LM} \mathbf{A} \cdot d\mathbf{r} + \int_{MP} \mathbf{A} \cdot d\mathbf{r}$$

The following approximations hold

$$\int_{PQ} \mathbf{A} \cdot d\mathbf{r} = (\mathbf{A} \text{ at } P) \cdot (h_2 \, \Delta u_2 \mathbf{e}_2) \tag{1}$$

Then

$$= (A_1 \mathbf{e}_1 + A_2 \mathbf{e}_2 + A_3 \mathbf{e}_3) \cdot (h_2 \, \Delta u_2 \mathbf{e}_2) = A_2 h_2 \, \Delta u_2$$

$$\int_{ML} \mathbf{A} \cdot d\mathbf{r} = A_2 h_2 \, \Delta u_2 + \frac{\partial}{\partial u_3}(A_2 h_2 \, \Delta u_2) \, \Delta u_3$$

or

$$\int_{LM} \mathbf{A} \cdot d\mathbf{r} = -A_2 h_2 \, \Delta u_2 - \frac{\partial}{\partial u_3}(A_2 h_2 \, \Delta u_2) \, \Delta u_3 \tag{2}$$

Similarly,

$$\int_{PM} \mathbf{A} \cdot d\mathbf{r} = (\mathbf{A} \text{ at } P) \cdot (h_3 \Delta u_3 \mathbf{e}_3) = A_3 h_3 \, \Delta u_3$$

or

$$\int_{MP} \mathbf{A} \cdot d\mathbf{r} = -A_3 h_3 \, \Delta u_3 \tag{3}$$

and

$$\int_{QL} \mathbf{A} \cdot d\mathbf{r} = A_3 h_3 \, \Delta u_3 + \frac{\partial}{\partial u_2}(A_3 h_3 \, \Delta u_3) \, \Delta u_2 \tag{4}$$

Adding (1), (2), (3), and (4), we have

$$\oint_{C_1} \mathbf{A} \cdot d\mathbf{r} = \frac{\partial}{\partial u_2}(A_3 h_3 \, \Delta u_3) \, \Delta u_2 - \frac{\partial}{\partial u_3}(A_2 h_2 \, \Delta u_2) \, \Delta u_3$$

$$= \left[\frac{\partial}{\partial u_2}(A_3 h_3) - \frac{\partial}{\partial u_3}(A_2 h_2) \right] \Delta u_2 \, \Delta u_3$$

apart from infinitesimals of order higher than $\Delta u_2 \, \Delta u_3$.

Dividing by the area of S_1 equal to $h_2 h_3 \, \Delta u_2 \, \Delta u_3$ and taking the limit as Δu_2 and Δu_3 approach zero,

$$(\text{curl } \mathbf{A}) \cdot \mathbf{e}_1 = \frac{1}{h_2 h_3} \left[\frac{\partial}{\partial u_2}(A_3 h_3) - \frac{\partial}{\partial u_3}(A_2 h_2) \right]$$

Similarly, by choosing areas S_2 and S_3 perpendicular to $\mathbf{e}_2$ and $\mathbf{e}_3$ at P, respectively, we find $(\text{curl } \mathbf{A}) \cdot \mathbf{e}_2$ and $(\text{curl } \mathbf{A}) \cdot \mathbf{e}_3$. This leads to the required result

$$\begin{aligned}
\text{curl } \mathbf{A} &= \frac{\mathbf{e}_1}{h_2 h_3} \left[\frac{\partial}{\partial u_2}(A_3 h_3) - \frac{\partial}{\partial u_3}(A_2 h_2) \right] \\
&+ \frac{\mathbf{e}_2}{h_3 h_1} \left[\frac{\partial}{\partial u_3}(A_1 h_1) - \frac{\partial}{\partial u_1}(A_3 h_3) \right] \\
&+ \frac{\mathbf{e}_3}{h_1 h_2} \left[\frac{\partial}{\partial u_1}(A_2 h_2) - \frac{\partial}{\partial u_2}(A_1 h_1) \right] \\
&= \frac{1}{h_1 h_2 h_3}
\begin{vmatrix}
h_1 \mathbf{e}_1 & h_2 \mathbf{e}_2 & h_3 \mathbf{e}_3 \\
\dfrac{\partial}{\partial u_1} & \dfrac{\partial}{\partial u_2} & \dfrac{\partial}{\partial u_3} \\
h_1 A_1 & h_2 A_2 & h_3 A_3
\end{vmatrix}
\end{aligned}$$

The result could also have been derived by choosing P as the center of area S_1; the calculation would then proceed as in Problem 6.36.

7.27. Express in cylindrical coordinates the quantities　(a) $\nabla \Phi$,　(b) $\nabla \cdot \mathbf{A}$,　(c) $\nabla \times \mathbf{A}$,　(d) $\nabla^2 \Phi$.

Solution

For cylindrical coordinates (ρ, ϕ, z),

$$u_1 = \rho, \quad u_2 = \phi, \quad u_3 = z; \qquad \mathbf{e}_1 = \mathbf{e}_\rho, \quad \mathbf{e}_2 = \mathbf{e}_\phi, \quad \mathbf{e}_3 = \mathbf{e}_z;$$

and

$$h_1 = h_\rho = 1, \quad h_2 = h_\phi = \rho, \quad h_3 = h_z = 1$$

(a) $$\begin{aligned}
\nabla \Phi &= \frac{1}{h_1} \frac{\partial \Phi}{\partial u_1} \mathbf{e}_1 + \frac{1}{h_2} \frac{\partial \Phi}{\partial u_2} \mathbf{e}_2 + \frac{1}{h_3} \frac{\partial \Phi}{\partial u_3} \mathbf{e}_3 \\
&= \frac{1}{1} \frac{\partial \Phi}{\partial \rho} \mathbf{e}_\rho + \frac{1}{\rho} \frac{\partial \Phi}{\partial \phi} \mathbf{e}_\phi + \frac{1}{1} \frac{\partial \Phi}{\partial z} \mathbf{e}_z \\
&= \frac{\partial \Phi}{\partial \rho} \mathbf{e}_\rho + \frac{1}{\rho} \frac{\partial \Phi}{\partial \phi} \mathbf{e}_\phi + \frac{\partial \Phi}{\partial z} \mathbf{e}_z
\end{aligned}$$

(b) $\nabla \cdot \mathbf{A} = \dfrac{1}{h_1 h_2 h_3} \left[\dfrac{\partial}{\partial u_1} (h_2 h_3 A_1) + \dfrac{\partial}{\partial u_2} (h_3 h_1 A_2) + \dfrac{\partial}{\partial u_3} (h_1 h_2 A_3) \right]$

$= \dfrac{1}{(1)(\rho)(1)} \left[\dfrac{\partial}{\partial \rho} \big((\rho)(1) A_\rho \big) + \dfrac{\partial}{\partial \phi} \big((1)(1) A_\phi \big) + \dfrac{\partial}{\partial z} \big((1)(\rho) A_z \big) \right]$

$= \dfrac{1}{\rho} \left[\dfrac{\partial}{\partial \rho} (\rho A_\rho) + \dfrac{\partial A_\phi}{\partial \phi} + \dfrac{\partial}{\partial z} (\rho A_z) \right]$

where $\mathbf{A} = A_\rho \mathbf{e}_1 + A_\phi \mathbf{e}_2 + A_z \mathbf{e}_3$, that is, $A_1 = A_\rho$, $A_2 = A_\phi$, $A_3 = A_z$.

(c) $\nabla \times \mathbf{A} = \dfrac{1}{h_1 h_2 h_3} \begin{vmatrix} h_1 \mathbf{e}_1 & h_2 \mathbf{e}_2 & h_3 \mathbf{e}_3 \\ \dfrac{\partial}{\partial u_1} & \dfrac{\partial}{\partial u_2} & \dfrac{\partial}{\partial u_3} \\ h_1 A_1 & h_2 A_2 & h_3 A_3 \end{vmatrix} = \dfrac{1}{\rho} \begin{vmatrix} \mathbf{e}_\rho & \rho \mathbf{e}_\phi & \mathbf{e}_z \\ \dfrac{\partial}{\partial \rho} & \dfrac{\partial}{\partial \phi} & \dfrac{\partial}{\partial z} \\ A_\rho & \rho A_\phi & A_z \end{vmatrix}$

$= \dfrac{1}{\rho} \left[\left(\dfrac{\partial A_z}{\partial \phi} - \dfrac{\partial}{\partial z} (\rho A_\phi) \right) \mathbf{e}_\rho + \left(\rho \dfrac{\partial A_\rho}{\partial z} - \rho \dfrac{\partial A_z}{\partial \rho} \right) \mathbf{e}_\phi + \left(\dfrac{\partial}{\partial \rho} (\rho A_\phi) - \dfrac{\partial A_\rho}{\partial \phi} \right) \mathbf{e}_z \right]$

(d) $\nabla^2 \Phi = \dfrac{1}{h_1 h_2 h_3} \left[\dfrac{\partial}{\partial u_1} \left(\dfrac{h_2 h_3}{h_1} \dfrac{\partial \Phi}{\partial u_1} \right) + \dfrac{\partial}{\partial u_2} \left(\dfrac{h_3 h_1}{h_2} \dfrac{\partial \Phi}{\partial u_2} \right) + \dfrac{\partial}{\partial u_3} \left(\dfrac{h_1 h_2}{h_3} \dfrac{\partial \Phi}{\partial u_3} \right) \right]$

$= \dfrac{1}{(1)(\rho)(1)} \left[\dfrac{\partial}{\partial \rho} \left(\dfrac{(\rho)(1)}{(1)} \dfrac{\partial \Phi}{\partial \rho} \right) + \dfrac{\partial}{\partial \phi} \left(\dfrac{(1)(1)}{\rho} \dfrac{\partial \Phi}{\partial \phi} \right) + \dfrac{\partial}{\partial z} \left(\dfrac{(1)(\rho)}{(1)} \dfrac{\partial \Phi}{\partial z} \right) \right]$

$= \dfrac{1}{\rho} \dfrac{\partial}{\partial \rho} \left(\rho \dfrac{\partial \Phi}{\partial \rho} \right) + \dfrac{1}{\rho^2} \dfrac{\partial^2 \Phi}{\partial \phi^2} + \dfrac{\partial^2 \Phi}{\partial z^2}$

7.28. Express (a) $\nabla \times \mathbf{A}$ and (b) $\nabla^2 \psi$ in spherical coordinates.

Solution

Here $u_1 = r$, $u_2 = \theta$, $u_3 = \phi$; $\mathbf{e}_1 = \mathbf{e}_r$, $\mathbf{e}_2 = \mathbf{e}_\theta$, $\mathbf{e}_3 = \mathbf{e}_\phi$; $h_1 = h_r = 1$, $h_2 = h_\theta = r$, $h_3 = h_\phi = r \sin \theta$.

(a) $\nabla \times \mathbf{A} = \dfrac{1}{h_1 h_2 h_3} \begin{vmatrix} h_1 \mathbf{e}_1 & h_2 \mathbf{e}_2 & h_3 \mathbf{e}_3 \\ \dfrac{\partial}{\partial u_1} & \dfrac{\partial}{\partial u_2} & \dfrac{\partial}{\partial u_3} \\ h_1 A_1 & h_2 A_2 & h_3 A_3 \end{vmatrix} = \dfrac{1}{(1)(r)(r \sin \theta)} \begin{vmatrix} \mathbf{e}_r & r \mathbf{e}_\theta & r \sin \theta \, \mathbf{e}_\phi \\ \dfrac{\partial}{\partial r} & \dfrac{\partial}{\partial \theta} & \dfrac{\partial}{\partial \phi} \\ A_r & r A_\theta & r \sin \theta \, A_\phi \end{vmatrix}$

$= \dfrac{1}{r^2 \sin \theta} \left[\left\{ \dfrac{\partial}{\partial \theta} (r \sin \theta A_\phi) - \dfrac{\partial}{\partial \phi} (r A_\theta) \right\} \mathbf{e}_r \right.$

$\left. + \left\{ \dfrac{\partial A_r}{\partial \phi} - \dfrac{\partial}{\partial r} (r \sin \theta \, A_\phi) \right\} r \mathbf{e}_\theta + \left\{ \dfrac{\partial}{\partial r} (r A_\theta) - \dfrac{\partial A_r}{\partial \theta} \right\} r \sin \theta \, \mathbf{e}_\phi \right]$

(b) $\nabla^2 \psi = \dfrac{1}{h_1 h_2 h_3} \left[\dfrac{\partial}{\partial u_1} \left(\dfrac{h_2 h_3}{h_1} \dfrac{\partial \psi}{\partial u_1} \right) + \dfrac{\partial}{\partial u_2} \left(\dfrac{h_3 h_1}{h_2} \dfrac{\partial \psi}{\partial u_2} \right) + \dfrac{\partial}{\partial u_3} \left(\dfrac{h_1 h_2}{h_3} \dfrac{\partial \psi}{\partial u_3} \right) \right]$

$= \dfrac{1}{(1)(r)(r \sin \theta)} \left[\dfrac{\partial}{\partial r} \left(\dfrac{(r)(r \sin \theta)}{(1)} \dfrac{\partial \psi}{\partial r} \right) + \dfrac{\partial}{\partial \theta} \left(\dfrac{(r \sin \theta)(1)}{r} \dfrac{\partial \psi}{\partial \theta} \right) + \dfrac{\partial}{\partial \phi} \left(\dfrac{(1)(r)}{r \sin \theta} \dfrac{\partial \psi}{\partial \phi} \right) \right]$

$= \dfrac{1}{r^2 \sin \theta} \left[\sin \theta \dfrac{\partial}{\partial r} \left(r^2 \dfrac{\partial \psi}{\partial r} \right) + \dfrac{\partial}{\partial \theta} \left(\sin \theta \dfrac{\partial \psi}{\partial \theta} \right) + \dfrac{1}{\sin \theta} \dfrac{\partial^2 \psi}{\partial \phi^2} \right]$

$= \dfrac{1}{r^2} \dfrac{\partial}{\partial r} \left(r^2 \dfrac{\partial \psi}{\partial r} \right) + \dfrac{1}{r^2 \sin \theta} \dfrac{\partial}{\partial \theta} \left(\sin \theta \dfrac{\partial \psi}{\partial \theta} \right) + \dfrac{1}{r^2 \sin^2 \theta} \dfrac{\partial^2 \psi}{\partial \phi^2}$

7.29. Write Laplace's equation in parabolic cylindrical coordinates.

Solution

From Problem 7.8(b),

$$u_1 = u, \quad u_2 = v, \quad u_3 = z; \qquad h_1 = \sqrt{u^2 + v^2}, \quad h_2 = \sqrt{u^2 + v^2}, \quad h_3 = 1$$

Then

$$\nabla^2 \psi = \frac{1}{u^2 + v^2} \left[\frac{\partial}{\partial u} \left(\frac{\partial \psi}{\partial u} \right) + \frac{\partial}{\partial v} \left(\frac{\partial \psi}{\partial v} \right) + \frac{\partial}{\partial z} \left((u^2 + v^2) \frac{\partial \psi}{\partial z} \right) \right]$$

$$= \frac{1}{u^2 + v^2} \left(\frac{\partial^2 \psi}{\partial u^2} + \frac{\partial^2 \psi}{\partial v^2} \right) + \frac{\partial^2 \psi}{\partial z^2}$$

and Laplace's equation is $\nabla^2 \psi = 0$ or

$$\frac{\partial^2 \psi}{\partial u^2} + \frac{\partial^2 \psi}{\partial v^2} + (u^2 + v^2) \frac{\partial^2 \psi}{\partial z^2} = 0$$

7.30. Express the heat conduction equation $\partial U / \partial t = \kappa \nabla^2 U$ in elliptic cylindrical coordinates.

Solution

Here $u_1 = u, u_2 = v, u_3 = z; h_1 = h_2 = a\sqrt{\sinh^2 u + \sin^2 v}, h_3 = 1$. Then

$$\nabla^2 U = \frac{1}{a^2(\sinh^2 u + \sin^2 v)} \left[\frac{\partial}{\partial u} \left(\frac{\partial U}{\partial u} \right) + \frac{\partial}{\partial v} \left(\frac{\partial U}{\partial v} \right) + \frac{\partial}{\partial z} \left(a^2(\sinh^2 u + \sin^2 v) \frac{\partial U}{\partial z} \right) \right]$$

$$= \frac{1}{a^2(\sinh^2 u + \sin^2 v)} \left[\frac{\partial^2 U}{\partial u^2} + \frac{\partial^2 U}{\partial v^2} \right] + \frac{\partial^2 U}{\partial z^2}$$

and the heat conduction equation is

$$\frac{\partial U}{\partial t} = \kappa \left\{ \frac{1}{a^2(\sinh^2 u + \sin^2 v)} \left[\frac{\partial^2 U}{\partial u^2} + \frac{\partial^2 U}{\partial v^2} \right] + \frac{\partial^2 U}{\partial z^2} \right\}$$

Surface Curvilinear Coordinates

7.31. Show that the square of the element of arc length on the surface $\mathbf{r} = \mathbf{r}(u, v)$ can be written

$$ds^2 = E\,du^2 + 2F\,du\,dv + G\,dv^2$$

Solution

We have

$$d\mathbf{r} = \frac{\partial \mathbf{r}}{\partial u}\,du + \frac{\partial \mathbf{r}}{\partial v}\,dv$$

Then

$$ds^2 = d\mathbf{r} \cdot d\mathbf{r}$$

$$= \frac{\partial \mathbf{r}}{\partial u} \cdot \frac{\partial \mathbf{r}}{\partial u}\,du^2 + 2\frac{\partial \mathbf{r}}{\partial u} \cdot \frac{\partial \mathbf{r}}{\partial v}\,du\,dv + \frac{\partial \mathbf{r}}{\partial v} \cdot \frac{\partial \mathbf{r}}{\partial v}\,dv^2$$

$$= E\,du^2 + 2F\,du\,dv + G\,dv^2$$

7.32. Show that the element of surface area of the surface $\mathbf{r} = \mathbf{r}(u, v)$ is given by

$$dS = \sqrt{EG - F^2}\, du\, dv$$

Solution

The element of area is given by

$$dS = \left|\left(\frac{\partial \mathbf{r}}{\partial u} du\right) \times \left(\frac{\partial \mathbf{r}}{\partial v} dv\right)\right| = \left|\frac{\partial \mathbf{r}}{\partial u} \times \frac{\partial \mathbf{r}}{\partial v}\right| du\, dv = \sqrt{\left(\frac{\partial \mathbf{r}}{\partial u} \times \frac{\partial \mathbf{r}}{\partial v}\right) \cdot \left(\frac{\partial \mathbf{r}}{\partial u} \times \frac{\partial \mathbf{r}}{\partial v}\right)}\, du\, dv$$

The quantity under the square root sign is equal to (see Problem 2.48)

$$\left(\frac{\partial \mathbf{r}}{\partial u} \cdot \frac{\partial \mathbf{r}}{\partial u}\right)\left(\frac{\partial \mathbf{r}}{\partial v} \cdot \frac{\partial \mathbf{r}}{\partial v}\right) - \left(\frac{\partial \mathbf{r}}{\partial u} \cdot \frac{\partial \mathbf{r}}{\partial v}\right)\left(\frac{\partial \mathbf{r}}{\partial v} \cdot \frac{\partial \mathbf{r}}{\partial u}\right) = EG - F^2$$

and the result follows.

Miscellaneous Problems on General Coordinates

7.33. Let $\mathbf{A}$ be a given vector defined with respect to two general curvilinear coordinate systems (u_1, u_2, u_3) and $(\bar{u}_1, \bar{u}_2, \bar{u}_3)$. Find the relation between the contravariant components of the vector in the two coordinate systems.

Solution

Suppose the transformation equations from a rectangular (x, y, z) system to the (u_1, u_2, u_3) and $(\bar{u}_1, \bar{u}_2, \bar{u}_3)$ systems are given by

$$\begin{cases} x = x_1(u_1, u_2, u_3), & y = y_1(u_1, u_2, u_3), & z = z_1(u_1, u_2, u_3) \\ x = x_2(\bar{u}_1, \bar{u}_2, \bar{u}_3), & y = y_2(\bar{u}_1, \bar{u}_2, \bar{u}_3), & z = z_2(\bar{u}_1, \bar{u}_2, \bar{u}_3) \end{cases} \tag{1}$$

Then there exists a transformation directly from the (u_1, u_2, u_3) system to the $(\bar{u}_1, \bar{u}_2, \bar{u}_3)$ system defined by

$$u_1 = u_1(\bar{u}_1, \bar{u}_2, \bar{u}_3), \quad u_2 = u_2(\bar{u}_1, \bar{u}_2, \bar{u}_3), \quad u_3 = u_3(\bar{u}_1, \bar{u}_2, \bar{u}_3) \tag{2}$$

and conversely. From (1),

$$d\mathbf{r} = \frac{\partial \mathbf{r}}{\partial u_1} du_1 + \frac{\partial \mathbf{r}}{\partial u_2} du_2 + \frac{\partial \mathbf{r}}{\partial u_3} du_3 = \boldsymbol{\alpha}_1\, du_1 + \boldsymbol{\alpha}_2\, du_2 + \boldsymbol{\alpha}_3\, du_3$$

$$d\mathbf{r} = \frac{\partial \mathbf{r}}{\partial \bar{u}_1} d\bar{u}_1 + \frac{\partial \mathbf{r}}{\partial \bar{u}_2} d\bar{u}_2 + \frac{\partial \mathbf{r}}{\partial \bar{u}_3} d\bar{u}_3 = \bar{\boldsymbol{\alpha}}_1\, d\bar{u}_1 + \bar{\boldsymbol{\alpha}}_2\, d\bar{u}_2 + \bar{\boldsymbol{\alpha}}_3\, d\bar{u}_3$$

Then

$$\boldsymbol{\alpha}_1\, du_1 + \boldsymbol{\alpha}_2\, du_2 + \boldsymbol{\alpha}_3\, du_3 = \bar{\boldsymbol{\alpha}}_1\, d\bar{u}_1 + \bar{\boldsymbol{\alpha}}_2\, d\bar{u}_2 + \bar{\boldsymbol{\alpha}}_3\, d\bar{u}_3 \tag{3}$$

From (2),

$$du_1 = \frac{\partial u_1}{\partial \bar{u}_1} d\bar{u}_1 + \frac{\partial u_1}{\partial \bar{u}_2} d\bar{u}_2 + \frac{\partial u_1}{\partial \bar{u}_3} d\bar{u}_3$$

$$du_2 = \frac{\partial u_2}{\partial \bar{u}_1} d\bar{u}_1 + \frac{\partial u_2}{\partial \bar{u}_2} d\bar{u}_2 + \frac{\partial u_2}{\partial \bar{u}_3} d\bar{u}_3$$

$$du_3 = \frac{\partial u_3}{\partial \bar{u}_1} d\bar{u}_1 + \frac{\partial u_3}{\partial \bar{u}_2} d\bar{u}_2 + \frac{\partial u_3}{\partial \bar{u}_3} d\bar{u}_3$$

Substituting into (3) and equating coefficients of $d\bar{u}_1$, $d\bar{u}_2$, $d\bar{u}_3$ on both sides, we find

$$
\begin{cases}
\bar{\alpha}_1 = \alpha_1 \dfrac{\partial u_1}{\partial \bar{u}_1} + \alpha_2 \dfrac{\partial u_2}{\partial \bar{u}_1} + \alpha_3 \dfrac{\partial u_3}{\partial \bar{u}_1} \\[2mm]
\bar{\alpha}_2 = \alpha_1 \dfrac{\partial u_1}{\partial \bar{u}_2} + \alpha_2 \dfrac{\partial u_2}{\partial \bar{u}_2} + \alpha_3 \dfrac{\partial u_3}{\partial \bar{u}_2} \\[2mm]
\bar{\alpha}_3 = \alpha_1 \dfrac{\partial u_1}{\partial \bar{u}_3} + \alpha_2 \dfrac{\partial u_2}{\partial \bar{u}_3} + \alpha_3 \dfrac{\partial u_3}{\partial \bar{u}_3}
\end{cases}
\tag{4}
$$

Now **A** can be expressed in the two coordinate systems as

$$
\mathbf{A} = C_1\alpha_1 + C_2\alpha_2 + C_3\alpha_3 \quad \text{and} \quad \mathbf{A} = \overline{C}_1\bar{\alpha}_1 + \overline{C}_2\bar{\alpha}_2 + \overline{C}_3\bar{\alpha}_3
\tag{5}
$$

where C_1, C_2, C_3 and $\overline{C}_1$, $\overline{C}_2$, $\overline{C}_3$ are the contravariant components of **A** in the two systems. Substituting (4) into (5),

$$
C_1\alpha_1 + C_2\alpha_2 + C_3\alpha_3 = \overline{C}_1\bar{\alpha}_1 + \overline{C}_2\bar{\alpha}_2 + \overline{C}_3\bar{\alpha}_3
$$

$$
= \left(\overline{C}_1 \frac{\partial u_1}{\partial \bar{u}_1} + \overline{C}_2 \frac{\partial u_1}{\partial \bar{u}_2} + \overline{C}_3 \frac{\partial u_1}{\partial \bar{u}_3} \right)\alpha_1 + \left(\overline{C}_1 \frac{\partial u_2}{\partial \bar{u}_1} + \overline{C}_2 \frac{\partial u_2}{\partial \bar{u}_2} + \overline{C}_3 \frac{\partial u_2}{\partial \bar{u}_3} \right)\alpha_2
$$

$$
+ \left(\overline{C}_1 \frac{\partial u_3}{\partial \bar{u}_1} + \overline{C}_2 \frac{\partial u_3}{\partial \bar{u}_2} + \overline{C}_3 \frac{\partial u_3}{\partial \bar{u}_3} \right)\alpha_3
$$

Then

$$
\begin{cases}
C_1 = \overline{C}_1 \dfrac{\partial u_1}{\partial \bar{u}_1} + \overline{C}_2 \dfrac{\partial u_1}{\partial \bar{u}_2} + \overline{C}_3 \dfrac{\partial u_1}{\partial \bar{u}_3} \\[2mm]
C_2 = \overline{C}_1 \dfrac{\partial u_2}{\partial \bar{u}_1} + \overline{C}_2 \dfrac{\partial u_2}{\partial \bar{u}_2} + \overline{C}_3 \dfrac{\partial u_2}{\partial \bar{u}_3} \\[2mm]
C_3 = \overline{C}_1 \dfrac{\partial u_3}{\partial \bar{u}_1} + \overline{C}_2 \dfrac{\partial u_3}{\partial \bar{u}_2} + \overline{C}_3 \dfrac{\partial u_3}{\partial \bar{u}_3}
\end{cases}
\tag{6}
$$

or in shorter notation

$$
C_p = \overline{C}_1 \frac{\partial u_p}{\partial \bar{u}_1} + \overline{C}_2 \frac{\partial u_p}{\partial \bar{u}_2} + \overline{C}_3 \frac{\partial u_p}{\partial \bar{u}_3} \qquad p = 1, 2, 3
\tag{7}
$$

and in even shorter notation

$$
C_p = \sum_{q=1}^{3} \overline{C}_q \frac{\partial u_p}{\partial \bar{u}_q} \qquad p = 1, 2, 3
\tag{8}
$$

Similarly, by interchanging the coordinates, we see that

$$
\overline{C}_p = \sum_{q=1}^{3} C_q \frac{\partial \bar{u}_p}{\partial u_q} \qquad p = 1, 2, 3
\tag{9}
$$

The above results lead us to adopt the following definition. If three quantities C_1, C_2, C_3 of a co-ordinate system (u_1, u_2, u_3) are related to three other quantities $\overline{C}_1$, $\overline{C}_2$, $\overline{C}_3$ of another coordinate system $(\bar{u}_1, \bar{u}_2, \bar{u}_3)$ by the transformation equations (6), (7), (8) or (9), then the quantities are called *components of a contravariant vector* or a *contravariant tensor of the first rank*.

7.34. Work Problem 7.33 for the covariant components of **A**.

Solution

Write the covariant components of **A** in the systems (u_1, u_2, u_3) and $(\bar{u}_1, \bar{u}_2, \bar{u}_3)$ as c_1, c_2, c_3 and $\bar{c}_1$, $\bar{c}_2$, $\bar{c}_3$, respectively. Then

$$
\mathbf{A} = c_1 \nabla u_1 + c_2 \nabla u_2 + c_3 \nabla u_3 = \bar{c}_1 \nabla \bar{u}_1 + \bar{c}_2 \nabla \bar{u}_2 + \bar{c}_3 \nabla \bar{u}_3
\tag{1}
$$

Now since $\overline{u}_p = \overline{u}_p(u_1, u_2, u_3)$ with $p = 1, 2, 3$,

$$\begin{cases} \dfrac{\partial \overline{u}_p}{\partial x} = \dfrac{\partial \overline{u}_p}{\partial u_1}\dfrac{\partial u_1}{\partial x} + \dfrac{\partial \overline{u}_p}{\partial u_2}\dfrac{\partial u_2}{\partial x} + \dfrac{\partial \overline{u}_p}{\partial u_3}\dfrac{\partial u_3}{\partial x} \\[2mm] \dfrac{\partial \overline{u}_p}{\partial y} = \dfrac{\partial \overline{u}_p}{\partial u_1}\dfrac{\partial u_1}{\partial y} + \dfrac{\partial \overline{u}_p}{\partial u_2}\dfrac{\partial u_2}{\partial y} + \dfrac{\partial \overline{u}_p}{\partial u_3}\dfrac{\partial u_3}{\partial y} \qquad p = 1, 2, 3 \\[2mm] \dfrac{\partial \overline{u}_p}{\partial z} = \dfrac{\partial \overline{u}_p}{\partial u_1}\dfrac{\partial u_1}{\partial z} + \dfrac{\partial \overline{u}_p}{\partial u_2}\dfrac{\partial u_2}{\partial z} + \dfrac{\partial \overline{u}_p}{\partial u_3}\dfrac{\partial u_3}{\partial z} \end{cases} \tag{2}$$

Also,

$$c_1 \nabla u_1 + c_2 \nabla u_2 + c_3 \nabla u_3 = \left(c_1 \dfrac{\partial u_1}{\partial x} + c_2 \dfrac{\partial u_2}{\partial x} + c_3 \dfrac{\partial u_3}{\partial x} \right)\mathbf{i}$$

$$+ \left(c_1 \dfrac{\partial u_1}{\partial y} + c_2 \dfrac{\partial u_2}{\partial y} + c_3 \dfrac{\partial u_3}{\partial y} \right)\mathbf{j} + \left(c_1 \dfrac{\partial u_1}{\partial z} + c_2 \dfrac{\partial u_2}{\partial z} + c_3 \dfrac{\partial u_3}{\partial z} \right)\mathbf{k} \tag{3}$$

and

$$\overline{c}_1 \nabla \overline{u}_1 + \overline{c}_2 \nabla \overline{u}_2 + \overline{c}_3 \nabla \overline{u}_3 = \left(\overline{c}_1 \dfrac{\partial \overline{u}_1}{\partial x} + \overline{c}_2 \dfrac{\partial \overline{u}_2}{\partial x} + \overline{c}_3 \dfrac{\partial \overline{u}_3}{\partial x} \right)\mathbf{i}$$

$$+ \left(\overline{c}_1 \dfrac{\partial \overline{u}_1}{\partial y} + \overline{c}_2 \dfrac{\partial \overline{u}_2}{\partial y} + \overline{c}_3 \dfrac{\partial \overline{u}_3}{\partial y} \right)\mathbf{j} + \left(\overline{c}_1 \dfrac{\partial \overline{u}_1}{\partial z} + \overline{c}_2 \dfrac{\partial \overline{u}_2}{\partial z} + \overline{c}_3 \dfrac{\partial \overline{u}_3}{\partial z} \right)\mathbf{k} \tag{4}$$

Equating coefficients of $\mathbf{i}, \mathbf{j}, \mathbf{k}$ in (3) and (4),

$$\begin{cases} c_1 \dfrac{\partial u_1}{\partial x} + c_2 \dfrac{\partial u_2}{\partial x} + c_3 \dfrac{\partial u_3}{\partial x} = \overline{c}_1 \dfrac{\partial \overline{u}_1}{\partial x} + \overline{c}_2 \dfrac{\partial \overline{u}_2}{\partial x} + \overline{c}_3 \dfrac{\partial \overline{u}_3}{\partial x} \\[2mm] c_1 \dfrac{\partial u_1}{\partial y} + c_2 \dfrac{\partial u_2}{\partial y} + c_3 \dfrac{\partial u_3}{\partial y} = \overline{c}_1 \dfrac{\partial \overline{u}_1}{\partial y} + \overline{c}_2 \dfrac{\partial \overline{u}_2}{\partial y} + \overline{c}_3 \dfrac{\partial \overline{u}_3}{\partial y} \\[2mm] c_1 \dfrac{\partial u_1}{\partial z} + c_2 \dfrac{\partial u_2}{\partial z} + c_3 \dfrac{\partial u_3}{\partial z} = \overline{c}_1 \dfrac{\partial \overline{u}_1}{\partial z} + \overline{c}_2 \dfrac{\partial \overline{u}_2}{\partial z} + \overline{c}_3 \dfrac{\partial \overline{u}_3}{\partial z} \end{cases} \tag{5}$$

Substituting equations (2) with $p = 1, 2, 3$ in any of the equations (5) and equating coefficients of

$$\dfrac{\partial u_1}{\partial x}, \dfrac{\partial u_2}{\partial x}, \dfrac{\partial u_3}{\partial x}, \dfrac{\partial u_1}{\partial y}, \dfrac{\partial u_2}{\partial y}, \dfrac{\partial u_3}{\partial y}, \dfrac{\partial u_1}{\partial z}, \dfrac{\partial u_2}{\partial z}, \dfrac{\partial u_3}{\partial z}$$

on each side, we find

$$\begin{cases} c_1 = \overline{c}_1 \dfrac{\partial \overline{u}_1}{\partial u_1} + \overline{c}_2 \dfrac{\partial \overline{u}_2}{\partial u_1} + \overline{c}_3 \dfrac{\partial \overline{u}_3}{\partial u_1} \\[2mm] c_2 = \overline{c}_1 \dfrac{\partial \overline{u}_1}{\partial u_2} + \overline{c}_2 \dfrac{\partial \overline{u}_2}{\partial u_2} + \overline{c}_3 \dfrac{\partial \overline{u}_3}{\partial u_2} \\[2mm] c_3 = \overline{c}_1 \dfrac{\partial \overline{u}_1}{\partial u_3} + \overline{c}_2 \dfrac{\partial \overline{u}_2}{\partial u_3} + \overline{c}_3 \dfrac{\partial \overline{u}_3}{\partial u_3} \end{cases} \tag{6}$$

which can be written

$$c_p = \overline{c}_1 \dfrac{\partial \overline{u}_1}{\partial u_p} + \overline{c}_2 \dfrac{\partial \overline{u}_2}{\partial u_p} + \overline{c}_3 \dfrac{\partial \overline{u}_3}{\partial u_p} \qquad p = 1, 2, 3 \tag{7}$$

or

$$c_p = \sum_{q=1}^{3} \overline{c}_q \dfrac{\partial \overline{u}_q}{\partial u_p} \qquad p = 1, 2, 3 \tag{8}$$

Similarly, we can show that

$$\bar{c}_p = \sum_{q=1}^{3} c_q \frac{\partial u_q}{\partial \bar{u}_p} \qquad p = 1, 2, 3 \tag{9}$$

The above results lead us to adopt the following definition. If three quantities c_1, c_2, c_3 of a co-ordinate system (u_1, u_2, u_3) are related to three other quantities $\bar{c}_1, \bar{c}_2, \bar{c}_3$ of another coordinate system $(\bar{u}_1, \bar{u}_2, \bar{u}_3)$ by the transformation equations (6), (7), (8), or (9), then the quantities are called *components of a covariant vector* or a *covariant tensor of the first rank*.

In generalizing the concepts in this Problem and in Problem 7.33 to higher dimensional spaces, and in generalizing the concept of vector, we are led to *tensor analysis* which we treat in Chapter 8. In the process of generalization, it is convenient to use a concise notation in order to express fundamental ideas in compact form. It should be remembered, however, that despite the notation used, the basic ideas treated in Chapter 8 are intimately connected with those treated in this chapter.

7.35. (a) Prove that, in general, coordinates (u_1, u_2, u_3),

$$g = \begin{vmatrix} g_{11} & g_{12} & g_{13} \\ g_{21} & g_{22} & g_{23} \\ g_{31} & g_{32} & g_{33} \end{vmatrix} = \left(\frac{\partial \mathbf{r}}{\partial u_1} \cdot \frac{\partial \mathbf{r}}{\partial u_2} \times \frac{\partial \mathbf{r}}{\partial u_3} \right)^2$$

where g_{pq} are the coefficients of $du_p \, du_q$ in ds^2 (Problem 7.17).

(b) Show that the volume element in general coordinates is $\sqrt{g} \, du_1 \, du_2 \, du_3$.

Solution

(a) From Problem 7.17,

$$g_{pq} = \boldsymbol{\alpha}_p \cdot \boldsymbol{\alpha}_q = \frac{\partial \mathbf{r}}{\partial u_p} \cdot \frac{\partial \mathbf{r}}{\partial u_q} = \frac{\partial x}{\partial u_p} \frac{\partial x}{\partial u_q} + \frac{\partial y}{\partial u_p} \frac{\partial y}{\partial u_q} + \frac{\partial z}{\partial u_p} \frac{\partial z}{\partial u_q} \qquad p, q = 1, 2, 3 \tag{1}$$

Then, using the following theorem on multiplication of determinants,

$$\begin{vmatrix} a_1 & a_2 & a_3 \\ b_1 & b_2 & b_3 \\ c_1 & c_2 & c_3 \end{vmatrix} \begin{vmatrix} A_1 & B_1 & C_1 \\ A_2 & B_2 & C_2 \\ A_3 & B_3 & C_3 \end{vmatrix} = \begin{vmatrix} a_1A_1 + a_2A_2 + a_3A_3 & a_1B_1 + a_2B_2 + a_3B_3 & a_1C_1 + a_2C_2 + a_3C_3 \\ b_1A_1 + b_2A_2 + b_3A_3 & b_1B_1 + b_2B_2 + b_3B_3 & b_1C_1 + b_2C_2 + b_3C_3 \\ c_1A_1 + c_2A_2 + c_3A_3 & c_1B_1 + c_2B_2 + c_3B_3 & c_1C_1 + c_2C_2 + c_3C_3 \end{vmatrix}$$

we have

$$\left(\frac{\partial \mathbf{r}}{\partial u_1} \cdot \frac{\partial \mathbf{r}}{\partial u_2} \times \frac{\partial \mathbf{r}}{\partial u_3} \right)^2 = \begin{vmatrix} \dfrac{\partial x}{\partial u_1} & \dfrac{\partial y}{\partial u_1} & \dfrac{\partial z}{\partial u_1} \\ \dfrac{\partial x}{\partial u_2} & \dfrac{\partial y}{\partial u_2} & \dfrac{\partial z}{\partial u_2} \\ \dfrac{\partial x}{\partial u_3} & \dfrac{\partial y}{\partial u_3} & \dfrac{\partial z}{\partial u_3} \end{vmatrix}^2$$

$$= \begin{vmatrix} \dfrac{\partial x}{\partial u_1} & \dfrac{\partial y}{\partial u_1} & \dfrac{\partial z}{\partial u_1} \\ \dfrac{\partial x}{\partial u_2} & \dfrac{\partial y}{\partial u_2} & \dfrac{\partial z}{\partial u_2} \\ \dfrac{\partial x}{\partial u_3} & \dfrac{\partial y}{\partial u_3} & \dfrac{\partial z}{\partial u_3} \end{vmatrix} \begin{vmatrix} \dfrac{\partial x}{\partial u_1} & \dfrac{\partial x}{\partial u_2} & \dfrac{\partial x}{\partial u_3} \\ \dfrac{\partial y}{\partial u_1} & \dfrac{\partial y}{\partial u_2} & \dfrac{\partial y}{\partial u_3} \\ \dfrac{\partial z}{\partial u_1} & \dfrac{\partial z}{\partial u_2} & \dfrac{\partial z}{\partial u_3} \end{vmatrix} = \begin{vmatrix} g_{11} & g_{12} & g_{13} \\ g_{21} & g_{22} & g_{23} \\ g_{31} & g_{32} & g_{33} \end{vmatrix}$$

(b) The volume element is given by

$$dV = \left| \left(\frac{\partial \mathbf{r}}{\partial u_1} du_1 \right) \cdot \left(\frac{\partial \mathbf{r}}{\partial u_2} du_2 \right) \times \left(\frac{\partial \mathbf{r}}{\partial u_3} du_3 \right) \right| = \left| \frac{\partial \mathbf{r}}{\partial u_1} \cdot \frac{\partial \mathbf{r}}{\partial u_2} \times \frac{\partial \mathbf{r}}{\partial u_3} \right| du_1 \, du_2 \, du_3$$

$$= \sqrt{g} \, du_1 \, du_2 \, du_3 \quad \text{by part (a).}$$

Note that $\sqrt{g}$ is the absolute value of the Jacobian of x, y, z with respect to u_1, u_2, u_3 (see Problem 7.13).

SUPPLEMENTARY PROBLEMS

7.36. Describe and sketch the coordinate surfaces and coordinate curves for (a) elliptic cylindrical, (b) bipolar, and (c) parabolic cylindrical coordinates.

7.37. Determine the transformation from (a) spherical to rectangular coordinates, (b) spherical to cylindrical coordinates.

7.38. Express each of the following loci in spherical coordinates: (a) the sphere $x^2 + y^2 + z^2 = 9$; (b) the cone $z^2 = 3(x^2 + y^2)$; (c) the paraboloid $z = x^2 + y^2$; (d) the plane $z = 0$; (e) the plane $y = x$.

7.39. Suppose ρ, ϕ, z are cylindrical coordinates. Describe each of the following loci and write the equation of each locus in rectangular coordinates: (a) $\rho = 4$, $z = 0$; (b) $\rho = 4$; (c) $\phi = \pi/2$; (d) $\phi = \pi/3$, $z = 1$.

7.40. Suppose u, v, z are elliptic cylindrical coordinates where $a = 4$. Describe each of the following loci and write the equation of each locus in rectangular coordinates:
(a) $v = \pi/4$; (b) $u = 0$, $z = 0$; (c) $u = \ln 2$, $z = 2$; (d) $v = 0$, $z = 0$.

7.41. Suppose u, v, z are parabolic cylindrical coordinates. Graph the curves or regions described by each of the following: (a) $u = 2$, $z = 0$; (b) $v = 1$, $z = 2$; (c) $1 \leq u \leq 2$, $2 \leq v \leq 3$, $z = 0$; (d) $1 < u < 2$, $2 < v < 3$, $z = 0$.

7.42. (a) Find the unit vectors $\mathbf{e}_r$, $\mathbf{e}_\theta$, and $\mathbf{e}_\phi$ of a spherical coordinate system in terms of $\mathbf{i}$, $\mathbf{j}$, and $\mathbf{k}$.
(b) Solve for $\mathbf{i}$, $\mathbf{j}$, and $\mathbf{k}$ in terms of $\mathbf{e}_r$, $\mathbf{e}_\theta$, and $\mathbf{e}_\phi$.

7.43. Represent the vector $\mathbf{A} = 2y\mathbf{i} - z\mathbf{j} + 3x\mathbf{k}$ in spherical coordinates and determine A_r, A_θ, and A_ϕ.

7.44. Prove that a spherical coordinate system is orthogonal.

7.45. Prove that (a) parabolic cylindrical, (b) elliptic cylindrical, and (c) oblate spheroidal coordinate systems are orthogonal.

7.46. Prove $\dot{\mathbf{e}}_r = \dot{\theta}\mathbf{e}_\theta + \sin\theta\,\dot{\phi}\mathbf{e}_\phi$, $\dot{\mathbf{e}}_\theta = -\dot{\theta}\mathbf{e}_r + \cos\theta\,\dot{\phi}\mathbf{e}_\phi$, $\dot{\mathbf{e}}_\phi = -\sin\theta\,\dot{\phi}\mathbf{e}_r - \cos\theta\,\dot{\phi}\mathbf{e}_\theta$.

7.47. Express the velocity $\mathbf{v}$ and acceleration $\mathbf{a}$ of a particle in spherical coordinates.

7.48. Find the square of the element of arc length and the corresponding scale factors in (a) paraboloidal, (b) elliptic cylindrical, and (c) oblate spheroidal coordinates.

7.49. Find the volume element dV in (a) paraboloidal, (b) elliptic cylindrical, and (c) bipolar coordinates.

7.50. Find (a) the scale factors and (b) the volume element dV for prolate spheroidal coordinates.

7.51. Derive expressions for the scale factors in (a) ellipsoidal and (b) bipolar coordinates.

7.52. Find the elements of area of a volume element in (a) cylindrical, (b) spherical, and (c) paraboloidal coordinates.

7.53. Prove that a necessary and sufficient condition that a curvilinear coordinate system be orthogonal is that $g_{pq} = 0$ for $p \neq q$.

7.54. Find the Jacobian $J\left(\dfrac{x, y, z}{u_1, u_2, u_3}\right)$ for (a) cylindrical, (b) spherical, (c) parabolic cylindrical,
(d) elliptic cylindrical, and (e) prolate spheroidal coordinates.

7.55. Evaluate $\iiint_V \sqrt{x^2 + y^2}\, dx\, dy\, dz$, where V is the region bounded by $z = x^2 + y^2$ and $z = 8 - (x^2 + y^2)$. *Hint:* Use cylindrical coordinates.

7.56. Find the volume of the smaller of the two regions bounded by the sphere $x^2 + y^2 + z^2 = 16$ and the cone $z^2 = x^2 + y^2$.

7.57. Use spherical coordinates to find the volume of the smaller of the two regions bounded by a sphere of radius a and a plane intersecting the sphere at a distance h from its center.

7.58. (a) Describe the coordinate surfaces and coordinate curves for the system

$$x^2 - y^2 = 2u_1 \cos u_2, \qquad xy = u_1 \sin u_2, \qquad z = u_3$$

(b) Show that the system is orthogonal. (c) Determine $J\left(\dfrac{x,\ y,\ z}{u_1,\ u_2,\ u_3}\right)$ for the system. (d) Show that u_1 and u_2 are related to the cylindrical coordinates ρ and ϕ and determine the relationship.

7.59. Find the moment of inertia of the region bounded by $x^2 - y^2 = 2$, $x^2 - y^2 = 4$, $xy = 1$, $xy = 2$, $z = 1$ and $z = 3$ with respect to the z-axis if the density is constant and equal to κ. *Hint:* Let $x^2 - y^2 = 2u$, $xy = v$.

7.60. Find $\partial \mathbf{r}/\partial u_1$, $\partial \mathbf{r}/\partial u_2$, $\partial \mathbf{r}/\partial u_3$, ∇u_1, ∇u_2, ∇u_3 in (a) cylindrical, (b) spherical, and (c) parabolic cylindrical coordinates. Show that $\mathbf{e}_1 = \mathbf{E}_1$, $\mathbf{e}_2 = \mathbf{E}_2$, $\mathbf{e}_3 = \mathbf{E}_3$ for these systems.

7.61. Given the coordinate transformation $u_1 = xy$, $2u_2 = x^2 + y^2$, $u_3 = z$. (a) Show that the coordinate system is not orthogonal. (b) Find $J\left(\dfrac{x,\ y,\ z}{u_1,\ u_2,\ u_3}\right)$. (c) Find ds^2.

7.62. Find $\nabla \Phi$, div $\mathbf{A}$ and curl $\mathbf{A}$ in parabolic cylindrical coordinates.

7.63. Express (a) $\nabla \psi$ and (b) $\nabla \cdot \mathbf{A}$ in spherical coordinates.

7.64. Find $\nabla^2 \psi$ in oblate spheroidal coordinates.

7.65. Write the equation $(\partial^2 \Phi/\partial x^2) + (\partial^2 \Phi/\partial y^2) = \Phi$ in elliptic cylindrical coordinates.

7.66. Express Maxwell's equation $\nabla \times \mathbf{E} = -(1/c)(\partial \mathbf{H}/\partial t)$ in prolate spheroidal coordinates.

7.67. Express Schroedinger's equation of quantum mechanics $\nabla^2 \psi + (8\pi^2 m/h^2)[E - V(x,\ y,\ z)]\psi = 0$ in parabolic cylindrical coordinates where m, h, and E are constants.

7.68. Write Laplace's equation in paraboloidal coordinates.

7.69. Express the heat equation $\partial U/\partial t = \kappa \nabla^2 U$ in spherical coordinates if U is independent of (a) ϕ, (b) ϕ and θ, (c) r and t, (d) ϕ, θ, and t.

7.70. Find the element of arc length on a sphere of radius a.

7.71. Prove that in any orthogonal curvilinear coordinate system, div curl $\mathbf{A} = 0$ and curl grad $\Phi = \mathbf{0}$.

7.72. Prove that the surface area of a given region R of the surface $\mathbf{r} = \mathbf{r}(u,\ v)$ is $\iint_R \sqrt{EG - F^2}\ du\ dv$. Use this to determine the surface area of a sphere.

7.73. Prove that a vector of length p, which is everywhere normal to the surface $\mathbf{r} = \mathbf{r}(u,\ v)$, is given by

$$\mathbf{A} = \pm p\left(\frac{\partial \mathbf{r}}{\partial u} \times \frac{\partial \mathbf{r}}{\partial v}\right) \Big/ \sqrt{EG - F^2}$$

7.74. (a) Describe the plane transformation $x = x(u,\ v)$, $y = y(u,\ v)$.
(b) Under what conditions will the u, v coordinate lines be orthogonal?

7.75. Let (x, y) be coordinates of a point P in a rectangular xy-plane and (u, v) the coordinates of a point Q in a rectangular uv-plane. If $x = x(u,\ v)$, and $y = y(u,\ v)$, we say that there is a *correspondence* or *mapping* between points P and Q.
(a) If $x = 2u + v$ and $y = u - 2v$, show that the lines in the xy-plane correspond to lines in the uv-plane.
(b) What does the square bounded by $x = 0$, $x = 5$, $y = 0$, and $y = 5$ correspond to in the uv-plane?
(c) Compute the Jacobian $J\left(\dfrac{x,\ y}{u,\ v}\right)$ and show that this is related to the ratios of the areas of the square and its image in the uv-plane.

7.76. Let $x = \frac{1}{2}(u^2 - v^2)$, and $y = uv$. Determine the image (or images) in the uv-plane of a square bounded by $x = 0$, $x = 1$, $y = 0$, $y = 1$ in the xy-plane.

7.77. Show that under suitable conditions on F and G,

$$\int_0^\infty \int_0^\infty e^{-s(x+y)} F(x)G(y)\, dx\, dy = \int_0^\infty e^{-st} \left\{ \int_0^t F(u)G(t-u)\, du \right\} dt$$

Hint: Use the transformation $x + y = t$, $x = v$ from the xy-plane to the vt-plane. The result is important in the theory of Laplace transforms.

7.78. (a) Let $x = 3u_1 + u_2 - u_3$, $y = u_1 + 2u_2 + 2u_3$, $z = 2u_1 - u_2 - u_3$. Find the volume of the cube bounded by $x = 0$, $x = 15$, $y = 0$, $y = 10$, $z = 0$ and $z = 5$, and the image of this cube in the $u_1 u_2 u_3$ rectangular coordinate system.

(b) Relate the ratio of these volumes to the Jacobian of the transformation.

7.79. Let (x, y, z) and (u_1, u_2, u_3) be the rectangular and curvilinear coordinates of a point, respectively.

(a) If $x = 3u_1 + u_2 - u_3$, $y = u_1 + 2u_2 + 2u_3$, $z = 2u_1 - u_2 - u_3$, is the system $u_1 u_2 u_3$ orthogonal?

(b) Find ds^2 and g for the system.

(c) What is the relation between this and the preceding problem?

7.80. Let $x = u_1^2 + 2$, $y = u_1 + u_2$, $z = u_3^2 - u_1$. Find (a) g and (b) the Jacobian $J = \dfrac{\partial(x, y, z)}{\partial(u_1, u_2, u_3)}$. Verify that $J^2 = g$.

ANSWERS TO SUPPLEMENTARY PROBLEMS

7.36. (a) $u = c_1$ and $v = c_2$ are elliptic and hyperbolic cylinders respectively, having z-axis as common axis. $z = c_3$ are planes. See Fig. 7-7.

(b) $u = c_1$ and $v = c_2$ are circular cylinders whose intersections with the xy-plane are circles with centers on the y- and x-axis, respectively, and intersecting at right angles. The cylinders $u = c_1$ all pass through the points $(-a, 0, 0)$ and $(a, 0, 0)$. $z = c_3$ are planes. See Fig. 7-8.

(c) $u = c_1$ and $v = c_2$ are parabolic cylinders whose traces on the xy-plane are intersecting mutually perpendicular coaxial parabolas with vertices on the x-axis but on opposite sides of the origin. $z = c_3$ are planes. See Fig. 7-6.

The coordinate curves are the intersections of the coordinate surfaces.

7.37. (a) $r = \sqrt{x^2 + y^2 + z^2}$, $\theta = \arctan \dfrac{\sqrt{x^2 + y^2}}{z}$, $\phi = \arctan \dfrac{y}{x}$.

(b) $r = \sqrt{\rho^2 + z^2}$, $\theta = \arctan \dfrac{\rho}{z}$, $\phi = \phi$.

7.38. (a) $r = 3$.

(b) $\theta = \pi/6$.

(c) $r \sin^2 \theta = \cos \theta$.

(d) $\theta = \pi/2$.

(e) The plane $y = x$ is made up of the two half planes $\phi = \pi/4$ and $\phi = 5\pi/4$.

7.39. (a) Circle in the xy-plane $x^2 + y^2 = 16$, $z = 0$.

(b) Cylinder $x^2 + y^2 = 16$ whose axis coincides with z-axis.

(c) The yz-plane where $y \geq 0$.

(d) The straight line $y = \sqrt{3}x$, $z = 1$ where $x \geq 0$, $y \geq 0$.

7.40. (a) Hyperbolic cylinder $x^2 - y^2 = 8$.

(b) The line joining points $(-4, 0, 0)$ and $(4, 0, 0)$, that is, $x = t$, $y = 0$, $z = 0$ where $-4 \leq t \leq 4$.

(c) Ellipse $\dfrac{x^2}{25} + \dfrac{y^2}{9} = 1$, $z = 2$. (d) The portion of the x-axis defined by $x \geq 4$, $y = 0$, $z = 0$.

7.41. (a) Parabola $y^2 = -8(x - 2)$, $z = 0$. (b) Parabola $y^2 = 2x + 1$, $z = 2$. (c) Region in xy-plane bounded by parabolas $y^2 = -2(x - 1/2)$, $y^2 = -8(x - 2)$, $y^2 = 8(x + 2)$ and $y^2 = 18(x + 9/2)$ *including* the boundary.

(d) Same as (c) but excluding the boundary.

7.42. (a) $\mathbf{e}_r = \sin\theta\,\cos\phi\,\mathbf{i} + \sin\theta\,\sin\phi\,\mathbf{j} + \cos\theta\,\mathbf{k}$, $\mathbf{e}_\theta = \cos\theta\,\cos\phi\,\mathbf{i} + \cos\theta\,\sin\phi\,\mathbf{j} - \sin\theta\,\mathbf{k}$
$\mathbf{e}_\phi = -\sin\phi\,\mathbf{i} + \cos\phi\,\mathbf{j}$.

(b) $\mathbf{i} = \sin\theta\,\cos\phi\,\mathbf{e}_r + \cos\theta\,\cos\phi\,\mathbf{e}_\theta - \sin\phi\,\mathbf{e}_\phi$, $\mathbf{j} = \sin\theta\,\sin\phi\,\mathbf{e}_r + \cos\theta\,\sin\phi\,\mathbf{e}_\theta + \cos\phi\,\mathbf{e}_\phi$
$\mathbf{k} = \cos\theta\,\mathbf{e}_r - \sin\theta\,\mathbf{e}_\theta$.

7.43. $\mathbf{A} = A_r\mathbf{e}_r + A_\theta\mathbf{e}_\theta + A_\phi\mathbf{e}_\phi$ where

$$A_r = 2r\sin^2\theta\,\sin\phi\,\cos\phi - r\sin\theta\,\cos\theta\,\sin\phi + 3r\sin\theta\,\cos\theta\,\cos\phi$$

$$A_\theta = 2r\sin\theta\,\cos\theta\,\sin\phi\,\cos\phi - r\cos^2\theta\,\sin\phi - 3r\sin^2\theta\,\cos\phi$$

$$A_\phi = -2r\sin\theta\,\sin^2\phi - r\cos\theta\,\cos\phi.$$

7.47. $\mathbf{v} = v_r\mathbf{e}_r + v_\theta\mathbf{e}_\theta + v_\phi\mathbf{e}_\phi$ where $v_r = \dot{r}$, $v_\theta = r\dot{\theta}$, $v_\phi = r\sin\theta\,\dot{\phi}$

$\mathbf{a} = a_r\mathbf{e}_r + a_\theta\mathbf{e}_\theta + a_\phi\mathbf{e}_\phi$ where $a_r = \ddot{r} - r\dot{\theta}^2 - r\sin^2\theta\,\dot{\phi}^2$

$$a_\theta = \frac{1}{r}\frac{d}{dt}(r^2\dot{\theta}) - r\sin\theta\,\cos\theta\,\dot{\phi}^2,$$

$$a_\phi = \frac{1}{r\sin\theta}\frac{d}{dt}(r^2\sin^2\theta\,\dot{\phi}).$$

7.48. (a) $ds^2 = (u^2 + v^2)(du^2 + dv^2) + u^2v^2\,d\phi^2$, $h_u = h_v = \sqrt{u^2 + v^2}$, $h_\phi = uv$.

(b) $ds^2 = a^2(\sinh^2 u + \sin^2 v)(du^2 + dv^2) + dz^2$, $h_u = h_v = a\sqrt{\sinh^2 u + \sin^2 v}$, $h_z = 1$.

(c) $ds^2 = a^2(\sinh^2\xi + \sin^2\eta)(d\xi^2 + d\eta^2) + a^2\cosh^2\xi\,\cos^2\eta\,d\phi^2$,
$h_\xi = h_\eta = a\sqrt{\sinh^2\xi + \sin^2\eta}$, $h_\phi = a\cosh\xi\,\cos\eta$.

7.49. (a) $uv(u^2 + v^2)\,du\,dv\,d\phi$, (b) $a^2(\sinh^2 u + \sin^2 v)\,du\,dv\,dz$, (c) $\dfrac{a^2\,du\,dv\,dz}{(\cosh v - \cos u)^2}$.

7.50. (a) $h_\xi = h_\eta = a\sqrt{\sinh^2\xi + \sin^2\eta}$, $h_\phi = a\sinh\xi\,\sin\eta$.

(b) $a^3(\sinh^2\xi + \sin^2\eta)\sinh\xi\,\sin\eta\,d\xi\,d\eta\,d\phi$.

7.52. (a) $\rho\,d\rho\,d\phi$, $\rho\,d\phi\,dx$, $d\rho\,dz$, (b) $r\sin\theta\,dr\,d\phi$, $r^2\sin\theta\,d\theta\,d\phi$, $r\,dr\,d\theta$,
(c) $(u^2 + v^2)du\,dv$, $uv\sqrt{u^2 + v^2}\,du\,d\phi$, $uv\sqrt{u^2 + v^2}\,dv\,d\phi$

7.54. (a) ρ, (b) $r^2\sin\theta$, (c) $u^2 + v^2$, (d) $a^2(\sinh^2 u + \sin^2 v)$, (e) $a^3(\sinh^2\xi + \sin^2\eta)\sinh\xi\,\sin\eta$

7.55. $\dfrac{256\pi}{15}$, **7.56.** $\dfrac{64\pi(2 - \sqrt{2})}{3}$, **7.57.** $\dfrac{\pi}{3}(2a^3 - 3a^2h + h^3)$, **7.58.** (c) $\tfrac{1}{2}$; (d) $u_1 = \tfrac{1}{2}\rho^2$, $u_2 = 2\phi$

7.59. 2κ

7.60. (a) $\dfrac{\partial\mathbf{r}}{\partial\rho} = \cos\phi\,\mathbf{i} + \sin\phi\,\mathbf{j}$, $\nabla\rho = \dfrac{x\mathbf{i} + y\mathbf{j}}{\sqrt{x^2 + y^2}} = \cos\phi\,\mathbf{i} + \sin\phi\,\mathbf{j}$,

$\dfrac{\partial\mathbf{r}}{\partial\phi} = -\rho\sin\phi\,\mathbf{i} + \rho\cos\phi\,\mathbf{j}$, $\nabla\phi = \dfrac{-\sin\phi\,\mathbf{i} + \cos\phi\,\mathbf{j}}{\rho}$, $\dfrac{\partial\mathbf{r}}{\partial z} = \mathbf{k}$, $\nabla z = \mathbf{k}$

(b) $\dfrac{\partial \mathbf{r}}{\partial r} = \sin\theta\ \cos\phi\,\mathbf{i} + \sin\theta\ \sin\phi\,\mathbf{j} + \cos\theta\,\mathbf{k}$

$\dfrac{\partial \mathbf{r}}{\partial \theta} = r\cos\theta\ \cos\phi\,\mathbf{i} + r\cos\theta\ \sin\phi\,\mathbf{j} - r\sin\theta\,\mathbf{k}$

$\dfrac{\partial \mathbf{r}}{\partial \phi} = -r\sin\theta\ \sin\theta\,\mathbf{i} + r\sin\theta\ \cos\phi\,\mathbf{j}$

$\nabla_r = \dfrac{x\mathbf{i} + y\mathbf{j} + z\mathbf{k}}{\sqrt{x^2 + y^2 + z^2}} = \sin\theta\ \cos\phi\,\mathbf{i} + \sin\theta\sin\phi\,\mathbf{j} + \cos\theta\,\mathbf{k}$

$\nabla_\theta = \dfrac{xz\mathbf{i} + yz\mathbf{j} - (x^2 + y^2)\mathbf{k}}{(x^2 + y^2 + z^2)\sqrt{x^2 + y^2}} = \dfrac{\cos\theta\cos\phi\,\mathbf{i} + \cos\theta\sin\phi\,\mathbf{j} - \sin\theta\,\mathbf{k}}{r}$

$\nabla_\phi = \dfrac{-y\mathbf{i} + x\mathbf{j}}{x^2 + y^2} = \dfrac{-\sin\phi\,\mathbf{i} + \cos\phi\,\mathbf{j}}{r\sin\theta}$

(c) $\dfrac{\partial \mathbf{r}}{\partial u} = u\mathbf{i} + v\mathbf{j}, \quad \dfrac{\partial \mathbf{r}}{\partial v} = -v\mathbf{i} + u\mathbf{j}, \quad \dfrac{\partial \mathbf{r}}{\partial x} = \mathbf{k}$

$\nabla_u = \dfrac{u\mathbf{i} + v\mathbf{j}}{u^2 + v^2}, \quad \nabla_v = \dfrac{-v\mathbf{i} + u\mathbf{j}}{u^2 + v^2}, \quad \nabla_z = \mathbf{k}$

7.61. (b) $\dfrac{1}{y^2 - x^2}$, (c) $ds^2 = \dfrac{(x^2 + y^2)(du_1^2 + du_2^2) - 4xy\,du_1 du_2}{(x^2 - y^2)^2} + du_3^2 = \dfrac{u^2(du_1^2 + du_2^2) - 2u_1\,du_1 du_2}{2(u_2^2 - u_1^2)} + du_3^2$

7.62. $\nabla\Phi = \dfrac{1}{\sqrt{u^2 + v^2}}\dfrac{\partial \Phi}{\partial u}\mathbf{e}_u + \dfrac{1}{\sqrt{u^2 + v^2}}\dfrac{\partial \Phi}{\partial v}\mathbf{e}_v + \dfrac{\partial \Phi}{\partial z}\mathbf{e}_z$

$\operatorname{div}\mathbf{A} = \dfrac{1}{u^2 + v^2}\left[\dfrac{\partial}{\partial u}\left(\sqrt{u^2 + v^2}A_u\right) + \dfrac{\partial}{\partial v}\left(\sqrt{u^2 + v^2}A_v\right)\right] + \dfrac{\partial A_z}{\partial z}$

$\operatorname{curl}\mathbf{A} = \dfrac{1}{u^2 + v^2}\left[\left\{\dfrac{\partial A_z}{\partial v} - \dfrac{\partial}{\partial z}\left(\sqrt{u^2 + v^2}A_v\right)\right\}\sqrt{u^2 + v^2}\,\mathbf{e}_u \right.$

$\left. + \left\{\dfrac{\partial}{\partial z}\left(\sqrt{u^2 + v^2}A_u\right) - \dfrac{\partial A_z}{\partial u}\right\}\sqrt{u^2 + v^2}\,\mathbf{e}_v \right.$

$\left. + \left\{\dfrac{\partial}{\partial u}\left(\sqrt{u^2 + v^2}A_v\right) - \dfrac{\partial}{\partial v}\left(\sqrt{u^2 + v^2}A_u\right)\right\}\mathbf{e}_z\right]$

7.63. (a) $\nabla\psi = \dfrac{\partial \psi}{\partial r}\mathbf{e}_r + \dfrac{1}{r}\dfrac{\partial \psi}{\partial \theta}\mathbf{e}_\theta + \dfrac{1}{r\sin\theta}\dfrac{\partial \psi}{\partial \phi}\mathbf{e}_\phi$

(b) $\nabla\cdot\mathbf{A} = \dfrac{1}{r^2}\dfrac{\partial}{\partial r}(r^2 A_r) + \dfrac{1}{r\sin\theta}\dfrac{\partial}{\partial \theta}(\sin\theta A_\theta) + \dfrac{1}{r\sin\theta}\dfrac{\partial A_\phi}{\partial \phi}$

7.64. $\nabla^2\psi = \dfrac{1}{a^2\cosh\xi(\sinh^2\xi + \sin^2\eta)}\dfrac{\partial}{\partial \xi}\left(\cosh\xi\dfrac{\partial \psi}{\partial \xi}\right)$

$+ \dfrac{1}{a^2\cos\eta(\sinh^2\xi + \sin^2\eta)}\dfrac{\partial}{\partial \eta}\left(\cos\eta\dfrac{\partial \psi}{\partial \eta}\right) + \dfrac{1}{a^2\cosh^2\xi\cos^2\eta}\dfrac{\partial^2\psi}{\partial \phi^2}$

7.65. $\dfrac{\partial^2\Phi}{\partial u^2} + \dfrac{\partial^2\Phi}{\partial v^2} = a^2(\sinh^2 u + \sin^2 v)\Phi$

7.66. $\dfrac{1}{aRS^2}\left[\left\{\dfrac{\partial}{\partial\eta}(RE_\phi)-\dfrac{\partial}{\partial\phi}(SE_\eta)\right\}Se_\xi\right.$

$\left.+\left\{\dfrac{\partial}{\partial\phi}(SE_\xi)-\dfrac{\partial}{\partial\xi}(RE_\phi)\right\}Se_\eta+\left\{\dfrac{\partial}{\partial\xi}(SE_\eta)-\dfrac{\partial}{\partial\eta}(SE_\xi)\right\}Re_\phi\right]$

$=-\dfrac{1}{c}\dfrac{\partial H_\xi}{\partial t}e_\xi-\dfrac{1}{c}\dfrac{\partial H_\eta}{\partial t}e_\eta-\dfrac{1}{c}\dfrac{\partial H_\phi}{\partial t}e_\phi$

where $R\equiv\sinh\xi\sin\eta$ and $S\equiv\sqrt{\sinh^2\xi+\sin^2\eta}$.

7.67. $\dfrac{1}{u^2+v^2}\left[\dfrac{\partial^2\psi}{\partial u^2}+\dfrac{\partial^2\psi}{\partial u^2}\right]+\dfrac{\partial^2\psi}{\partial z^2}+\dfrac{8\pi^2 m}{h^2}[E-W(u,v,z)]\psi=0$, where $W(u,v,z)=V(x,y,z)$.

7.68. $uv^2\dfrac{\partial}{\partial u}\left(u\dfrac{\partial\psi}{\partial u}\right)+u^2v\dfrac{\partial}{\partial v}\left(v\dfrac{\partial\psi}{\partial v}\right)+(u^2+v^2)\dfrac{\partial^2\psi}{\partial\phi^2}=0$

7.69. (a) $\dfrac{\partial U}{\partial t}=\kappa\left[\dfrac{1}{r^2}\dfrac{\partial}{\partial r}\left(r^2\dfrac{\partial U}{\partial r}\right)+\dfrac{1}{r^2\sin\theta}\dfrac{\partial}{\partial\theta}\left(\sin\theta\dfrac{\partial U}{\partial\theta}\right)\right]$ (b) $\dfrac{\partial U}{\partial t}=\kappa\left[\dfrac{1}{r^2}\dfrac{\partial}{\partial r}\left(r^2\dfrac{\partial U}{\partial r}\right)\right]$

(c) $\sin\theta\dfrac{\partial}{\partial\theta}\left(\sin\theta\dfrac{\partial U}{\partial\theta}\right)+\dfrac{\partial^2 U}{\partial\phi^2}=0$ (d) $\dfrac{d}{dr}\left(r^2\dfrac{dU}{dr}\right)=0$

7.70. $ds^2=a^2[d\theta^2+\sin^2\theta\,d\phi^2]$, **7.74.** (b) $\dfrac{\partial x}{\partial u}\dfrac{\partial x}{\partial v}+\dfrac{\partial y}{\partial u}\dfrac{\partial y}{\partial v}=0$

7.78. (a) 750, 75; (b) Jacobian $=10$

7.79. (a) No, (b) $ds^2=14du_1^2+6du_2^2+6du_3^2+6du_1du_2-6du_1du_3+8du_2du_3$, $g=100$

7.80. (a) $g=16u_1^2u_3^2$, (b) $J=4u_1u_3$

CHAPTER 8

Tensor Analysis

8.1 Introduction

Physical laws, if they are to be valid, must be independent of any particular coordinate system used to describe them mathematically. A study of the consequence of this requirement leads to *tensor analysis*, which is of great use in general relativity theory, differential geometry, mechanics, elasticity, hydrodynamics, electromagnetic theory, and numerous other fields of science and engineering.

8.2 Spaces of *N* Dimensions

A point in three-dimensional space is a set of three numbers, called coordinates, determined by specifying a particular coordinate system or frame of reference. For example (x, y, z), (ρ, ϕ, z), (r, θ, ϕ) are coordinates of a point in rectangular, cylindrical, and spherical coordinate systems, respectively. A point in N-dimensional space is, by analogy, a set of N numbers denoted by $(x^1, x^2, \ldots, x^N)$ where $1, 2, \ldots, N$ are taken not as exponents but as *superscripts*, a policy which will prove useful.

The fact that we cannot visualize points in spaces of dimension higher than three has of course nothing whatsoever to do with their existence.

8.3 Coordinate Transformations

Let $(x^1, x^2, \ldots, x^N)$ and $(\bar{x}^1, \bar{x}^2, \ldots, \bar{x}^N)$ be coordinates of a point in two different frames of reference. Suppose there exists N independent relations between the coordinates of the two systems having the form

$$
\begin{aligned}
\bar{x}^1 &= \bar{x}^1(x^1, x^2, \ldots, x^N) \\
\bar{x}^2 &= \bar{x}^2(x^1, x^2, \ldots, x^N) \\
&\ \ \vdots \qquad\qquad\quad \vdots \\
\bar{x}^N &= \bar{x}^N(x^1, x^2, \ldots, x^N)
\end{aligned}
\tag{1}
$$

which we can indicate briefly by

$$
\bar{x}^k = \bar{x}^k(x^1, x^2, \ldots, x^N) \quad k = 1, 2, \ldots, N
\tag{2}
$$

where it is supposed that the functions involved are single-valued, continuous, and have continuous derivatives. Then, conversely to each set of coordinates $(\bar{x}^1, \bar{x}^2, \ldots, \bar{x}^N)$, there will correspond a unique set $(x^1, x^2, \ldots, x^N)$ given by

$$
x^k = x^k(\bar{x}^1, \bar{x}^2, \ldots, \bar{x}^N) \quad k = 1, 2, \ldots, N
\tag{3}
$$

The relations (2) or (3) define a *transformation of coordinates* from one frame of reference to another.

Summation Convention

Consider the expression $a_1x^1 + a_2x^2 + \cdots + a_Nx^N$. This can be written using the short notation $\sum_{j=1}^{N} a_jx^j$. An even shorter notation is simply to write it as a_jx^j where we adopt the convention that whenever an index (subscript or superscript) is repeated in a given term, we are to sum over that index from 1 to N unless otherwise specified. This is called the *summation convention*. Clearly, instead of using the index j, we could have used another letter, say p, and the sum could be written a_px^p. Any index that is repeated in a given term, so that the summation convention applies, is called a *dummy index* or *umbral index*.

An index occurring only once in a given term is called a *free index* and can stand for any of the numbers $1, 2, \ldots, N$ such as k in equation (2) or (3), each of which represents N equations.

8.4　Contravariant and Covariant Vectors

Suppose N quantities $A^1, A^2, \ldots, A^N$ in a coordinate system $(x^1, x^2, \ldots, x^N)$ are related to N other quantities $\overline{A}^1, \overline{A}^2, \ldots, \overline{A}^N$ in another coordinate system $(\overline{x}^1, \overline{x}^2, \ldots, \overline{x}^N)$ by the transformation equations

$$\overline{A}^p = \sum_{q=1}^{N} \frac{\partial \overline{x}^p}{\partial x^q} A^q \qquad p = 1, 2, \ldots, N$$

which by the conventions adopted can simply be written as

$$\overline{A}^p = \frac{\partial \overline{x}^p}{\partial x^q} A^q$$

Then they are called components of a *contravariant vector* or *contravariant tensor of the first rank or first order*. To provide motivation for this and later transformations, see Problems 7.33 and 7.34.

On the other hand, suppose N quantities $A_1, A_2, \ldots, A_N$ in a coordinate system $(x^1, x^2, \ldots, x^N)$ are related to N other quantities $\overline{A}_1, \overline{A}_2, \ldots, \overline{A}_N$ in another coordinate system $(\overline{x}^1, \overline{x}^2, \ldots, \overline{x}^N)$ by the transformation equations

$$\overline{A}_p = \sum_{q=1}^{N} \frac{\partial x^q}{\partial \overline{x}^p} A_q \qquad p = 1, 2, \ldots, N$$

or

$$\overline{A}_p = \frac{\partial x^q}{\partial \overline{x}^p} A_q$$

Then they are called components of a *covariant vector* or *covariant tensor of the first rank or first order*.

Note that a superscript is used to indicate contravariant components whereas a subscript is used to indicate covariant components; an exception occurs in the notation for coordinates.

Instead of speaking of a tensor whose components are A^p or A_p, we shall often refer simply to the tensor A^p or A_p. No confusion should arise from this.

8.5　Contravariant, Covariant, and Mixed Tensors

Suppose N^2 quantities A^{qs} in a coordinate system $(x^1, x^2, \ldots, x^N)$ are related to N^2 other quantities $\overline{A}^{pr}$ in another coordinate system $(\overline{x}^1, \overline{x}^2, \ldots, \overline{x}^N)$ by the transformation equations

$$\overline{A}^{pr} = \sum_{s=1}^{N} \sum_{q=1}^{N} \frac{\partial \overline{x}^p}{\partial x^q} \frac{\partial \overline{x}^r}{\partial x^s} A^{qs} \qquad p, r = 1, 2, \ldots, N$$

or

$$\overline{A}^{pr} = \frac{\partial \overline{x}^p}{\partial x^q} \frac{\partial \overline{x}^r}{\partial x^s} A^{qs}$$

by the adopted conventions, they are called *contravariant components of a tensor of the second rank* or rank two.

The N^2 quantities A_{qs} are called *covariant components of a tensor of the second rank* if

$$\overline{A}_{pr} = \frac{\partial x^q}{\partial \overline{x}^p} \frac{\partial x^s}{\partial \overline{x}^r} A_{qs}$$

Similarly, the N^2 quantities A_s^q are called *components of a mixed tensor of the second rank* if

$$\overline{A}_r^p = \frac{\partial \overline{x}^p}{\partial x^q} \frac{\partial x^s}{\partial \overline{x}^r} A_s^q$$

Kronecker Delta

The Kronecker delta, denoted by δ_k^j, is defined as follows:

$$\delta_k^j = \begin{cases} 0 & \text{if } j \neq k \\ 1 & \text{if } j = k \end{cases}$$

As its notation indicates, it is a mixed tensor of the second rank.

8.6 Tensors of Rank Greater Than Two, Tensor Fields

Tensors of rank three or more are easily defined. Specifically, for example, A_{kl}^{qst} are the components of a mixed tensor of rank 5, contravariant of order 3 and covariant of order 2, where they transform according to the relations

$$\overline{A}_{ij}^{prm} = \frac{\partial \overline{x}^p}{\partial x^q} \frac{\partial \overline{x}^r}{\partial x^s} \frac{\partial \overline{x}^m}{\partial x^t} \frac{\partial x^k}{\partial \overline{x}^i} \frac{\partial x^l}{\partial \overline{x}^j} A_{kl}^{qst}$$

Scalars or Invariants

Suppose ϕ is a function of the coordinates x^k, and let $\overline{\phi}$ denote the functional value under a transformation to a new set of coordinates $\overline{x}^k$. Then ϕ is called a *scalar* or *invariant* with respect to the coordinate transformation if $\phi = \overline{\phi}$. A scalar or invariant is also called a *tensor of rank zero*.

Tensor Fields

If to each point of a region in N-dimensional space there corresponds a definite tensor, we say that a *tensor field* has been defined. This is a *vector field* or a *scalar field* according as the tensor is of rank one or zero. It should be noted that a tensor or tensor field is not just the set of its components in one special coordinate system but *all the possible sets* under *any* transformation of coordinates.

Symmetric and Skew-Symmetric Tensors

A tensor is called *symmetric with respect to two contravariant or two covariant indices* if its components remain unaltered upon interchange of the indices. Thus, if $A_{qs}^{mpr} = A_{qs}^{pmr}$, the tensor is symmetric in m and p. If a tensor is symmetric with respect to *any* two contravariant *and any* two covariant indices, it is called *symmetric*.

A tensor is called *skew-symmetric with respect to two contravariant or two covariant indices* if its components change sign upon interchange of the indices. Thus, if $A_{qs}^{mpr} = -A_{qs}^{pmr}$, the tensor is skew-symmetric in m and p. If a tensor is skew-symmetric with respect to *any* two contravariant *and any* two covariant indices, it is called *skew-symmetric*.

8.7   Fundamental Operations with Tensors

The following operations apply.

1. **Addition.** The *sum* of two or more tensors of the same rank and type (i.e. same number of contravariant indices and same number of covariant indices) is also a tensor of the same rank and type. Thus, if A_q^{mp} and B_q^{mp} are tensors, then $C_q^{mp} = A_q^{mp} + B_q^{mp}$ is also a tensor. Addition of tensors is commutative and associative.

2. **Subtraction.** The *difference* of two tensors of the same rank and type is also a tensor of the same rank and type. Thus, if A_q^{mp} and B_q^{mp} are tensors, then $D_q^{mp} = A_q^{mp} - B_q^{mp}$ is also a tensor.

3. **Outer Multiplication.** The *product* of two tensors is a tensor whose rank is the sum of the ranks of the given tensors. This product, which involves ordinary multiplication of the components of the tensor, is called the *outer product*. For example, $A_q^{pr} B_s^m = C_{qs}^{prm}$ is the outer product of A_q^{pr} and B_s^m. However, note that not every tensor can be written as a product of two tensors of lower rank. For this reason, division of tensors is not always possible.

4. **Contraction.** If one contravariant and one covariant index of a tensor are set equal, the result indicates that a summation over the equal indices is to be taken according to the summation convention. This resulting sum is a tensor of rank two less than that of the original tensor. The process is called *contraction*. For example, in the tensor of rank 5, A_{qs}^{mpr}, set $r = s$ to obtain $A_{qr}^{mpr} = B_q^{mp}$, a tensor of rank 3. Further, by setting $p = q$, we obtain $B_p^{mp} = C^m$, a tensor of rank 1.

5. **Inner Multiplication.** By the process of outer multiplication of two tensors followed by a contraction, we obtain a new tensor called an *inner product* of the given tensors. The process is called *inner multiplication*. For example, given the tensors A_q^{mp} and B_{st}^r, the outer product is $A_q^{mp} B_{st}^r$. Letting $q = r$, we obtain the inner product $A_r^{mp} B_{st}^r$. Letting $q = r$ and $p = s$, another inner product $A_r^{mp} B_{pt}^r$ is obtained. Inner and outer multiplication of tensors is commutative and associative.

6. **Quotient Law.** Suppose it is not known whether a quantity X is a tensor or not. If an inner product of X with an arbitrary tensor is itself a tensor, then X is also a tensor. This is called the *quotient law*.

8.8   Matrices

A *matrix A* of order m by n is an array of quantities a_{pq}, called *elements*, which are arranged in m rows and n columns and generally denoted by

$$
\begin{bmatrix} a_{11} & a_{12} & \cdots & a_{1n} \\ a_{21} & a_{22} & \cdots & a_{2n} \\ \vdots & \vdots & & \vdots \\ a_{m1} & a_{m2} & \cdots & a_{mn} \end{bmatrix}
\quad \text{or} \quad
\begin{pmatrix} a_{11} & a_{12} & \cdots & a_{1n} \\ a_{21} & a_{22} & \cdots & a_{2n} \\ \vdots & \vdots & & \vdots \\ a_{m1} & a_{m2} & \cdots & a_{mn} \end{pmatrix}
$$

or in abbreviated form by $[a_{pq}]$ or (a_{pq}), $p = 1, \ldots, m$; $q = 1, \ldots, n$. We use the former notation, $[a_{pq}]$, unless otherwise stated or implied. If $m = n$, the matrix is a *square matrix* of order u or simply *order m*. If $m = 1$, it is a *row matrix* or *row vector*; if $n = 1$, it is a *column matrix* or *column vector*.

The diagonal of a square matrix containing the elements $a_{11}, a_{22}, \ldots, a_{mm}$ is called the *principal* or *main diagonal*. A square matrix whose elements are equal to one in the principal diagonal and zero elsewhere is called a *unit matrix* and is denoted by I. A *null matrix*, denoted by O, is a matrix whose elements are all zero.

Matrix Algebra

Suppose $A = [a_{pq}]$ and $B = [b_{pq}]$ are matrices having the same order (m by n). Then the following definitions apply:

(1) $A = B$ if $a_{pq} = b_{pq}$ for all p and q.

(2) The sum S and difference D of A and B are the matrices defined by

$$S = A + B = [a_{pq} + b_{pq}], \quad D = A - B = [a_{pq} - b_{pq}].$$

That is, the sum $S = A + B$ [difference $D = A - B$] is obtained by adding (subtracting) corresponding elements of A and B.

(3) The product of a scalar λ by a matrix $A = [a_{pq}]$, denoted by λA, is the matrix $[\lambda a_{pq}]$ where each element of A is multiplied by λ.

(4) The transpose of a matrix A is a matrix A^T, which is formed from A by interchanging its rows and columns. Thus, if $A = [a_{pq}]$, then $A^T = [a_{qp}]$.

Matrix Multiplication

Now suppose A and B are two matrices such that the number of columns of A is equal to the number of rows of B, say A is an $m \times p$ matrix and B is a $p \times n$ matrix. Then the product of A and B is defined and the product, denoted by AB, is the matrix whose ij-entry is obtained by multiplying the elements of row i of A by the corresponding elements of column j of B and then adding. Thus, if $A = [a_{ik}]$ and $B = [b_{kj}]$, then $AB = [c_{ij}]$ where

$$c_{ij} = a_{i1}b_{1j} + a_{i2}b_{2j} + \cdots + a_{ip}b_{pj} = \sum_{k=1}^{p} a_{ik}b_{kj}$$

Matrices whose product is defined are called *conformable*.

Determinants

Consider an n-square matrix $A = [a_{ij}]$. The *determinant* of A is denoted by $|A|$, $\det A$, $|a_{ij}|$, or $\det [a_{ij}]$. The reader may be familiar with the definition of $\det A$ when $n \leq 3$. The general definition of $\det A$ follows:

$$|A| = \sum_{\sigma \in S_n} (\text{sgn } \sigma) a_{1\sigma(1)} a_{2\sigma(2)} \cdots a_{n\sigma(n)}$$

Here S_n consists of all permutations σ of $\{1, 2, \ldots, n\}$, and sign $\sigma = \pm 1$ according as σ is an even or odd permutation.

One main property of the determinant follows.

PROPOSITION 8.1: Let $P = AB$ where A and B are n-square matrices. Then

$$\det P = (\det A)(\det B)$$

Inverses

The inverse of a square matrix A is a matrix, denoted by A^{-1}, such that

$$AA^{-1} = A^{-1}A = I$$

where I is the unit matrix. A necessary and sufficient condition that A^{-1} exists is that $\det A \neq 0$. If $\det A = 0$, then A is called <u>singular</u>, otherwise A is called <u>nonsingular</u>.

8.9   Line Element and Metric Tensor

The differential of arc length ds in rectangular coordinates (x, y, z) is obtained from $ds^2 = dx^2 + dy^2 + dz^2$. By transforming to general curvilinear coordinates (see Problem 7.17), this becomes

$$ds^2 = \sum_{p=1}^{3} \sum_{q=1}^{3} g_{pq} \, du_p \, du_q$$

Such spaces are called *three-dimensional Euclidean spaces*.

A generalization to N-dimensional space with coordinates $(x^1, x^2, \ldots, x^N)$ is immediate. We define the *line element ds* in this space to be given by the quadratic form, called the *metric form* or *metric*,

$$ds^2 = \sum_{p=1}^{N} \sum_{q=1}^{N} g_{pq} \, dx^p \, dx^q$$

or, using the summation convention,

$$ds^2 = g_{pq} \, dx^p \, dx^q$$

In the special case where there exists a transformation of coordinates from x^j to $\bar{x}^k$ such that the metric form is transformed into $(d\bar{x}^1)^2 + (d\bar{x}^2)^2 + \cdots + (d\bar{x}^N)^2$ or $d\bar{x}^k d\bar{x}^k$, then the space is called *N-dimensional Euclidean space*. In the general case, however, the space is called *Riemannian*.

The quantities g_{pq} are the components of a covariant tensor of rank two called the *metric tensor* or *fundamental tensor*. We can, and always will, choose this tensor to be symmetric (see Problem 8.29).

Conjugate or Reciprocal Tensors

Let $g = |g_{pq}|$ denote the determinant with elements g_{pq} and suppose $g \neq 0$. Define g^{pq} by

$$g^{pq} = \frac{\text{cofactor of } g_{pq}}{g}$$

Then g^{pq} is a symmetric contravariant tensor of rank two called the *conjugate* or *reciprocal tensor* of g_{pq} (see Problem 8.34). It can be shown (see Problem 8.33) that

$$g^{pq} g_{rq} = \delta_r^p$$

8.10   Associated Tensors

Given a tensor, we can derive other tensors by raising or lowering indices. For example, given the tensor A_{pq}, by raising the index p, we obtain the tensor $A^p_{.q}$, the dot indicating the original position of the moved index. By raising the index q also, we obtain $A^{pq}_{..}$. Where no confusion can arise, we shall often omit the dots; thus $A^{pq}_{..}$ can be written A^{pq}. These derived tensors can be obtained by forming inner products of the given tensor with the metric tensor g_{pq} or its conjugate g^{pq}. Thus, for example

$$A^p_{.q} = g^{rp} A_{rq}, \qquad A^{pq} = g^{rp} g^{sq} A_{rs}, \qquad A^p_{.rs} = g_{rq} A^{pq}_{..s}, \qquad A^{qm \cdot tk}_{..n} = g^{pk} g_{sn} g^{rm} A^{q \cdot st}_{.r \cdot p}$$

These become clear if we interpret multiplication by g^{rp} as meaning: let $r = p$ (or $p = r$) in whatever follows and *raise* this index. Similarly, we interpret multiplication by g_{rq} as meaning: let $r = q$ (or $q = r$) in whatever follows and lower this index.

All tensors obtained from a given tensor by forming inner products with the metric tensor and its conjugate are called *associated tensors* of the given tensor. For example A^m and A_m are associated tensors, the first are contravariant and the second covariant components. The relation between them is given by

$$A_p = g_{pq}A^q \quad \text{or} \quad A^p = g^{pq}A_q$$

For rectangular coordinates $g_{pq} = 1$ if $p = q$, and 0 if $p \neq q$, so that $A_p = A^p$, which explains why no distinction was made between contravariant and covariant components of a vector in earlier chapters.

8.11   Christoffel's Symbols

The following symbols

$$[pq, r] = \frac{1}{2}\left(\frac{\partial g_{pr}}{\partial x^q} + \frac{\partial g_{qr}}{\partial x^p} - \frac{\partial g_{pq}}{\partial x^r}\right)$$

$$\begin{Bmatrix} s \\ pq \end{Bmatrix} = g^{sr}[pq, r]$$

are called the *Christoffel symbols of the first and second kind*, respectively. Other symbols used instead of $\begin{Bmatrix} s \\ pq \end{Bmatrix}$ are $\{pq, s\}$ and Γ_{pq}^s. The latter symbol suggests, however, a tensor character, which is not true in general.

Transformation Laws of Christoffel's Symbols

Suppose we denote by a bar a symbol in a coordinate system $\bar{x}^k$. Then

$$\overline{[jk, m]} = [pq, r]\frac{\partial x^p}{\partial \bar{x}^j}\frac{\partial x^q}{\partial \bar{x}^k}\frac{\partial x^r}{\partial \bar{x}^m} + g_{pq}\frac{\partial x^p}{\partial \bar{x}^m}\frac{\partial^2 x^q}{\partial \bar{x}^j \partial \bar{x}^k}$$

$$\overline{\begin{Bmatrix} n \\ jk \end{Bmatrix}} = \begin{Bmatrix} s \\ pq \end{Bmatrix}\frac{\partial \bar{x}^n}{\partial x^s}\frac{\partial x^p}{\partial \bar{x}^j}\frac{\partial x^q}{\partial \bar{x}^k} + \frac{\partial \bar{x}^n}{\partial x^q}\frac{\partial^2 x^q}{\partial \bar{x}^j \partial \bar{x}^k}$$

are the laws of transformation of the Christoffel symbols showing that they are not tensors unless the second terms on the right are zero.

8.12   Length of a Vector, Angle between Vectors, Geodesics

The quantity $A^p B_p$, which is the inner product of A^p and B_p, is a scalar analogous to the scalar product in rectangular coordinates. We define the length L of the vector A^p or A_q as given by

$$L^2 = A^p A_p = g^{pq}A_p A_q = g_{pq}A^p A^q$$

We can define the angle θ between A^p and B_p as given by

$$\cos\theta = \frac{A^p B_p}{\sqrt{(A^p A_p)(B^p B_p)}}$$

Geodesics

The distance s between two points t_1 and t_2 on a curve $x^r = x^r(t)$ in a Riemannian space is given by

$$s = \int_{t_1}^{t_2} \sqrt{g_{pq}\frac{dx^p}{dt}\frac{dx^q}{dt}}\, dt$$

That curve in the space, which makes the distance a minimum, is called a *geodesic* of the space. By use of the *calculus of variations* (see Problems 8.50 and 8.51), the geodesics are found from the differential equation

$$\frac{d^2 x^r}{ds^2} + \left\{ \begin{matrix} r \\ pq \end{matrix} \right\} \frac{dx^p}{ds} \frac{dx^q}{ds} = 0$$

where s is the arc length parameter. As examples, the geodesics on a plane are straight lines whereas the geodesics on a sphere are arcs of great circles.

Physical Components

The physical components of a vector A^p or A_p, denoted by A_u, A_v, A_w, are the projections of the vector on the tangents to the coordinate curves and are given in the case of orthogonal coordinates by

$$A_u = \sqrt{g_{11}} A^1 = \frac{A_1}{\sqrt{g_{11}}}, \quad A_v = \sqrt{g_{22}} A^2 = \frac{A_2}{\sqrt{g_{22}}}, \quad A_w = \sqrt{g_{33}} A^3 = \frac{A_3}{\sqrt{g_{33}}}$$

Similarly, the physical components of a tensor A^{pq} or A_{pq} are given by

$$A_{uu} = g_{11} A^{11} = \frac{A_{11}}{g_{11}}, \quad A_{uv} = \sqrt{g_{11} g_{22}} A^{12} = \frac{A_{12}}{\sqrt{g_{11} g_{22}}}, \quad A_{uw} = \sqrt{g_{11} g_{33}} A^{13} = \frac{A_{13}}{\sqrt{g_{11} g_{33}}}, \quad \text{etc.}$$

8.13 Covariant Derivative

The covariant derivative of a tensor A_p with respect to x^q is denoted by $A_{p,q}$ and is defined by

$$A_{p,q} \equiv \frac{\partial A_p}{\partial x^q} - \left\{ \begin{matrix} s \\ pq \end{matrix} \right\} A_s$$

a covariant tensor of rank two.

The covariant derivative of a tensor A^p with respect to x^q is denoted by $A^p_{,q}$ and is defined by

$$A^p_{,q} \equiv \frac{\partial A^p}{\partial x^q} + \left\{ \begin{matrix} p \\ qs \end{matrix} \right\} A^s$$

a mixed tensor of rank two.

For rectangular systems, the Christoffel symbols are zero and the covariant derivatives are the usual partial derivatives. Covariant derivatives of tensors are also tensors (see Problem 8.52).

The above results can be extended to covariant derivatives of higher rank tensors. Thus

$$A^{p_1 \cdots p_m}_{r_1 \cdots r_n, \, q} \equiv \frac{\partial A^{p_1 \cdots p_m}_{r_1 \cdots r_n}}{\partial x^q}$$

$$- \left\{ \begin{matrix} s \\ r_1 q \end{matrix} \right\} A^{p_1 \cdots p_m}_{s r_2 \cdots r_n} - \left\{ \begin{matrix} s \\ r_2 q \end{matrix} \right\} A^{p_1 \cdots p_m}_{r_1 s r_3 \cdots r_n} - \cdots - \left\{ \begin{matrix} s \\ r_n q \end{matrix} \right\} A^{p_1 \cdots p_m}_{r_1 \cdots r_{n-1} s}$$

$$+ \left\{ \begin{matrix} p_1 \\ qs \end{matrix} \right\} A^{s p_2 \cdots p_m}_{r_1 \cdots r_n} + \left\{ \begin{matrix} p_2 \\ qs \end{matrix} \right\} A^{p_1 s p_3 \cdots p_m}_{r_1 \cdots r_n} + \cdots + \left\{ \begin{matrix} p_m \\ qs \end{matrix} \right\} A^{p_1 \cdots p_{m-1} s}_{r_1 \cdots r_n}$$

is the covariant derivative of $A^{p_1 \cdots p_m}_{r_1 \cdots r_n}$ with respect to x^q.

The rules of covariant differentiation for sums and products of tensors are the same as those for ordinary differentiation. In performing the differentiations, the tensors g_{pq}, g^{pq}, and δ^p_q may be treated as constants since their covariant derivatives are zero (see Problem 8.54). Since covariant derivatives express rates of change of physical quantities independent of any frames of reference, they are of great importance in expressing physical laws.

8.14 Permutation Symbols and Tensors

The symbol e_{pqr} is defined by the following relations:

$$e_{123} = e_{231} = e_{312} = +1, \quad e_{213} = e_{132} = e_{321} = -1, \quad e_{pqr} = 0$$

if two or more indices are equal.

The symbol e^{pqr} is defined in the same manner. The symbols e_{pqr} and e^{pqr} are called <u>permutation symbols</u> in three-dimensional space.

Further, let us define

$$\epsilon_{pqr} = \frac{1}{\sqrt{g}} e_{pqr}, \quad \epsilon^{pqr} = \sqrt{g}\, e^{pqr}$$

It can be shown that ϵ_{pqr} and ϵ^{pqr} are respectively, covariant and contravariant tensors, called *permutation tensors* in three-dimensional space. Generalizations to higher dimensions are possible.

8.15 Tensor Form of Gradient, Divergence, and Curl

1. **Gradient.** If Φ is a scalar or invariant, the gradient of Φ is defined by

$$\nabla \Phi = \text{grad } \Phi = \Phi_{,p} = \frac{\partial \Phi}{\partial x^p}$$

    where $\Phi_{,p}$ is the covariant derivative of Φ with respect to x^p.

2. **Divergence.** The divergence of A^p is the contraction of its covariant derivative with respect to x^q, i.e. the contraction of $A^p_{,q}$. Then

$$\text{div } A^p = A^p_{,p} = \frac{1}{\sqrt{g}} \frac{\partial}{\partial x^k} \left(\sqrt{g}\, A^k \right)$$

3. **Curl.** The curl of A_p is $A_{p,q} - A_{q,p} = \dfrac{\partial A_p}{\partial x^q} - \dfrac{\partial A_q}{\partial x^p}$, a tensor of rank two. The curl is also defined as $-\epsilon^{pqr} A_{p,q}$.

4. **Laplacian.** The Laplacian of Φ is the divergence of grad Φ or

$$\nabla^2 \Phi = \text{div } \Phi_{,p} = \frac{1}{\sqrt{g}} \frac{\partial}{\partial x^j} \left(\sqrt{g}\, g^{jk} \frac{\partial \Phi}{\partial x^k} \right)$$

    In case $g < 0$, $\sqrt{g}$ must be replaced by $\sqrt{-g}$. Both cases $g > 0$ and $g < 0$ can be included by writing $\sqrt{|g|}$ in place of $\sqrt{g}$.

8.16 Intrinsic or Absolute Derivative

The intrinsic or absolute derivative of A_p along a curve $x^q = x^q(t)$, denoted by $\dfrac{\delta A_p}{\delta t}$, is defined as the inner product of the covariant derivative of A_p and $\dfrac{dx^q}{dt}$, that is $A_{p,q} \dfrac{dx^q}{dt}$, and is given by

$$\frac{\delta A_p}{\delta t} \equiv \frac{dA_p}{dt} - \left\{ \begin{matrix} r \\ pq \end{matrix} \right\} A_r \frac{dx^q}{dt}$$

Similarly, we define

$$\frac{\delta A^p}{\delta t} \equiv \frac{dA^p}{dt} + \left\{ \begin{matrix} p \\ qr \end{matrix} \right\} A^r \frac{dx^q}{dt}$$

The vectors A_p or A^p are said to *move parallelly* along a curve if their intrinsic derivatives along the curve are zero, respectively.

Intrinsic derivatives of higher rank tensors are similarly defined.

8.17   Relative and Absolute Tensors

A tensor $A^{p_1 \cdots p_m}_{r_1 \cdots r_n}$ is called a *relative tensor of weight w* if its components transform according to the equation

$$\overline{A}^{q_1 \cdots q_m}_{s_1 \cdots s_n} = \left|\frac{\partial x}{\partial \overline{x}}\right|^w A^{p_1 \cdots p_m}_{r_1 \cdots r_n} \frac{\partial \overline{x}^{q_1}}{\partial x^{p_1}} \cdots \frac{\partial \overline{x}^{q_m}}{\partial x^{p_m}} \frac{\partial x^{r_1}}{\partial \overline{x}^{s_1}} \cdots \frac{\partial x^{r_n}}{\partial \overline{x}^{s_n}}$$

where $J = \left|\dfrac{\partial x}{\partial \overline{x}}\right|$ is the Jacobian of the transformation. If $w = 0$, the tensor is called *absolute* and is the type of tensor with which we have been dealing above. If $w = 1$, the relative tensor is called a *tensor density*. The operations of addition, multiplication, etc., of relative tensors are similar to those of absolute tensors. See, for example, Problem 8.64.

SOLVED PROBLEMS

Summation Convention

8.1. Write each of the following using the summation convention.

(a) $d\phi = \dfrac{\partial \phi}{\partial x^1} dx^1 + \dfrac{\partial \phi}{\partial x^2} dx^2 + \cdots + \dfrac{\partial \phi}{\partial x^N} dx^N$,   (d) $ds^2 = g_{11}(dx^1)^2 + g_{22}(dx^2)^2 + g_{33}(dx^3)^2$,

(b) $\dfrac{d\overline{x}^k}{dt} = \dfrac{\partial \overline{x}^k}{\partial x^1}\dfrac{dx^1}{dt} + \dfrac{\partial \overline{x}^k}{\partial x^2}\dfrac{dx^2}{dt} + \cdots + \dfrac{\partial \overline{x}^k}{\partial x^N}\dfrac{dx^N}{dt}$,   (e) $\displaystyle\sum_{p=1}^{3}\sum_{q=1}^{3} g_{pq}\, dx^p dx^q$

(c) $(x^1)^2 + (x^2)^2 + (x^3)^2 + \cdots + (x^N)^2$

Solution

(a) $d\phi = \dfrac{\partial \phi}{\partial x^j} dx^j$,   (b) $\dfrac{d\overline{x}^k}{dt} = \dfrac{\partial \overline{x}^k}{\partial x^m}\dfrac{dx^m}{dt}$,   (c) $x^k x^k$

(d) $ds^2 = g_{kk}\, dx^k dx^k$, $N = 3$,   (e) $g_{pq}\, dx^p\, dx^q$, $N = 3$

8.2. Write the terms in each of the following indicated sums.

(a) $a_{jk}x^k$,   (b) $A_{pq}A^{qr}$,   (c) $\overline{g}_{rs} = g_{jk} \dfrac{\partial x^j}{\partial \overline{x}^r}\dfrac{\partial x^k}{\partial \overline{x}^s}$, $N = 3$

Solution

(a) $\displaystyle\sum_{k=1}^{N} a_{jk}x^k = a_{j1}x^1 + a_{j2}x^2 + \cdots + a_{jN}x^N$,   (b) $\displaystyle\sum_{q=1}^{N} A_{pq}A^{qr} = A_{p1}A^{1r} + A_{p2}A^{2r} + \cdots + A_{pN}A^{Nr}$

(c) $\displaystyle\overline{g}_{rs} = \sum_{j=1}^{3}\sum_{k=1}^{3} g_{jk} \dfrac{\partial x^j}{\partial \overline{x}^r}\dfrac{\partial x^k}{\partial \overline{x}^s}$

$$= \sum_{j=1}^{3}\left(g_{j1} \frac{\partial x^j}{\partial \overline{x}^r}\frac{\partial x^1}{\partial \overline{x}^s} + g_{j2} \frac{\partial x^j}{\partial \overline{x}^r}\frac{\partial x^2}{\partial \overline{x}^s} + g_{j3} \frac{\partial x^j}{\partial \overline{x}^r}\frac{\partial x^3}{\partial \overline{x}^s} \right)$$

$$= g_{11} \frac{\partial x^1}{\partial \overline{x}^r}\frac{\partial x^1}{\partial \overline{x}^s} + g_{21} \frac{\partial x^2}{\partial \overline{x}^r}\frac{\partial x^1}{\partial \overline{x}^s} + g_{31} \frac{\partial x^3}{\partial \overline{x}^r}\frac{\partial x^1}{\partial \overline{x}^s}$$

$$+ g_{12} \frac{\partial x^1}{\partial \overline{x}^r}\frac{\partial x^2}{\partial \overline{x}^s} + g_{22} \frac{\partial x^2}{\partial \overline{x}^r}\frac{\partial x^2}{\partial \overline{x}^s} + g_{32} \frac{\partial x^3}{\partial \overline{x}^r}\frac{\partial x^2}{\partial \overline{x}^s}$$

$$+ g_{13} \frac{\partial x^1}{\partial \overline{x}^r}\frac{\partial x^3}{\partial \overline{x}^s} + g_{23} \frac{\partial x^2}{\partial \overline{x}^r}\frac{\partial x^3}{\partial \overline{x}^s} + g_{33} \frac{\partial x^3}{\partial \overline{x}^r}\frac{\partial x^3}{\partial \overline{x}^s}$$

8.3. Suppose x^k, $k = 1, 2, \ldots, N$, are rectangular coordinates. What locus, if any, is represented by each of the following equations for $N = 2$, 3, and $N \geq 4$. Assume that the functions are single-valued, have continuous derivatives and are independent, when necessary.

(a) $a_k x^k = 1$, where a_k are constants

(b) $x^k x^k = 1$

(c) $x^k = x^k(u)$

(d) $x^k = x^k(u, v)$

Solution

(a) For $N = 2$, $a_1 x^1 + a_2 x^2 = 1$, a line in two dimensions, that is, a line in a plane.

For $N = 3$, $a_1 x^1 + a_2 x^2 + a_3 x^3 = 1$, a plane in three dimensions.

For $N \geq 4$, $a_1 x^1 + a_2 x^2 + \cdots + a_N x^N = 1$ is a *hyperplane*.

(b) For $N = 2$, $(x^1)^2 + (x^2)^2 = 1$, a circle of unit radius in the plane.

For $N = 3$, $(x^1)^2 + (x^2)^2 + (x^3)^2 = 1$, a sphere of unit radius.

For $N \geq 4$, $(x^1)^2 + (x^2)^2 + \cdots + (x^N)^2 = 1$, a *hypersphere* of unit radius.

(c) For $N = 2$, $x^1 = x^1(u)$, $x^2 = x^2(u)$, a plane curve with parameter u.

For $N = 3$, $x^1 = x^1(u)$, $x^2 = x^2(u)$, $x^3 = x^3(u)$, a three-dimensional space curve.

For $N \geq 4$, an N-dimensional space curve.

(d) For $N = 2$, $x^1 = x^1(u, v)$, $x^2 = x^2(u, v)$ is a transformation of coordinates from (u, v) to (x^1, x^2).

For $N = 3$, $x^1 = x^1(u, v)$, $x^2 = x^2(u, v)$, $x^3 = x^3(u, v)$ is a three-dimensional surface with parameters u and v.

For $N \geq 4$, a *hypersurface*.

Contravariant and Covariant Vectors and Tensors

8.4. Write the law of transformation for the tensors (a) A^i_{jk}, (b) B^{mn}_{ijk}, (c) C^m.

Solution

(a) $\overline{A}^p_{qr} = \dfrac{\partial \overline{x}^p}{\partial x^i} \dfrac{\partial x^j}{\partial \overline{x}^q} \dfrac{\partial x^k}{\partial \overline{x}^r} A^i_{jk}$

As an aid for remembering the transformation, note that the relative positions of indices p, q, r on the left side of the transformation are the same as those on the right side. Since these indices are associated with the $\overline{x}$ coordinates and since indices i, j, k are associated, respectively, with indices p, q, r, the required transformation is easily written.

(b) $\overline{B}^{pq}_{rst} = \dfrac{\partial \overline{x}^p}{\partial x^m} \dfrac{\partial \overline{x}^q}{\partial x^n} \dfrac{\partial x^i}{\partial \overline{x}^r} \dfrac{\partial x^j}{\partial \overline{x}^s} \dfrac{\partial x^k}{\partial \overline{x}^t} B^{mn}_{ijk}$, (c) $\overline{C}^p = \dfrac{\partial \overline{x}^p}{\partial x^m} C^m$

8.5. A quantity $A(j, k, l, m)$, which is a function of coordinates x^i, transforms to another coordinate system $\overline{x}^i$ according to the rule

$$\overline{A}(p, q, r, s) = \dfrac{\partial x^j}{\partial \overline{x}^p} \dfrac{\partial \overline{x}^q}{\partial x^k} \dfrac{\partial \overline{x}^r}{\partial x^l} \dfrac{\partial \overline{x}^s}{\partial x^m} A(j, k, l, m)$$

(a) Is the quantity a tensor? (b) If so, write the tensor in suitable notation and (c) give the contravariant and covariant order and rank.

Solution

(a) Yes, (b) A^{klm}_j, (c) Contravariant of order 3, covariant of order 1 and rank $3 + 1 = 4$

8.6. Determine whether each of the following quantities is a tensor. If so, state whether it is contravariant or covariant and give its rank: (a) dx^k, (b) $\dfrac{\partial \phi(x^1, \ldots, x^N)}{\partial x^k}$.

Solution

(a) Assume the transformation of coordinates $\bar{x}^j = \bar{x}^j(x^1, \ldots, x^N)$. Then $d\bar{x}^j = \dfrac{\partial \bar{x}^j}{\partial x^k} dx^k$ and so dx^k is a contravariant tensor of rank one or a contravariant vector. Note that the location of the index k is appropriate.

(b) Under the transformation $x^k = x^k(\bar{x}^1, \ldots, \bar{x}^N)$, ϕ is a function of x^k and hence $\bar{x}^j$ such that $\phi(x^1, \ldots, x^N) = \bar{\phi}(\bar{x}^1, \ldots, \bar{x}^N)$, that is, ϕ is a scalar or invariant (tensor of rank zero). By the chain rule for partial differentiation, $\dfrac{\partial \bar{\phi}}{\partial \bar{x}^j} = \dfrac{\partial \phi}{\partial \bar{x}^j} = \dfrac{\partial \phi}{\partial x^k} \dfrac{\partial x^k}{\partial \bar{x}^j} = \dfrac{\partial x^k}{\partial \bar{x}^j} \dfrac{\partial \phi}{\partial x^k}$ and $\dfrac{\partial \phi}{\partial x^k}$ transforms like $\bar{A}_j = \dfrac{\partial x^k}{\partial \bar{x}^j} A_k$. Then $\dfrac{\partial \phi}{\partial x^k}$

is a covariant tensor of rank one or a covariant vector.

Note that in $\dfrac{\partial \phi}{\partial x^k}$ the index appears in the denominator and thus acts like a subscript which indicates its

covariant character. We refer to the tensor $\dfrac{\partial \phi}{\partial x^k}$ or equivalently, the tensor with components $\dfrac{\partial \phi}{\partial x^k}$, as the

gradient of ϕ, written grad ϕ or $\nabla \phi$.

8.7. A covariant tensor has components xy, $2y - z^2$, xz in rectangular coordinates. Find its covariant components in spherical coordinates.

Solution

Let A_j denote the covariant components in rectangular coordinates $x^1 = x$, $x^2 = y$, $x^3 = z$. Then

$$A_1 = xy = x^1 x^2, \quad A_2 = 2y - z^2 = 2x^2 - (x^3)^2, \quad A_3 = x^1 x^3$$

where care must be taken to distinguish between superscripts and exponents.

Let $\bar{A}_k$ denote the covariant components in spherical coordinates $\bar{x}^1 = r$, $\bar{x}^2 = \theta$, $\bar{x}^3 = \phi$. Then

$$\bar{A}_k = \frac{\partial x^j}{\partial \bar{x}^k} A_j \tag{1}$$

The transformation equations between coordinate systems are

$$x^1 = \bar{x}^1 \sin \bar{x}^2 \cos \bar{x}^3, \quad x^2 = \bar{x}^1 \sin \bar{x}^2 \sin \bar{x}^3, \quad x^3 = \bar{x}^1 \cos \bar{x}^2$$

Then equations (1) yield the required covariant components

$$\bar{A}_1 = \frac{\partial x^1}{\partial \bar{x}^1} A_1 + \frac{\partial x^2}{\partial \bar{x}^1} A_2 + \frac{\partial x^3}{\partial \bar{x}^1} A_3$$

$$= (\sin \bar{x}^2 \cos \bar{x}^3)(x^1 x^2) + (\sin \bar{x}^2 \sin \bar{x}^3)(2x^2 - (x^3)^2) + (\cos \bar{x}^2)(x^1 x^3)$$

$$= (\sin \theta \cos \phi)(r^2 \sin^2 \theta \sin \phi \cos \phi)$$

$$\quad + (\sin \theta \sin \phi)(2r \sin \theta \sin \phi - r^2 \cos^2 \theta)$$

$$\quad + (\cos \theta)(r^2 \sin \theta \cos \theta \cos \phi)$$

$$\bar{A}_2 = \frac{\partial x^1}{\partial \bar{x}^2} A_1 + \frac{\partial x^2}{\partial \bar{x}^2} A_2 + \frac{\partial x^3}{\partial \bar{x}^2} A_3$$

$$= (r \cos \theta \cos \phi)(r^2 \sin^2 \theta \sin \phi \cos \phi)$$

$$\quad + (r \cos \theta \sin \phi)(2r \sin \theta \sin \phi - r^2 \cos^2 \theta)$$

$$\quad + (-r \sin \theta)(r^2 \sin \theta \cos \theta \cos \phi)$$

$$\overline{A}_3 = \frac{\partial x^1}{\partial \overline{x}^3} A_1 + \frac{\partial x^2}{\partial \overline{x}^3} A_2 + \frac{\partial x^3}{\partial \overline{x}^3} A_3$$

$$= (-r \sin \theta \sin \phi)(r^2 \sin^2 \theta \sin \phi \cos \phi)$$

$$+ (r \sin \theta \cos \phi)(2r \sin \theta \sin \phi - r^2 \cos^2 \theta)$$

$$+ (0)(r^2 \sin \theta \cos \theta \cos \phi)$$

8.8. Show that $\dfrac{\partial A_p}{\partial x^q}$ is not a tensor even though A_p is a covariant tensor of rank one.

Solution

By hypothesis, $\overline{A}_j = \dfrac{\partial x^p}{\partial \overline{x}^j} A_p$. Differentiating with respect to $\overline{x}^k$.

$$\frac{\partial \overline{A}_j}{\partial \overline{x}^k} = \frac{\partial x^p}{\partial \overline{x}^j} \frac{\partial A_p}{\partial \overline{x}^k} + \frac{\partial^2 x^p}{\partial \overline{x}^k \partial \overline{x}^j} A_p = \frac{\partial x^p}{\partial \overline{x}^j} \frac{\partial A_p}{\partial x^q} \frac{\partial x^q}{\partial \overline{x}^k} + \frac{\partial^2 x^p}{\partial \overline{x}^k \partial \overline{x}^j} A_p = \frac{\partial x^p}{\partial \overline{x}^j} \frac{\partial x^q}{\partial \overline{x}^k} \frac{\partial A_p}{\partial x^q} + \frac{\partial^2 x^p}{\partial \overline{x}^k \partial \overline{x}^j} A_p$$

Since the second term on the right is present, $\dfrac{\partial A_p}{\partial x^q}$ does not transform as a tensor should. Later, we shall show how the addition of a suitable quantity to $\dfrac{\partial A_p}{\partial x^q}$ causes the result to be a tensor (Problem 8.52).

8.9. Show that the velocity of a fluid at any point is a contravariant tensor of rank one.

Solution

The velocity of a fluid at any point has components $\dfrac{dx^k}{dt}$ in the coordinate system x^k. In the coordinate system $\overline{x}^j$, the velocity is $\dfrac{d\overline{x}^j}{dt}$. But, by the chain rule,

$$\frac{d\overline{x}^j}{dt} = \frac{\partial \overline{x}^j}{\partial x^k} \frac{dx^k}{dt}$$

and it follows that the velocity is a contravariant tensor of rank one or a contravariant vector.

The Kronecker Delta

8.10. Evaluate (a) $\delta_q^p A_s^{qr}$, (b) $\delta_q^p \delta_r^q$.

Solution

Since $\delta_q^p = 1$ if $p = q$ and 0 if $p \neq q$, we have

(a) $\delta_q^p A_s^{qr} = A_s^{pr}$,   (b) $\delta_q^p \delta_r^q = \delta_r^p$

8.11. Show that $\dfrac{\partial x^p}{\partial x^q} = \delta_q^p$.

Solution

If $p = q$, $\dfrac{\partial x^p}{\partial x^q} = 1$ since $x^p = x^q$. If $p \neq q$, $\dfrac{\partial x^p}{\partial x^q} = 0$ since x^p and x^q are independent.

Then $\dfrac{\partial x^p}{\partial x^q} = \delta_q^p$.

8.12. Prove that $\dfrac{\partial x^p}{\partial \overline{x}^q}\dfrac{\partial \overline{x}^q}{\partial x^r} = \delta_r^p$.

Solution

Coordinates x^p are functions of coordinates $\overline{x}^q$, which are in turn functions of coordinates x^r. Then, by the chain rule and Problem 8.11,

$$\frac{\partial x^p}{\partial x^r} = \frac{\partial x^p}{\partial \overline{x}^q}\frac{\partial \overline{x}^q}{\partial x^r} = \delta_r^p$$

8.13. Let $\overline{A}^p = \dfrac{\partial \overline{x}^p}{\partial x^q} A^q$. Prove that $A^q = \dfrac{\partial x^q}{\partial \overline{x}^p}\overline{A}^p$.

Solution

Multiply equation $\overline{A}^p = \dfrac{\partial \overline{x}^p}{\partial x^q} A^q$ by $\dfrac{\partial x^r}{\partial \overline{x}^p}$.

Then $\dfrac{\partial x^r}{\partial \overline{x}^p}\overline{A}^p = \dfrac{\partial x^r}{\partial \overline{x}^p}\dfrac{\partial \overline{x}^p}{\partial x^q} A^q = \delta_q^r A^q = A^r$ by Problem 8.12. Placing $r = q$, the result follows. This indicates that in the transformation equations for the tensor components, the quantities with bars and quantities without bars can be interchanged, a result which can be proved in general.

8.14. Prove that δ_q^p is a mixed tensor of the second rank.

Solution

If δ_q^p is a mixed tensor of the second rank, it must transform according to the rule

$$\overline{\delta}_k^j = \frac{\partial \overline{x}^j}{\partial x^p}\frac{\partial x^q}{\partial \overline{x}^k}\delta_q^p$$

The right side equals $\dfrac{\partial \overline{x}^j}{\partial x^p}\dfrac{\partial x^p}{\partial \overline{x}^k} = \delta_k^j$ by Problem 8.12. Since $\overline{\delta}_k^j = \delta_k^j = 1$ if $j = k$, and 0 if $j \neq k$, it follows that δ_q^p is a mixed tensor of rank two, justifying the notation used.

   Note that we sometimes use $\delta_{pq} = 1$ if $p = q$ and 0 if $p \neq q$, as the Kronecker delta. This is, however, *not* a covariant tensor of the second rank as the notation would seem to indicate.

Fundamental Operations with Tensors

8.15. Suppose A_r^{pq} and B_r^{pq} are tensors. Prove that their sum and difference are tensors.

Solution

By hypothesis A_r^{pq} and B_r^{pq} are tensors, so that

$$\overline{A}_l^{jk} = \frac{\partial \overline{x}^j}{\partial x^p}\frac{\partial \overline{x}^k}{\partial x^q}\frac{\partial x^r}{\partial \overline{x}^l} A_r^{pq} \quad \text{and} \quad \overline{B}_l^{jk} = \frac{\partial \overline{x}^j}{\partial x^p}\frac{\partial \overline{x}^k}{\partial x^q}\frac{\partial x^r}{\partial \overline{x}^l} B_r^{pq}$$

Adding,

$$(\overline{A}_l^{jk} + \overline{B}_l^{jk}) = \frac{\partial \overline{x}^j}{\partial x^p}\frac{\partial \overline{x}^k}{\partial x^q}\frac{\partial x^r}{\partial \overline{x}^l}(A_r^{pq} + B_r^{pq})$$

Subtracting,

$$(\overline{A}_l^{jk} - \overline{B}_l^{jk}) = \frac{\partial \overline{x}^j}{\partial x^p}\frac{\partial \overline{x}^k}{\partial x^q}\frac{\partial x^r}{\partial \overline{x}^l}(A_r^{pq} - B_r^{pq})$$

Then $A_r^{pq} + B_r^{pq}$ and $A_r^{pq} - B_r^{pq}$ are tensors of the same rank and type as A_r^{pq} and B_r^{pq}.

8.16. Suppose A_r^{pq} and B_t^s are tensors. Prove that $C_{rt}^{pqs} = A_r^{pq} B_t^s$ is also a tensor.

Solution

We must prove that C_{rt}^{pqs} is a tensor whose components are formed by taking the products of components of tensors A_r^{pq} and B_t^s. Since A_r^{pq} and B_t^s are tensors,

$$\overline{A}_l^{jk} = \frac{\partial \overline{x}^j}{\partial x^p} \frac{\partial \overline{x}^k}{\partial x^q} \frac{\partial x^r}{\partial \overline{x}^l} A_r^{pq} \quad \text{and} \quad \overline{B}_n^m = \frac{\partial \overline{x}^m}{\partial x^s} \frac{\partial x^t}{\partial \overline{x}^n} B_t^s$$

Multiplying,

$$\overline{A}_l^{jk} \overline{B}_n^m = \frac{\partial \overline{x}^j}{\partial x^p} \frac{\partial \overline{x}^k}{\partial x^q} \frac{\partial x^r}{\partial \overline{x}^l} \frac{\partial \overline{x}^m}{\partial x^s} \frac{\partial x^t}{\partial \overline{x}^n} A_r^{pq} B_t^s$$

which shows that $A_r^{pq} B_t^s$ is a tensor of rank 5, with contravariant indices p, q, s and covariant indices r, t, thus warranting the notation C_{rt}^{pqs}. We call $C_{rt}^{pqs} = A_r^{pq} B_t^s$ the *outer product* of A_r^{pq} and B_t^s.

8.17. Let A_{rst}^{pq} be a tensor. (a) Choose $p = t$ and show that A_{rsp}^{pq}, where the summation convention is employed, is a tensor. What is its rank? (b) Choose $p = t$ and $q = s$ and show similarly that A_{rqp}^{pq} is a tensor. What is its rank?

Solution

(a)   Since A_{rst}^{pq} is a tensor,

$$\overline{A}_{lmn}^{jk} = \frac{\partial \overline{x}^j}{\partial x^p} \frac{\partial \overline{x}^k}{\partial x^q} \frac{\partial x^r}{\partial \overline{x}^l} \frac{\partial x^s}{\partial \overline{x}^m} \frac{\partial x^t}{\partial \overline{x}^n} A_{rst}^{pq} \tag{1}$$

We must show that A_{rsp}^{pq} is a tensor. Place the corresponding indices j and n equal to each other and sum over this index. Then

$$\overline{A}_{lmj}^{jk} = \frac{\partial \overline{x}^j}{\partial x^p} \frac{\partial \overline{x}^k}{\partial x^q} \frac{\partial x^r}{\partial \overline{x}^l} \frac{\partial x^s}{\partial \overline{x}^m} \frac{\partial x^t}{\partial \overline{x}^j} A_{rst}^{pq} = \frac{\partial x^t}{\partial \overline{x}^j} \frac{\partial \overline{x}^j}{\partial x^p} \frac{\partial \overline{x}^k}{\partial x^q} \frac{\partial x^r}{\partial \overline{x}^l} \frac{\partial x^s}{\partial \overline{x}^m} A_{rst}^{pq}$$

$$= \delta_p^t \frac{\partial \overline{x}^k}{\partial x^q} \frac{\partial x^r}{\partial \overline{x}^l} \frac{\partial x^s}{\partial \overline{x}^m} A_{rst}^{pq} = \frac{\partial \overline{x}^k}{\partial x^q} \frac{\partial x^r}{\partial \overline{x}^l} \frac{\partial x^s}{\partial \overline{x}^m} A_{rsp}^{pq}$$

and so A_{rsp}^{pq} is a tensor of rank 3 and can be denoted by B_{rs}^q. The process of placing a contravariant index equal to a covariant index in a tensor and summing is called *contraction*. By such a process a tensor is formed whose rank is two less than the rank of the original tensor.

(b)   We must show that A_{rqp}^{pq} is a tensor. Placing $j = n$ and $k = m$ in equation (1) of part (a) and summing over j and k, we have

$$\overline{A}_{lkj}^{jk} = \frac{\partial \overline{x}^j}{\partial x^p} \frac{\partial \overline{x}^k}{\partial x^q} \frac{\partial x^r}{\partial \overline{x}^l} \frac{\partial x^s}{\partial \overline{x}^k} \frac{\partial x^t}{\partial \overline{x}^j} A_{rst}^{pq} = \frac{\partial x^t}{\partial \overline{x}^j} \frac{\partial \overline{x}^j}{\partial x^p} \frac{\partial x^s}{\partial \overline{x}^k} \frac{\partial \overline{x}^k}{\partial x^q} \frac{\partial x^r}{\partial \overline{x}^l} A_{rst}^{pq}$$

$$= \delta_p^t \delta_q^s \frac{\partial x^r}{\partial \overline{x}^l} A_{rst}^{pq} = \frac{\partial x^r}{\partial \overline{x}^l} A_{rqp}^{pq}$$

which shows that A_{rqp}^{pq} is a tensor of rank one and can be denoted by C_r. Note that by contracting twice, the rank was reduced by 4.

8.18. Prove that the contraction of the tensor A_q^p is a scalar or invariant.

Solution

We have

$$\overline{A}_k^j = \frac{\partial \overline{x}^j}{\partial x^p} \frac{\partial x^q}{\partial \overline{x}^k} A_q^p$$

Putting $j = k$ and summing,

$$\overline{A}_j^j = \frac{\partial \overline{x}^j}{\partial x^p}\frac{\partial x^q}{\partial \overline{x}^j}A_q^p = \delta_p^q A_q^p = A_p^p$$

Then $\overline{A}_j^j = A_p^p$ and it follows that A_p^p must be an invariant. Since A_q^p is a tensor of rank two and contraction with respect to a single index lowers the rank by two, we are led to define an invariant as a tensor of rank zero.

8.19. Show that the contraction of the outer product of the tensors A^p and B_q is an invariant.

Solution

Since A^p and B_q are tensors, $\overline{A}^j = \dfrac{\partial \overline{x}^j}{\partial x^p}A^p$, $\overline{B}_k = \dfrac{\partial x^q}{\partial \overline{x}^k}B_q$. Then

$$\overline{A}^j \overline{B}_k = \frac{\partial \overline{x}^j}{\partial x^p}\frac{\partial x^q}{\partial \overline{x}^k}A^p B_q$$

By contraction (putting $j = k$ and summing)

$$\overline{A}^j \overline{B}_j = \frac{\partial \overline{x}^j}{\partial x^p}\frac{\partial x^q}{\partial \overline{x}^j}A^p B_q = \delta_p^q A^p B_q = A^p B_p$$

and so $A^p B_p$ is an invariant. The process of multiplying tensors (outer multiplication) and then contracting is called *inner multiplication* and the result is called an *inner product*. Since $A^p B_p$ is a scalar, it is often called the *scalar product* of the vectors A^p and B_q.

8.20. Show that any inner product of the tensors A_r^p and B_t^{qs} is a tensor of rank three.

Solution

Outer product of A_r^p and $B_t^{qs} = A_r^p B_t^{qs}$.

Let us contract with respect to indices p and t, that is, let $p = t$ and sum. We must show that the resulting inner product, represented by $A_r^p B_p^{qs}$, is a tensor of rank three.

By hypothesis, A_r^p and B_t^{qs} are tensors; then

$$\overline{A}_k^j = \frac{\partial \overline{x}^j}{\partial x^p}\frac{\partial x^r}{\partial \overline{x}^k}A_r^p, \quad \overline{B}_n^{lm} = \frac{\partial \overline{x}^l}{\partial x^q}\frac{\partial \overline{x}^m}{\partial x^s}\frac{\partial x^t}{\partial \overline{x}^n}B_t^{qs}$$

Multiplying, letting $j = n$ and summing, we have

$$\overline{A}_k^j \overline{B}_j^{lm} = \frac{\partial \overline{x}^j}{\partial x^p}\frac{\partial x^r}{\partial \overline{x}^k}\frac{\partial \overline{x}^l}{\partial x^q}\frac{\partial \overline{x}^m}{\partial x^s}\frac{\partial x^t}{\partial \overline{x}^j}A_r^p B_t^{qs}$$

$$= \delta_p^t \frac{\partial x^r}{\partial \overline{x}^k}\frac{\partial \overline{x}^l}{\partial x^q}\frac{\partial \overline{x}^m}{\partial x^s}A_r^p B_t^{qs}$$

$$= \frac{\partial x^r}{\partial \overline{x}^k}\frac{\partial \overline{x}^l}{\partial x^q}\frac{\partial \overline{x}^m}{\partial x^s}A_r^p B_p^{qs}$$

showing that $A_r^p B_p^{qs}$ is a tensor of rank three. By contracting with respect to q and r or s and r in the product $A_r^p B_t^{qs}$, we can similarly show that any inner product is a tensor of rank three.

Another Method. The outer product of two tensors is a tensor whose rank is the sum of the ranks of the given tensors. Then $A_r^p B_t^{qs}$ is a tensor of rank $3 + 2 = 5$. Since a contraction results in a tensor whose rank is two less than that of the given tensor, it follows that any contraction of $A_r^p B_t^{qs}$ is a tensor of rank $5 - 2 = 3$.

8.21. Let $X(p,\ q,\ r)$ be a quantity such that $X(p,\ q,\ r)B_r^{qn} = 0$ for an arbitrary tensor B_r^{qn}. Prove that $X(p,\ q,\ r) = 0$ identically.

Solution

Since B_r^{qn} is an arbitrary tensor, choose one particular component (say the one with $q = 2$, $r = 3$) not equal to zero, while all other components are zero. Then $X(p,\ 2,\ 3)B_3^{2n} = 0$, so that $X(p,\ 2,\ 3) = 0$ since $B_3^{2n} \neq 0$. By similar reasoning with all possible combinations of q and r, we have $X(p,\ q,\ r) = 0$ and the result follows.

8.22. Suppose in the coordinate system x^i, a quantity $A(p, q, r)$ is $A(p, q, r)B_r^{qs} = C_p^s$ where B_r^{qs} is an arbitrary tensor and C_p^s is a tensor. Prove that $A(p, q, r)$ is a tensor.

Solution

In the transformed coordinates $\bar{x}^i$, $\bar{A}(j, k, l)\bar{B}_l^{km} = \bar{C}_j^m$.

Then $\bar{A}(j, k, l)\dfrac{\partial \bar{x}^k}{\partial x^q}\dfrac{\partial \bar{x}^m}{\partial x^s}\dfrac{\partial x^r}{\partial \bar{x}^l}B_r^{qs} = \dfrac{\partial \bar{x}^m}{\partial x^s}\dfrac{\partial x^p}{\partial \bar{x}^j}C_p^s = \dfrac{\partial \bar{x}^m}{\partial x^s}\dfrac{\partial x^p}{\partial \bar{x}^j}A(p, q, r)B_r^{qs}$

or

$$\frac{\partial \bar{x}^m}{\partial x^s}\left[\frac{\partial \bar{x}^k}{\partial x^q}\frac{\partial x^r}{\partial \bar{x}^l}\bar{A}(j, k, l) - \frac{\partial x^p}{\partial \bar{x}^j}A(p, q, r)\right]B_r^{qs} = 0$$

Inner multiplication by $\dfrac{\partial x^n}{\partial \bar{x}^m}$ (i.e. multiplying by $\dfrac{\partial x^n}{\partial \bar{x}^t}$ and then contracting with $t = m$) yields

$$\delta_s^n\left[\frac{\partial \bar{x}^k}{\partial x^q}\frac{\partial x^r}{\partial \bar{x}^l}\bar{A}(j, k, l) - \frac{\partial x^p}{\partial \bar{x}^j}A(p, q, r)\right]B_r^{qs} = 0$$

or

$$\left[\frac{\partial \bar{x}^k}{\partial x^q}\frac{\partial x^r}{\partial \bar{x}^l}\bar{A}(j, k, l) - \frac{\partial x^p}{\partial \bar{x}^j}A(p, q, r)\right]B_r^{qn} = 0.$$

Since B_r^{qn} is an arbitrary, tensor, we have by Problem 8.21,

$$\frac{\partial \bar{x}^k}{\partial x^q}\frac{\partial x^r}{\partial \bar{x}^l}\bar{A}(j, k, l) - \frac{\partial x^p}{\partial \bar{x}^j}A(p, q, r) = 0$$

Inner multiplication by $\dfrac{\partial x^q}{\partial \bar{x}^m}\dfrac{\partial \bar{x}^n}{\partial x^r}$ yields

$$\delta_m^k\delta_l^n\bar{A}(j, k, l) - \frac{\partial x^p}{\partial \bar{x}^j}\frac{\partial x^q}{\partial \bar{x}^m}\frac{\partial \bar{x}^n}{\partial x^r}A(p, q, r) = 0$$

or

$$\bar{A}(j, m, n) = \frac{\partial x^p}{\partial \bar{x}^j}\frac{\partial x^q}{\partial \bar{x}^m}\frac{\partial \bar{x}^n}{\partial x^r}A(p, q, r)$$

which shows that $A(p, q, r)$ is a tensor and justifies use of the notation A_{pq}^r.

In this problem we have established a special case of the *quotient law* which states that if an inner product of a quantity X with an arbitrary tensor B is a tensor C, then X is a tensor.

Symmetric and Skew-Symmetric Tensors

8.23. Suppose a tensor A_{st}^{pqr} is symmetric (skew-symmetric) with respect to indices p and q in one coordinate system. Show that it remains symmetric (skew-symmetric) with respect to p and q in any coordinate system.

Solution

Since only indices p and q are involved, we shall prove the results for B^{pq}. If B^{pq} is symmetric, $B^{pq} = B^{qp}$. Then

$$\bar{B}^{jk} = \frac{\partial \bar{x}^j}{\partial x^p}\frac{\partial \bar{x}^k}{\partial x^q}B^{pq} = \frac{\partial \bar{x}^k}{\partial x^q}\frac{\partial \bar{x}^j}{\partial x^p}B^{qp} = \bar{B}^{kj}$$

and B^{pq} remains symmetric in the $\bar{x}^i$ coordinate system.

If B^{pq} is skew-symmetric, $B^{pq} = -B^{qp}$. Then

$$\bar{B}^{jk} = \frac{\partial \bar{x}^j}{\partial x^p}\frac{\partial \bar{x}^k}{\partial x^q}B^{pq} = -\frac{\partial \bar{x}^k}{\partial x^q}\frac{\partial \bar{x}^j}{\partial x^p}B^{qp} = -\bar{B}^{kj}$$

and B^{pq} remains skew-symmetric in the $\bar{x}^i$ coordinate system.

The above results are, of course, valid for other symmetric (skew-symmetric) tensors.

8.24. Show that every tensor can be expressed as the sum of two tensors, one of which is symmetric and the other skew-symmetric in a pair of covariant or contravariant indices.

Solution

Consider, for example, the tensor B^{pq}. We have

$$B^{pq} = \tfrac{1}{2}(B^{pq} + B^{qp}) + \tfrac{1}{2}(B^{pq} - B^{qp})$$

But $R^{pq} = \tfrac{1}{2}(B^{pq} + B^{qp}) = R^{qp}$ is symmetric, and $S^{pq} = \tfrac{1}{2}(B^{pq} - B^{qp}) = -S^{qp}$ is skew-symmetric. By similar reasoning, the result is seen to be true for any tensor.

8.25. Let $\Phi = a_{jk}A^jA^k$. Show that we can always write $\Phi = b_{jk}A^jA^k$ where b_{jk} is symmetric.

Solution

$$\Phi = a_{jk}A^jA^k = a_{kj}A^kA^j = a_{kj}A^jA^k$$

Then

$$2\Phi \doteq a_{jk}A^jA^k + a_{kj}A^jA^k = (a_{jk} + a_{kj})A^jA^k \quad \text{and} \quad \Phi = \tfrac{1}{2}(a_{jk} + a_{kj})A^jA^k = b_{jk}A^jA^k$$

where $b_{jk} = \tfrac{1}{2}(a_{jk} + a_{kj}) = b_{kj}$ is symmetric.

Matrices

8.26. Write the sum $S = A + B$, difference $D = A - B$, and products $P = AB$, $Q = BA$ of the matrices

$$A = \begin{bmatrix} 3 & 1 & -2 \\ 4 & -2 & 3 \\ -2 & 1 & -1 \end{bmatrix}, \quad B = \begin{bmatrix} 2 & 0 & -1 \\ -4 & 1 & 2 \\ 1 & -1 & 0 \end{bmatrix}$$

Solution

$$S = A + B = \begin{bmatrix} 3+2 & 1+0 & -2-1 \\ 4-4 & -2+1 & 3+2 \\ -2+1 & 1-1 & -1+0 \end{bmatrix} = \begin{bmatrix} 5 & 1 & -3 \\ 0 & -1 & 5 \\ -1 & 0 & -1 \end{bmatrix}$$

$$D = A - B = \begin{bmatrix} 3-2 & 1-0 & -2+1 \\ 4+4 & -2-1 & 3-2 \\ -2-1 & 1+1 & -1-0 \end{bmatrix} = \begin{bmatrix} 1 & 1 & -1 \\ 8 & -3 & 1 \\ -3 & 2 & -1 \end{bmatrix}$$

$$P = AB = \begin{bmatrix} (3)(2)+(1)(-4)+(-2)(1) & (3)(0)+(1)(1)+(-2)(-1) & (3)(-1)+(1)(2)+(-2)(0) \\ (4)(2)+(-2)(-4)+(3)(1) & (4)(0)+(-2)(1)+(3)(-1) & (4)(-1)+(-2)(2)+(3)(0) \\ (-2)(2)+(1)(-4)+(-1)(1) & (-2)(0)+(1)(1)+(-1)(-1) & (-2)(-1)+(1)(2)+(-1)(0) \end{bmatrix}$$

$$= \begin{bmatrix} 0 & 3 & -1 \\ 19 & -5 & -8 \\ -9 & 2 & 4 \end{bmatrix}$$

$$Q = BA = \begin{bmatrix} 8 & 1 & -3 \\ -12 & -4 & 9 \\ -1 & 3 & -5 \end{bmatrix}$$

This shows that $AB \neq BA$, that is, multiplication of matrices is not commutative in general.

8.27. Let $A = \begin{bmatrix} 2 & 1 \\ -1 & 3 \end{bmatrix}$ and $B = \begin{bmatrix} -1 & 2 \\ 3 & -2 \end{bmatrix}$. Show that $(A+B)(A-B) \neq A^2 - B^2$.

Solution

$A + B = \begin{bmatrix} 1 & 3 \\ 2 & 1 \end{bmatrix}$, $A - B = \begin{bmatrix} 3 & -1 \\ -4 & 5 \end{bmatrix}$. Then $(A+B)(A-B) = \begin{bmatrix} 1 & 3 \\ 2 & 1 \end{bmatrix}\begin{bmatrix} 3 & -1 \\ -4 & 5 \end{bmatrix} = \begin{bmatrix} -9 & 14 \\ 2 & 3 \end{bmatrix}$.

$A^2 = \begin{bmatrix} 2 & 1 \\ -1 & 3 \end{bmatrix}\begin{bmatrix} 2 & 1 \\ -1 & 3 \end{bmatrix} = \begin{bmatrix} 3 & 5 \\ -5 & 8 \end{bmatrix}$, $B^2 = \begin{bmatrix} -1 & 2 \\ 3 & -2 \end{bmatrix}\begin{bmatrix} -1 & 2 \\ 3 & -2 \end{bmatrix} = \begin{bmatrix} 7 & -6 \\ -9 & 10 \end{bmatrix}$.

Then $A^2 - B^2 = \begin{bmatrix} -4 & 11 \\ 4 & -2 \end{bmatrix}$.

Therefore, $(A+B)(A-B) \neq A^2 - B^2$. However, $(A+B)(A-B) = A^2 - AB + BA - B^2$.

8.28. Express in matrix notation the transformation equations for (a) a covariant vector, (b) a contravariant tensor of rank two, assuming $N = 3$.

Solution

(a) The transformation equations $\overline{A}_p = \dfrac{\partial x^q}{\partial \overline{x}^p} A_q$ can be written

$$\begin{bmatrix} \overline{A}_1 \\ \overline{A}_2 \\ \overline{A}_3 \end{bmatrix} = \begin{bmatrix} \dfrac{\partial x^1}{\partial \overline{x}^1} & \dfrac{\partial x^2}{\partial \overline{x}^1} & \dfrac{\partial x^3}{\partial \overline{x}^1} \\ \dfrac{\partial x^1}{\partial \overline{x}^2} & \dfrac{\partial x^2}{\partial \overline{x}^2} & \dfrac{\partial x^3}{\partial \overline{x}^2} \\ \dfrac{\partial x^1}{\partial \overline{x}^3} & \dfrac{\partial x^2}{\partial \overline{x}^3} & \dfrac{\partial x^3}{\partial \overline{x}^3} \end{bmatrix} \begin{bmatrix} A_1 \\ A_2 \\ A_3 \end{bmatrix}$$

in terms of column vectors, or equivalently in terms of row vectors

$$[\overline{A}_1 \; \overline{A}_2 \; \overline{A}_3] = [A_1 \; A_2 \; A_3] \begin{bmatrix} \dfrac{\partial x^1}{\partial \overline{x}^1} & \dfrac{\partial x^1}{\partial \overline{x}^2} & \dfrac{\partial x^1}{\partial \overline{x}^3} \\ \dfrac{\partial x^2}{\partial \overline{x}^1} & \dfrac{\partial x^2}{\partial \overline{x}^2} & \dfrac{\partial x^2}{\partial \overline{x}^3} \\ \dfrac{\partial x^3}{\partial \overline{x}^1} & \dfrac{\partial x^3}{\partial \overline{x}^2} & \dfrac{\partial x^3}{\partial \overline{x}^3} \end{bmatrix}$$

(b) The transformation equations $\overline{A}^{pr} = \dfrac{\partial \overline{x}^p}{\partial x^q} \dfrac{\partial \overline{x}^r}{\partial x^s} A^{qs}$ can be written

$$\begin{bmatrix} \overline{A}^{11} & \overline{A}^{12} & \overline{A}^{13} \\ \overline{A}^{21} & \overline{A}^{22} & \overline{A}^{23} \\ \overline{A}^{31} & \overline{A}^{32} & \overline{A}^{33} \end{bmatrix} = \begin{bmatrix} \dfrac{\partial \overline{x}^1}{\partial x^1} & \dfrac{\partial \overline{x}^1}{\partial x^2} & \dfrac{\partial \overline{x}^1}{\partial x^3} \\ \dfrac{\partial \overline{x}^2}{\partial x^1} & \dfrac{\partial \overline{x}^2}{\partial x^2} & \dfrac{\partial \overline{x}^2}{\partial x^3} \\ \dfrac{\partial \overline{x}^3}{\partial x^1} & \dfrac{\partial \overline{x}^3}{\partial x^2} & \dfrac{\partial \overline{x}^3}{\partial x^3} \end{bmatrix} \begin{bmatrix} A^{11} & A^{12} & A^{13} \\ A^{21} & A^{22} & A^{23} \\ A^{31} & A^{32} & A^{33} \end{bmatrix} \begin{bmatrix} \dfrac{\partial \overline{x}^1}{\partial x^1} & \dfrac{\partial \overline{x}^2}{\partial x^1} & \dfrac{\partial \overline{x}^3}{\partial x^1} \\ \dfrac{\partial \overline{x}^1}{\partial x^2} & \dfrac{\partial \overline{x}^2}{\partial x^2} & \dfrac{\partial \overline{x}^3}{\partial x^2} \\ \dfrac{\partial \overline{x}^1}{\partial x^3} & \dfrac{\partial \overline{x}^2}{\partial x^3} & \dfrac{\partial \overline{x}^3}{\partial x^3} \end{bmatrix}$$

Extensions of these results can be made for $N > 3$. For higher rank tensors, however, the matrix notation fails.

The Line Element and Metric Tensor

8.29. Suppose $ds^2 = g_{jk}\, dx^j\, dx^k$ is an invariant. Show that g_{jk} is a symmetric covariant tensor of rank two.

Solution

By Problem 8.25, $\Phi = ds^2$, $A^j = dx^j$ and $A^k = dx^k$; it follows that g_{jk} can be chosen symmetric. Also, since ds^2 is an invariant,

$$\bar{g}_{pq}\, d\bar{x}^p\, d\bar{x}^q = g_{jk}\, dx^j\, dx^k = g_{jk}\, \frac{\partial x^j}{\partial \bar{x}^p}\, d\bar{x}^p\, \frac{\partial x^k}{\partial \bar{x}^q}\, d\bar{x}^q = g_{jk}\, \frac{\partial x^j}{\partial \bar{x}^p}\, \frac{\partial x^k}{\partial \bar{x}^q}\, d\bar{x}^p\, d\bar{x}^q$$

Then $\bar{g}_{pq} = g_{jk}\, \dfrac{\partial x^j}{\partial \bar{x}^p}\, \dfrac{\partial x^k}{\partial \bar{x}^q}$ and g_{jk} is a symmetric covariant tensor of rank two, called the *metric tensor*.

8.30. Determine the metric tensor in (a) cylindrical and (b) spherical coordinates.

Solution

(a) As in Problem 7.7, $ds^2 = d\rho^2 + \rho^2\, d\phi^2 + dz^2$.
  If $x^1 = \rho$, $x^2 = \phi$, $x^3 = z$, then $g_{11} = 1$, $g_{22} = \rho^2$, $g_{33} = 1$, $g_{12} = g_{21} = 0$, $g_{23} = g_{32} = 0$, $g_{31} = g_{13} = 0$.

In matrix form, the metric tensor can be written $\begin{bmatrix} g_{11} & g_{12} & g_{13} \\ g_{21} & g_{22} & g_{23} \\ g_{31} & g_{32} & g_{33} \end{bmatrix} = \begin{bmatrix} 1 & 0 & 0 \\ 0 & \rho^2 & 0 \\ 0 & 0 & 1 \end{bmatrix}$

(b) As in Problem 7.8(a), $ds^2 = dr^2 + r^2\, d\theta^2 + r^2 \sin^2 \theta\, d\phi^2$.

If $x^1 = r$, $x^2 = \theta$, $x^3 = \phi$, the metric tensor can be written $\begin{bmatrix} 1 & 0 & 0 \\ 0 & r^2 & 0 \\ 0 & 0 & r^2 \sin^2 \theta \end{bmatrix}$

In general, for orthogonal coordinates, $g_{jk} = 0$ for $j \neq k$.

8.31. (a) Express the determinant $g = \begin{vmatrix} g_{11} & g_{12} & g_{13} \\ g_{21} & g_{22} & g_{23} \\ g_{31} & g_{32} & g_{33} \end{vmatrix}$ in terms of the elements in the second row and their corresponding cofactors. (b) Show that $g_{jk}G(j, k) = g$ where $G(j, k)$ is the cofactor of g_{jk} in g and where summation is over k only.

Solution

(a) The cofactor of g_{jk} is the determinant obtained from g by (1) deleting the row and column in which g_{jk} appears and (2) associating the sign $(-1)^{j+k}$ to this determinant. Thus,

$$\text{Cofactor of } g_{21} = (-1)^{2+1}\begin{vmatrix} g_{12} & g_{13} \\ g_{32} & g_{33} \end{vmatrix}, \quad \text{Cofactor of } g_{22} = (-1)^{2+2}\begin{vmatrix} g_{11} & g_{13} \\ g_{31} & g_{33} \end{vmatrix},$$

$$\text{Cofactor of } g_{23} = (-1)^{2+3}\begin{vmatrix} g_{11} & g_{12} \\ g_{31} & g_{32} \end{vmatrix}$$

Denote these cofactors by $G(2, 1)$, $G(2, 2)$, and $G(2, 3)$ respectively. Then, by an elementary principle of determinants

$$g_{21}G(2, 1) + g_{22}G(2, 2) + g_{23}G(2, 3) = g$$

(b) By applying the result of (a) to any row or column, we have $g_{jk}G(j, k) = g$ where the summation is over k only. These results hold where $g = |g_{jk}|$ is an Nth order determinant.

8.32. (a) Prove that $g_{21}G(3, 1) + g_{22}G(3, 2) + g_{23}G(3, 3) = 0$.

(b) Prove that $g_{jk}G(p, k) = 0$ if $j \neq p$.

Solution

(a) Consider the determinant $\begin{vmatrix} g_{11} & g_{12} & g_{13} \\ g_{21} & g_{22} & g_{23} \\ g_{21} & g_{22} & g_{23} \end{vmatrix}$ which is zero since its last two rows are identical. Expanding according to elements of the last row we have

$$g_{21}G(3, 1) + g_{22}G(3, 2) + g_{23}G(3, 3) = 0$$

(b) By setting the corresponding elements of any two rows (or columns) equal, we can show, as in part (a), that $g_{jk}G(p, k) = 0$ if $j \neq p$. This result holds for Nth-order determinants as well.

8.33. Define $g^{jk} = \dfrac{G(j, k)}{g}$ where $G(j, k)$ is the cofactor of g_{jk} in the determinant $g = |g_{jk}| \neq 0$. Prove that $g_{jk}g^{pk} = \delta_j^p$.

Solution

By Problem 8.31, $g_{jk}\dfrac{G(j, k)}{g} = 1$ or $g_{jk}g^{jk} = 1$, where summation is over k only.

By Problem 8.32, $g_{jk}\dfrac{G(p, k)}{g} = 0$ or $g_{jk}g^{pk} = 0$ if $p \neq j$.

Then $g_{jk}g^{pk} (= 1$ if $p = j$, and 0 if $p \neq j) = \delta_j^p$.

We have used the notation g^{jk} although we have not yet shown that the notation is warranted (i.e. that g^{jk} is a contravariant tensor of rank two). This is established in Problem 8.34. Note that the cofactor has been written $G(j, k)$ and not G^{jk} since we can show that it is not a tensor in the usual sense. However, it can be shown to be a *relative tensor* of weight two which is contravariant and with this extension of the tensor concept the notation G^{jk} can be justified (see Supplementary Problem 8.152).

8.34. Prove that g^{jk} is a symmetric contravariant tensor of rank two.

Solution

Since g_{jk} is symmetric, $G(j, k)$ is symmetric and so $g^{jk} = G(j, k)/g$ is symmetric.

If B^p is an arbitrary contravariant vector, $B_q = g_{pq}B^p$ is an arbitrary covariant vector. Multiplying by g^{jq},

$$g^{jq}B_q = g^{jq}g_{pq}B^p = \delta_p^j B^p = B^j \quad \text{or} \quad g^{jq}B_q = B^j$$

Since B_q is an arbitrary vector, g^{jq} is a contravariant tensor of rank two, by application of the quotient law. The tensor g^{jk} is called the *conjugate metric tensor*.

8.35. Determine the conjugate metric tensor in (a) cylindrical and (b) spherical coordinates.

Solution

(a) From Problem 8.30(a), $g = \begin{vmatrix} 1 & 0 & 0 \\ 0 & \rho^2 & 0 \\ 0 & 0 & 1 \end{vmatrix} = \rho^2$

$$g^{11} = \frac{\text{cofactor of } g_{11}}{g} = \frac{1}{\rho^2}\begin{vmatrix} \rho^2 & 0 \\ 0 & 1 \end{vmatrix} = 1, \qquad g^{33} = \frac{\text{cofactor of } g_{33}}{g} = \frac{1}{\rho^2}\begin{vmatrix} 1 & 0 \\ 0 & \rho^2 \end{vmatrix} = 1$$

$$g^{22} = \frac{\text{cofactor of } g_{22}}{g} = \frac{1}{\rho^2}\begin{vmatrix} 1 & 0 \\ 0 & 1 \end{vmatrix} = \frac{1}{\rho^2}, \qquad g^{12} = \frac{\text{cofactor of } g_{12}}{g} = -\frac{1}{\rho^2}\begin{vmatrix} 0 & 0 \\ 0 & 1 \end{vmatrix} = 0$$

Similarly, $g^{jk} = 0$ if $j \neq k$. In matrix form, the conjugate metric tensor can be represented by

$$\begin{bmatrix} 1 & 0 & 0 \\ 0 & 1/\rho^2 & 0 \\ 0 & 0 & 1 \end{bmatrix}$$

(b) From Problem 8.30(b), $g = \begin{vmatrix} 1 & 0 & 0 \\ 0 & r^2 & 0 \\ 0 & 0 & r^2 \sin^2 \theta \end{vmatrix} = r^4 \sin^2 \theta$

As in part (a), we find $g^{11} = 1$, $g^{22} = \dfrac{1}{r^2}$, $g^{33} = \dfrac{1}{r^2 \sin^2 \theta}$ and $g^{jk} = 0$ for $j \neq k$, and in matrix form this can be written

$$\begin{bmatrix} 1 & 0 & 0 \\ 0 & 1/r^2 & 0 \\ 0 & 0 & 1/r^2 \sin^2 \theta \end{bmatrix}$$

8.36. Find (a) g and (b) g^{jk} corresponding to $ds^2 = 5(dx^1)^2 + 3(dx^2)^2 + 4(dx^3)^2 - 6\,dx^1 dx^2 + 4\,dx^2 dx^3$.

Solution

(a) $g_{11} = 5$, $g_{22} = 3$, $g_{33} = 4$, $g_{12} = g_{21} = -3$, $g_{23} = g_{32} = 2$, $g_{13} = g_{31} = 0$.

Then $g = \begin{vmatrix} 5 & -3 & 0 \\ -3 & 3 & 2 \\ 0 & 2 & 4 \end{vmatrix} = 4$.

(b) The cofactors $G(j, k)$ of g_{jk} are

$G(1, 1) = 8$, $G(2, 2) = 20$, $G(3, 3) = 6$, $G(1, 2) = G(2, 1) = 12$, $G(2, 3) = G(3, 2) = -10$, $G(1, 3) = G(3, 1) = -6$

Then $g^{11} = 2$, $g^{22} = 5$, $g^{33} = 3/2$, $g^{12} = g^{21} = 3$, $g^{23} = g^{32} = -5/2$, $g^{13} = g^{31} = -3/2$

Note that the product of the matrices (g_{jk}) and (g^{jk}) is the unit matrix $\mathbf{I}$, that is

$$\begin{bmatrix} 5 & -3 & 0 \\ -3 & 3 & 2 \\ 0 & 2 & 4 \end{bmatrix} \begin{bmatrix} 2 & 3 & -3/2 \\ 3 & 5 & -5/2 \\ -3/2 & -5/2 & 3/2 \end{bmatrix} = \begin{bmatrix} 1 & 0 & 0 \\ 0 & 1 & 0 \\ 0 & 0 & 1 \end{bmatrix}$$

Associated Tensors

8.37. Let $A_j = g_{jk}A^k$. Show that $A^k = g^{jk}A_j$.

Solution

Multiply $A_j = g_{jk}A^k$ by g^{jq}. Then $g^{jq}A_j = g^{jq}g_{jk}A^k = \delta_k^q A^k = A^q$, that is $A^q = g^{jq}A_j$ or $A^k = g^{jk}A_j$.

The tensors of rank one, A_j and A^k, are called *associated*. They represent the covariant and contravariant components of a vector.

8.38. (a) Show that $L^2 = g_{pq}A^pA^q$ is an invariant. (b) Show that $L^2 = g^{pq}A_pA_q$.

Solution

(a) Let A_j and A^k be the covariant and contravariant components of a vector. Then

$$\bar{A}_p = \frac{\partial x^j}{\partial \bar{x}^p}A_j, \quad \bar{A}^q = \frac{\partial \bar{x}^q}{\partial x^k}A^k$$

and

$$\overline{A}_p \overline{A}^p = \frac{\partial x^j}{\partial \overline{x}^p} \frac{\partial \overline{x}^p}{\partial \overline{x}^k} A_j A^k = \delta_k^j A_j A^k = A_j A^j$$

so that $A_j A^j$ is an invariant which we call L^2. Then we can write

$$L^2 = A_j A^j = g_{jk} A^k A^j = g_{pq} A^p A^q$$

(b) From (a), $L^2 = A_j A^j = A_j g^{kj} A_k = g^{jk} A_j A_k = g^{pq} A_p A_q$.

The scalar or invariant quantity $L = \sqrt{A_p A^p}$ is called the magnitude or length of the vector with covariant components A_p and contravariant components A^p.

8.39. (a) Suppose A^p and B^q are vectors. Show that $g_{pq} A^p B^q$ is an invariant.

(b) Show that $\dfrac{g_{pq} A^p B^q}{\sqrt{(A^p A_p)(B^q B_q)}}$ is an invariant.

Solution

(a) By Problem 8.38, $A^p B_p = A^p g_{pq} B^q = g_{pq} A^p B^q$ is an invariant.

(b) Since $A^p A_p$ and $B^q B_q$ are invariants, $\sqrt{(A^p A_p)(B^q B_q)}$ is an invariant and so $\dfrac{g_{pq} A^p B^q}{\sqrt{(A^p A_p)(B^q B_q)}}$ is an invariant. We define

$$\cos \theta = \frac{g_{pq} A^p B^q}{\sqrt{(A^p A_p)(B^q B_q)}}$$

as the *cosine of the angle between vectors* A^p and B^q. If $g_{pq} A^p B^q = A^p B_p = 0$, the vectors are called *orthogonal*.

8.40. Express the relationship between the associated tensors:

(a) A^{jkl} and A_{pqr}, (b) $A_{j\cdot l}^{\cdot k}$ and A^{qkr}, (c) $A_{\cdot q \cdot t}^{p \cdot rs \cdot}$ and $A_{jqk}^{\cdots sl}$.

Solution

(a) $A^{jkl} = g^{jp} g^{kq} g^{lr} A_{pqr}$ or $A_{pqr} = g_{jp} g_{kq} g_{lr} A^{jkl}$
(b) $A_{j\cdot l}^{\cdot k} = g_{jq} g_{lr} A^{qkr}$ or $A^{qkr} = g^{jq} g^{lr} A_{j\cdot l}^{\cdot k}$
(c) $A_{\cdot q \cdot t}^{p \cdot rs \cdot} = g^{pj} g^{rk} g_{tl} A_{jqk}^{\cdots sl}$ or $A_{jqk}^{\cdots sl} = g_{pj} g_{rk} g^{tl} A_{\cdot q \cdot t}^{p \cdot rs \cdot}$

8.41. Prove that the angles $\theta_{12}, \theta_{23},$ and θ_{31} between the coordinate curves in a three-dimensional coordinate system are given by

$$\cos \theta_{12} = \frac{g_{12}}{\sqrt{g_{11} g_{22}}}, \qquad \cos \theta_{23} = \frac{g_{23}}{\sqrt{g_{22} g_{33}}}, \qquad \cos \theta_{31} = \frac{g_{31}}{\sqrt{g_{33} g_{11}}}$$

Solution

Along the x^1 coordinate curve, $x^2 = $ constant and $x^3 = $ constant.

Then, from the metric form, $ds^2 = g_{11}(dx^1)^2$ or $\dfrac{dx^1}{ds} = \dfrac{1}{\sqrt{g_{11}}}$.

Thus, a unit tangent vector along the x^1 curve is $A_1^r = \dfrac{1}{\sqrt{g_{11}}} \delta_1^r$. Similarly, unit tangent vectors along the x^2 and x^3 coordinate curves are $A_2^r = \dfrac{1}{\sqrt{g_{22}}} \delta_2^r$ and $A_3^r = \dfrac{1}{\sqrt{g_{33}}} \delta_3^r$.

The cosine of the angle θ_{12} between A_1^r and A_2^r is given by

$$\cos\theta_{12} = g_{pq}A_1^p A_2^q = g_{pq}\frac{1}{\sqrt{g_{11}}}\frac{1}{\sqrt{g_{22}}}\delta_1^p\delta_2^q = \frac{g_{12}}{\sqrt{g_{11}g_{22}}}.$$

Similarly, we obtain the other results.

8.42. Prove that for an orthogonal coordinate system, $g_{12} = g_{23} = g_{31} = 0$.

Solution

This follows at once from Problem 8.41 by placing $\theta_{12} = \theta_{23} = \theta_{31} = 90°$. From the fact that $g_{pq} = g_{qp}$, it also follows that $g_{21} = g_{32} = g_{13} = 0$.

8.43. Prove that for an orthogonal coordinate system, $g_{11} = \dfrac{1}{g^{11}}$, $g_{22} = \dfrac{1}{g^{22}}$, $g_{33} = \dfrac{1}{g^{33}}$.

Solution

From Problem 8.33, $g^{pr}g_{rq} = \delta_q^p$.

If $p = q = 1$, $g^{1r}g_{r1} = 1$ or $g^{11}g_{11} + g^{12}g_{21} + g^{13}g_{31} = 1$.

Then, using Problem 8.42, $g_{11} = \dfrac{1}{g^{11}}$.

Similarly, if $p = q = 2$, $g_{22} = \dfrac{1}{g^{22}}$; and if $p = q = 3$, $g_{33} = \dfrac{1}{g^{33}}$.

Christoffel's Symbols

8.44. Prove (a) $[pq, r] = [qp, r]$, (b) $\left\{\begin{matrix} s \\ pq \end{matrix}\right\} = \left\{\begin{matrix} s \\ qp \end{matrix}\right\}$, (c) $[pq, r] = g_{rs}\left\{\begin{matrix} s \\ pq \end{matrix}\right\}$.

Solution

(a) $[pq, r] = \dfrac{1}{2}\left(\dfrac{\partial g_{pr}}{\partial x^q} + \dfrac{\partial g_{qr}}{\partial x^p} - \dfrac{\partial g_{pq}}{\partial x^r}\right) = \dfrac{1}{2}\left(\dfrac{\partial g_{qr}}{\partial x^p} + \dfrac{\partial g_{pr}}{\partial x^q} - \dfrac{\partial g_{qp}}{\partial x^r}\right) = [qp, r].$

(b) $\left\{\begin{matrix} s \\ pq \end{matrix}\right\} = g^{sr}[pq, r] = g^{sr}[qp, r] = \left\{\begin{matrix} s \\ qp \end{matrix}\right\}$

(c) $g_{ks}\left\{\begin{matrix} s \\ pq \end{matrix}\right\} = g_{ks}g^{sr}[pq, r] = \delta_k^r[pq, r] = [pq, k]$

or

$[pq, k] = g_{ks}\left\{\begin{matrix} s \\ pq \end{matrix}\right\}$; that is, $[pq, r] = g_{rs}\left\{\begin{matrix} s \\ pq \end{matrix}\right\}$.

Note that multiplying $[pq, r]$ by g^{sr} has the effect of replacing r by s, raising this index and replacing square brackets by braces to yield $\left\{\begin{matrix} s \\ pq \end{matrix}\right\}$. Similarly, multiplying $\left\{\begin{matrix} s \\ pq \end{matrix}\right\}$ by g_{rs} or g_{sr} has the effect of replacing s by r, lowering this index and replacing braces by square brackets to yield $[pq, r]$.

8.45. Prove

(a) $\dfrac{\partial g_{pq}}{\partial x^m} = [pm, q] + [qm, p]$, (b) $\dfrac{\partial g^{pq}}{\partial x^m} = -g^{pn}\left\{\begin{matrix} q \\ mn \end{matrix}\right\} - g^{qn}\left\{\begin{matrix} p \\ mn \end{matrix}\right\}$

(c) $\left\{\begin{matrix} p \\ pq \end{matrix}\right\} = \dfrac{\partial}{\partial x^q}\ln\sqrt{g}$

Solution

(a) $[pm, q] + [qm, p] = \dfrac{1}{2}\left(\dfrac{\partial g_{pq}}{\partial x^m} + \dfrac{\partial g_{mq}}{\partial x^p} - \dfrac{\partial g_{pm}}{\partial x^q}\right) + \dfrac{1}{2}\left(\dfrac{\partial g_{qp}}{\partial x^m} + \dfrac{\partial g_{mp}}{\partial x^q} - \dfrac{\partial g_{qm}}{\partial x^p}\right) = \dfrac{\partial g_{pq}}{\partial x^m}$

(b) $\dfrac{\partial}{\partial x^m}\left(g^{jk}g_{ij}\right) = \dfrac{\partial}{\partial x^m}\left(\delta_i^k\right) = 0$. Then

$$g^{jk}\frac{\partial g_{ij}}{\partial x^m} + \frac{\partial g^{jk}}{\partial x^m}g_{ij} = 0 \quad \text{or} \quad g_{ij}\frac{\partial g^{jk}}{\partial x^m} = -g^{jk}\frac{\partial g_{ij}}{\partial x^m}$$

Multiplying by g^{ir},

$$g^{ir}g_{ij}\frac{\partial g^{jk}}{\partial x^m} = -g^{ir}g^{jk}\frac{\partial g_{ij}}{\partial x^m}$$

that is

$$\delta_j^r \frac{\partial g^{jk}}{\partial x^m} = -g^{ir}g^{jk}([im, j] + [jm, i])$$

or

$$\frac{\partial g^{rk}}{\partial x^m} = -g^{ir}\begin{Bmatrix} k \\ im \end{Bmatrix} - g^{jk}\begin{Bmatrix} r \\ jm \end{Bmatrix}$$

and the result follows on replacing r, k, i, j by p, q, n, n, respectively.

(c) From Problem 8.31, $g = g_{jk}G(j, k)$ (sum over k only).

Since $G(j, k)$ does not contain g_{jk} explicitly, $\dfrac{\partial g}{\partial g_{jr}} = G(j, r)$. Then, summing over j and r,

$$\frac{\partial g}{\partial x^m} = \frac{\partial g}{\partial g_{jr}}\frac{\partial g_{jr}}{\partial x^m} = G(j, r)\frac{\partial g_{jr}}{\partial x^m}$$

$$= gg^{jr}\frac{\partial g_{jr}}{\partial x^m} = gg^{jr}([jm, r] + [rm, j])$$

$$= g\left(\begin{Bmatrix} j \\ jm \end{Bmatrix} + \begin{Bmatrix} r \\ rm \end{Bmatrix}\right) = 2g\begin{Bmatrix} j \\ jm \end{Bmatrix}$$

Thus

$$\frac{1}{2g}\frac{\partial g}{\partial x^m} = \begin{Bmatrix} j \\ jm \end{Bmatrix} \quad \text{or} \quad \begin{Bmatrix} j \\ jm \end{Bmatrix} = \frac{\partial}{\partial x^m}\ln\sqrt{g}$$

The result follows on replacing j by p and m by q.

8.46. Derive transformation laws for the Christoffel symbols of (a) the first kind, (b) the second kind.

Solution

(a) Since $\bar{g}_{jk} = \dfrac{\partial x^p}{\partial \bar{x}^j}\dfrac{\partial x^q}{\partial \bar{x}^k}g_{pq}$,

$$\frac{\partial \bar{g}_{jk}}{\partial \bar{x}^m} = \frac{\partial x^p}{\partial \bar{x}^j}\frac{\partial x^q}{\partial \bar{x}^k}\frac{\partial g_{pq}}{\partial x^r}\frac{\partial x^r}{\partial \bar{x}^m} + \frac{\partial x^p}{\partial \bar{x}^j}\frac{\partial^2 x^q}{\partial \bar{x}^m \partial \bar{x}^k}g_{pq} + \frac{\partial^2 x^p}{\partial \bar{x}^m \partial \bar{x}^j}\frac{\partial x^q}{\partial \bar{x}^k}g_{pq} \qquad (1)$$

By cyclic permutation of indices j, k, m and p, q, r,

$$\frac{\partial \bar{g}_{km}}{\partial \bar{x}^j} = \frac{\partial x^q}{\partial \bar{x}^k}\frac{\partial x^r}{\partial \bar{x}^m}\frac{\partial g_{qr}}{\partial x^p}\frac{\partial x^p}{\partial \bar{x}^j} + \frac{\partial x^q}{\partial \bar{x}^k}\frac{\partial^2 x^r}{\partial \bar{x}^j \partial \bar{x}^m}g_{qr} + \frac{\partial^2 x^q}{\partial \bar{x}^j \partial \bar{x}^k}\frac{\partial x^r}{\partial \bar{x}^m}g_{qr} \qquad (2)$$

$$\frac{\partial \bar{g}_{mj}}{\partial \bar{x}^k} = \frac{\partial x^r}{\partial \bar{x}^m}\frac{\partial x^p}{\partial \bar{x}^j}\frac{\partial g_{rp}}{\partial x^q}\frac{\partial x^q}{\partial \bar{x}^k} + \frac{\partial x^r}{\partial \bar{x}^m}\frac{\partial^2 x^p}{\partial \bar{x}^k \partial \bar{x}^j}g_{rp} + \frac{\partial^2 x^r}{\partial \bar{x}^k \partial \bar{x}^m}\frac{\partial x^p}{\partial \bar{x}^j}g_{rp} \qquad (3)$$

Subtracting (1) from the sum of (2) and (3) and multiplying by $\frac{1}{2}$, we obtain on using the definition of the Christoffel symbols of the first kind,

$$\overline{[jk, m]} = \frac{\partial x^p}{\partial \overline{x}^j} \frac{\partial x^q}{\partial \overline{x}^k} \frac{\partial x^r}{\partial \overline{x}^m} [pq, r] + \frac{\partial^2 x^p}{\partial \overline{x}^j \partial \overline{x}^k} \frac{\partial x^q}{\partial \overline{x}^m} g_{pq} \tag{4}$$

(b) Multiply (4) by $\overline{g}^{nm} = \frac{\partial \overline{x}^n}{\partial x^s} \frac{\partial \overline{x}^m}{\partial x^t} g^{st}$ to obtain

$$\overline{g}^{nm} \overline{[jk, m]} = \frac{\partial x^p}{\partial \overline{x}^j} \frac{\partial x^q}{\partial \overline{x}^k} \frac{\partial x^r}{\partial \overline{x}^m} \frac{\partial \overline{x}^n}{\partial x^s} \frac{\partial \overline{x}^m}{\partial x^t} g^{st} [pq, r] + \frac{\partial^2 x^p}{\partial \overline{x}^j \partial \overline{x}^k} \frac{\partial x^q}{\partial \overline{x}^m} \frac{\partial \overline{x}^n}{\partial x^s} \frac{\partial \overline{x}^m}{\partial x^t} g^{st} g_{pq}$$

Then

$$\overline{\begin{Bmatrix} n \\ jk \end{Bmatrix}} = \frac{\partial x^p}{\partial \overline{x}^j} \frac{\partial x^q}{\partial \overline{x}^k} \frac{\partial \overline{x}^n}{\partial x^s} \delta_t^r g^{st} [pq, r] + \frac{\partial^2 x^p}{\partial \overline{x}^j \partial \overline{x}^k} \frac{\partial \overline{x}^n}{\partial x^s} \delta_t^q g^{st} g_{pq}$$

$$= \frac{\partial x^p}{\partial \overline{x}^j} \frac{\partial x^q}{\partial \overline{x}^k} \frac{\partial \overline{x}^n}{\partial x^s} \begin{Bmatrix} s \\ pq \end{Bmatrix} + \frac{\partial^2 x^p}{\partial \overline{x}^j \partial \overline{x}^k} \frac{\partial \overline{x}^n}{\partial x^p}$$

since $\delta_t^r g^{st} [pq, r] = g^{sr} [pq, r] = \begin{Bmatrix} s \\ pq \end{Bmatrix}$ and $\delta_t^q g^{st} g_{pq} = g^{sq} g_{pq} = \delta_p^s$.

8.47. Prove $\dfrac{\partial^2 x^m}{\partial \overline{x}^j \partial \overline{x}^k} = \overline{\begin{Bmatrix} n \\ jk \end{Bmatrix}} \dfrac{\partial x^m}{\partial \overline{x}^n} - \dfrac{\partial x^p}{\partial \overline{x}^j} \dfrac{\partial x^q}{\partial \overline{x}^k} \begin{Bmatrix} m \\ pq \end{Bmatrix}$.

Solution

From Problem 8.46(b), $\overline{\begin{Bmatrix} n \\ jk \end{Bmatrix}} = \dfrac{\partial x^p}{\partial \overline{x}^j} \dfrac{\partial x^q}{\partial \overline{x}^k} \dfrac{\partial \overline{x}^n}{\partial x^s} \begin{Bmatrix} s \\ pq \end{Bmatrix} + \dfrac{\partial^2 x^p}{\partial \overline{x}^j \partial \overline{x}^k} \dfrac{\partial \overline{x}^n}{\partial x^p}$.

Multiplying by $\dfrac{\partial x^m}{\partial \overline{x}^n}$,

$$\overline{\begin{Bmatrix} n \\ jk \end{Bmatrix}} \frac{\partial x^m}{\partial \overline{x}^n} = \frac{\partial x^p}{\partial \overline{x}^j} \frac{\partial x^q}{\partial \overline{x}^k} \delta_s^m \begin{Bmatrix} s \\ pq \end{Bmatrix} + \frac{\partial^2 x^p}{\partial \overline{x}^j \partial \overline{x}^k} \delta_p^m$$

$$= \frac{\partial x^p}{\partial \overline{x}^j} \frac{\partial x^q}{\partial \overline{x}^k} \begin{Bmatrix} m \\ pq \end{Bmatrix} + \frac{\partial^2 x^m}{\partial \overline{x}^j \partial \overline{x}^k}$$

Solving for $\dfrac{\partial^2 x^m}{\partial \overline{x}^j \partial \overline{x}^k}$, the result follows.

8.48. Evaluate the Christoffel symbols of (a) the first kind, (b) the second kind, for spaces where $g_{pq} = 0$ if $p \neq q$.

Solution

(a) If $p = q = r$, $\quad [pq, r] = [pp, p] = \dfrac{1}{2}\left(\dfrac{\partial g_{pp}}{\partial x^p} + \dfrac{\partial g_{pp}}{\partial x^p} - \dfrac{\partial g_{pp}}{\partial x^p}\right) = \dfrac{1}{2}\dfrac{\partial g_{pp}}{\partial x^p}$.

If $p = q \neq r$, $\quad [pq, r] = [pp, r] = \dfrac{1}{2}\left(\dfrac{\partial g_{pr}}{\partial x^p} + \dfrac{\partial g_{pr}}{\partial x^p} - \dfrac{\partial g_{pp}}{\partial x^r}\right) = -\dfrac{1}{2}\dfrac{\partial g_{pp}}{\partial x^r}$.

If $p = r \neq q$, $\quad [pq, r] = [pq, p] = \dfrac{1}{2}\left(\dfrac{\partial g_{pp}}{\partial x^q} + \dfrac{\partial g_{qp}}{\partial x^p} - \dfrac{\partial g_{pq}}{\partial x^p}\right) = \dfrac{1}{2}\dfrac{\partial g_{pp}}{\partial x^q}$.

If p, q, r are distinct, $[pq, r] = 0$.

We have not used the summation convention here.

(b)   By Problem 8.43, $g^{jj} = \dfrac{1}{g_{jj}}$ (not summed). Then

$$\begin{Bmatrix} s \\ pq \end{Bmatrix} = g^{sr}[pq,\, r] = 0 \text{ if } r \neq s, \text{ and } = g^{ss}[pq,\, s] = \frac{[pq,\, s]}{g_{ss}} \text{ (not summed) if } r = s.$$

By (a):

If $p = q = s$, $\begin{Bmatrix} s \\ pq \end{Bmatrix} = \begin{Bmatrix} p \\ pp \end{Bmatrix} = \dfrac{[pp,\, p]}{g_{pp}} = \dfrac{1}{2g_{pp}}\dfrac{\partial g_{pp}}{\partial x^p} = \dfrac{1}{2}\dfrac{\partial}{\partial x^p}\ln g_{pp}.$

If $p = q \neq s$, $\begin{Bmatrix} s \\ pq \end{Bmatrix} = \begin{Bmatrix} s \\ pp \end{Bmatrix} = \dfrac{[pp,\, s]}{g_{ss}} = -\dfrac{1}{2g_{ss}}\dfrac{\partial g_{pp}}{\partial x^s}$

If $p = s \neq q$, $\begin{Bmatrix} s \\ pq \end{Bmatrix} = \begin{Bmatrix} p \\ pq \end{Bmatrix} = \dfrac{[pq,\, p]}{g_{pp}} = \dfrac{1}{2g_{pp}}\dfrac{\partial g_{pp}}{\partial x^q} = \dfrac{1}{2}\dfrac{\partial}{\partial x^q}\ln g_{pp}.$

If p, q, s are distinct, $\begin{Bmatrix} s \\ pq \end{Bmatrix} = 0.$

8.49. Determine the Christoffel symbols of the second kind in (a) rectangular, (b) cylindrical, and (c) spherical coordinates.

Solution

We can use the results of Problem 8.48, since for orthogonal coordinates $g_{pq} = 0$ if $p \neq q$.

(a)   In rectangular coordinates, $g_{pp} = 1$ so that $\begin{Bmatrix} s \\ pq \end{Bmatrix} = 0.$

(b)   In cylindrical coordinates, $x^1 = \rho,\ x^2 = \phi,\ x^3 = z$, we have by Problem 8.30(a), $g_{11} = 1$, $g_{22} = \rho^2$, $g_{33} = 1$. The only non-zero Christoffel symbols of the second kind can occur where $p = 2$. These are

$$\begin{Bmatrix} 1 \\ 22 \end{Bmatrix} = -\frac{1}{2g_{11}}\frac{\partial g_{22}}{\partial x^1} = -\frac{1}{2}\frac{\partial}{\partial \rho}(\rho^2) = -\rho,$$

$$\begin{Bmatrix} 2 \\ 21 \end{Bmatrix} = \begin{Bmatrix} 2 \\ 12 \end{Bmatrix} = \frac{1}{2g_{22}}\frac{\partial g_{22}}{\partial x^1} = \frac{1}{2\rho^2}\frac{\partial}{\partial \rho}(\rho^2) = \frac{1}{\rho}$$

(c)   In spherical coordinates, $x^1 = r,\ x^2 = \theta,\ x^3 = \phi$, we have by Problem 8.30(b), $g_{11} = 1$, $g_{22} = r^2$, $g_{33} = r^2 \sin^2\theta$. The only non-zero Christoffel symbols of the second kind can occur where $p = 2$ or 3. These are

$$\begin{Bmatrix} 1 \\ 22 \end{Bmatrix} = -\frac{1}{2g_{11}}\frac{\partial g_{22}}{\partial x^1} = -\frac{1}{2}\frac{\partial}{\partial r}(r^2) = -r$$

$$\begin{Bmatrix} 2 \\ 21 \end{Bmatrix} = \begin{Bmatrix} 2 \\ 12 \end{Bmatrix} = \frac{1}{2g_{22}}\frac{\partial g_{22}}{\partial x^1} = \frac{1}{2r^2}\frac{\partial}{\partial r}(r^2) = \frac{1}{r}$$

$$\begin{Bmatrix} 1 \\ 33 \end{Bmatrix} = -\frac{1}{2g_{11}}\frac{\partial g_{33}}{\partial x^1} = -\frac{1}{2}\frac{\partial}{\partial r}(r^2 \sin^2\theta) = -r\sin^2\theta$$

$$\begin{Bmatrix} 2 \\ 33 \end{Bmatrix} = -\frac{1}{2g_{22}}\frac{\partial g_{33}}{\partial x^2} = -\frac{1}{2r^2}\frac{\partial}{\partial \theta}(r^2 \sin^2\theta) = -\sin\theta\cos\theta$$

$$\begin{Bmatrix} 3 \\ 31 \end{Bmatrix} = \begin{Bmatrix} 3 \\ 13 \end{Bmatrix} = \frac{1}{2g_{33}}\frac{\partial g_{33}}{\partial x^1} = \frac{1}{2r^2\sin^2\theta}\frac{\partial}{\partial r}(r^2 \sin^2\theta) = \frac{1}{r}$$

$$\begin{Bmatrix} 3 \\ 32 \end{Bmatrix} = \begin{Bmatrix} 3 \\ 23 \end{Bmatrix} = \frac{1}{2g_{33}}\frac{\partial g_{33}}{\partial x^2} = \frac{1}{2r^2\sin^2\theta}\frac{\partial}{\partial \theta}(r^2 \sin^2\theta) = \cot\theta$$

Geodesics

8.50. Prove that a necessary condition that $I = \int_{t_1}^{t_2} F(t, x, \dot{x})\, dt$ be an extremum (maximum or minimum) is that $\dfrac{\partial F}{\partial x} - \dfrac{d}{dt}\left(\dfrac{\partial F}{\partial \dot{x}}\right) = 0$.

Solution

Let the curve which makes I an extremum be $x = X(t)$, $t_1 \leq t \leq t_2$. Then, $x = X(t) + \epsilon\eta(t)$, where ϵ is independent of t, is a neighboring curve through t_1 and t_2 so that $\eta(t_1) = \eta(t_2) = 0$. The value of I for the neighboring curve is

$$I(\epsilon) = \int_{t_1}^{t_2} F(t, X + \epsilon\eta, \dot{X} + \epsilon\dot{\eta})\, dt$$

This is an extremum for $\epsilon = 0$. A necessary condition that this be so is that $\dfrac{dI}{d\epsilon}\Big|_{\epsilon=0} = 0$. But by differentiation under the integral sign, assuming this valid,

$$\frac{dI}{d\epsilon}\Big|_{\epsilon=0} = \int_{t_1}^{t_2}\left(\frac{\partial F}{\partial x}\eta + \frac{\partial F}{\partial \dot{x}}\dot{\eta}\right) dt = 0$$

which can be written as

$$\int_{t_1}^{t_2} \frac{\partial F}{\partial x}\eta\, dt + \frac{\partial F}{\partial \dot{x}}\eta\Big|_{t_1}^{t_2} - \int_{t_1}^{t_2} \eta\frac{d}{dt}\left(\frac{\partial F}{\partial \dot{x}}\right) dt = \int_{t_1}^{t_2}\eta\left(\frac{\partial F}{\partial x} - \frac{d}{dt}\left(\frac{\partial F}{\partial \dot{x}}\right)\right) dt = 0$$

Since η is arbitrary, the integrand $\dfrac{\partial F}{\partial x} - \dfrac{d}{dt}\left(\dfrac{\partial F}{\partial \dot{x}}\right) = 0$.

The result is easily extended to the integral $\int_{t_1}^{t_2} F(t, x^1, \dot{x}^1, x^2, \dot{x}^2, \ldots, x^N, \dot{x}^N)\, dt$ and yields

$$\frac{\partial F}{\partial x^k} - \frac{d}{dt}\left(\frac{\partial F}{\partial \dot{x}^k}\right) = 0$$

called *Euler's* or *Lagrange's equations* (see also Problem 8.73).

8.51. Show that the geodesics in a Riemannian space are given by $\dfrac{d^2 x^r}{ds^2} + \left\{\begin{matrix} r \\ pq \end{matrix}\right\}\dfrac{dx^p}{ds}\dfrac{dx^q}{ds} = 0$.

Solution

We must determine the extremum of $\int_{t_1}^{t_2} \sqrt{g_{pq}\dot{x}^p\dot{x}^q}\, dt$ using Euler's equations (Problem 8.50) with $F = \sqrt{g_{pq}\dot{x}^p\dot{x}^q}$. We have

$$\frac{\partial F}{\partial x^k} = \frac{1}{2}(g_{pq}\dot{x}^p\dot{x}^q)^{-1/2}\frac{\partial g_{pq}}{\partial x^k}\dot{x}^p\dot{x}^q, \quad \frac{\partial F}{\partial \dot{x}^k} = \frac{1}{2}(g_{pq}\dot{x}^p\dot{x}^q)^{-1/2}2g_{pk}\dot{x}^p$$

Using $\dfrac{ds}{dt} = \sqrt{g_{pq}\dot{x}^p\dot{x}^q}$, Euler's equations can be written

$$\frac{d}{dt}\left(\frac{g_{pk}\dot{x}^p}{\dot{s}}\right) - \frac{1}{2\dot{s}}\frac{\partial g_{pq}}{\partial x^k}\dot{x}^p\dot{x}^q = 0$$

or

$$g_{pk}\ddot{x}^p + \frac{\partial g_{pk}}{\partial x^q}\dot{x}^p\dot{x}^q - \frac{1}{2}\frac{\partial g_{pq}}{\partial x^k}\dot{x}^p\dot{x}^q = \frac{g_{pk}\dot{x}^p\ddot{s}}{\dot{s}}$$

Writing $\dfrac{\partial g_{pk}}{\partial x^q}\dot{x}^p\dot{x}^q = \dfrac{1}{2}\left(\dfrac{\partial g_{pk}}{\partial x^q}+\dfrac{\partial g_{qk}}{\partial x^p}\right)\dot{x}^p\dot{x}^q$ this equation becomes

$$g_{pk}\ddot{x}^p + [pq, k]\dot{x}^p\dot{x}^q = \frac{g_{pk}\dot{x}^p\ddot{s}}{\dot{s}}$$

If we use arc length as parameter, $\dot{s}=1$, $\ddot{s}=0$ and the equation becomes

$$g_{pk}\frac{d^2x^p}{ds^2} + [pq, k]\frac{dx^p}{ds}\frac{dx^q}{ds} = 0$$

Multiplying by g^{rk}, we obtain

$$\frac{d^2x^r}{ds^2} + \left\{\begin{array}{c} r \\ pq \end{array}\right\}\frac{dx^p}{ds}\frac{dx^q}{ds} = 0$$

The Covariant Derivative

8.52. Suppose A_p and A^p are tensors. Show that (a) $A_{p,q} \equiv \dfrac{\partial A_p}{\partial x^q} - \left\{\begin{array}{c} s \\ pq \end{array}\right\}A_s$

and (b) $A^p_{,q} \equiv \dfrac{\partial A^p}{\partial x^q} + \left\{\begin{array}{c} p \\ qs \end{array}\right\}A^s$ are tensors.

Solution

(a) Since $\overline{A}_j = \dfrac{\partial x^r}{\partial \overline{x}^j}A_r$,

$$\frac{\partial \overline{A}_j}{\partial \overline{x}^k} = \frac{\partial x^r}{\partial \overline{x}^j}\frac{\partial A_r}{\partial x^t}\frac{\partial x^t}{\partial \overline{x}^k} + \frac{\partial^2 x^r}{\partial \overline{x}^j\partial \overline{x}^k}A_r \tag{1}$$

From Problem 8.47,

$$\frac{\partial^2 x^r}{\partial \overline{x}^j\partial \overline{x}^k} = \left\{\begin{array}{c} n \\ jk \end{array}\right\}\frac{\partial x^r}{\partial \overline{x}^n} - \frac{\partial x^i}{\partial \overline{x}^j}\frac{\partial x^l}{\partial \overline{x}^k}\left\{\begin{array}{c} r \\ il \end{array}\right\}$$

Substituting in (1),

$$\frac{\partial \overline{A}_j}{\partial \overline{x}^k} = \frac{\partial x^r}{\partial \overline{x}^j}\frac{\partial x^t}{\partial \overline{x}^k}\frac{\partial A_r}{\partial x^t} + \overline{\left\{\begin{array}{c} n \\ jk \end{array}\right\}\frac{\partial x^r}{\partial \overline{x}^n}}A_r - \frac{\partial x^i}{\partial \overline{x}^j}\frac{\partial x^l}{\partial \overline{x}^k}\left\{\begin{array}{c} r \\ il \end{array}\right\}A_r$$

$$= \frac{\partial x^p}{\partial \overline{x}^j}\frac{\partial x^q}{\partial \overline{x}^k}\frac{\partial A_p}{\partial x^q} + \overline{\left\{\begin{array}{c} n \\ jk \end{array}\right\}}\overline{A}_n - \frac{\partial x^p}{\partial \overline{x}^j}\frac{\partial x^q}{\partial \overline{x}^k}\left\{\begin{array}{c} s \\ pq \end{array}\right\}A_s$$

or

$$\frac{\partial \overline{A}_j}{\partial \overline{x}^k} - \overline{\left\{\begin{array}{c} n \\ jk \end{array}\right\}}\overline{A}_n = \frac{\partial x^p}{\partial \overline{x}^j}\frac{\partial x^q}{\partial \overline{x}^k}\left(\frac{\partial A_p}{\partial x^q} - \left\{\begin{array}{c} s \\ pq \end{array}\right\}A_s\right)$$

and $\dfrac{\partial A_p}{\partial x^q} - \left\{\begin{array}{c} s \\ pq \end{array}\right\}A_s$ is a covariant tensor of second rank, called the *covariant derivative* of A_p with *respect to* x^q and written $A_{p,q}$.

(b) Since $\overline{A}^j = \dfrac{\partial \overline{x}^j}{\partial x^r}A^r$,

$$\frac{\partial \overline{A}^j}{\partial \overline{x}^k} = \frac{\partial \overline{x}^j}{\partial x^r}\frac{\partial A^r}{\partial x^t}\frac{\partial x^t}{\partial \overline{x}^k} + \frac{\partial^2 \overline{x}^j}{\partial x^r\partial x^t}\frac{\partial x^t}{\partial \overline{x}^k}A^r \tag{2}$$

From Problem 8.47, interchanging x and $\overline{x}$ coordinates,

$$\frac{\partial^2 \overline{x}^j}{\partial x^r\partial x^t} = \left\{\begin{array}{c} n \\ rt \end{array}\right\}\frac{\partial \overline{x}^j}{\partial x^n} - \frac{\partial \overline{x}^i}{\partial x^r}\frac{\partial \overline{x}^l}{\partial x^t}\overline{\left\{\begin{array}{c} i \\ il \end{array}\right\}}$$

Substituting in (2),

$$\frac{\partial \bar{A}^j}{\partial \bar{x}^k} = \frac{\partial \bar{x}^j}{\partial x^r}\frac{\partial x^t}{\partial \bar{x}^k}\frac{\partial A^r}{\partial x^t} + \left\{ {n \atop rt} \right\}\frac{\partial \bar{x}^j}{\partial x^n}\frac{\partial x^t}{\partial \bar{x}^k}A^r - \frac{\partial \bar{x}^i}{\partial x^r}\frac{\partial \bar{x}^l}{\partial x^t}\frac{\partial x^t}{\partial \bar{x}^k}\overline{\left\{ {j \atop il} \right\}}A^r$$

$$= \frac{\partial \bar{x}^j}{\partial x^r}\frac{\partial x^t}{\partial \bar{x}^k}\frac{\partial A^r}{\partial x^t} + \left\{ {n \atop rt} \right\}\frac{\partial \bar{x}^j}{\partial x^n}\frac{\partial x^t}{\partial \bar{x}^k}A^r - \frac{\partial \bar{x}^i}{\partial x^r}\delta_k^l\overline{\left\{ {j \atop il} \right\}}A^r$$

$$= \frac{\partial \bar{x}^j}{\partial x^p}\frac{\partial x^q}{\partial \bar{x}^k}\frac{\partial A^p}{\partial x^q} + \left\{ {p \atop sq} \right\}\frac{\partial \bar{x}^j}{\partial x^p}\frac{\partial x^q}{\partial \bar{x}^k}A^s - \overline{\left\{ {j \atop ik} \right\}}\bar{A}^i$$

or

$$\frac{\partial \bar{A}^j}{\partial \bar{x}^k} + \overline{\left\{ {j \atop ki} \right\}}\bar{A}^i = \frac{\partial \bar{x}^j}{\partial x^p}\frac{\partial x^q}{\partial \bar{x}^k}\left(\frac{\partial A^p}{\partial x^q} + \left\{ {p \atop qs} \right\}A^s \right)$$

and $\dfrac{\partial A^p}{\partial x^q} + \left\{ {p \atop qs} \right\}A^s$ is a mixed tensor of second rank, called the *covariant derivative of A^p with respect to* x^q and written $A^p_{,q}$.

8.53. Write the covariant derivative with respect to x^q of each of the following tensors:

(a) A_{jk}, (b) A^{jk}, (c) A^j_k, (d) A^j_{kl}, (e) A^{jkl}_{mn}.

Solution

(a) $A_{jk,q} = \dfrac{\partial A_{jk}}{\partial x^q} - \left\{ {s \atop jq} \right\}A_{sk} - \left\{ {s \atop kq} \right\}A_{js}$ (b) $A^{jk}_{,q} = \dfrac{\partial A^{jk}}{\partial x^q} + \left\{ {j \atop qs} \right\}A^{sk} + \left\{ {k \atop qs} \right\}A^{js}$

(c) $A^j_{k,q} = \dfrac{\partial A^j_k}{\partial x^q} - \left\{ {s \atop kq} \right\}A^j_s + \left\{ {j \atop qs} \right\}A^s_k$ (d) $A^j_{kl,q} = \dfrac{\partial A^j_{kl}}{\partial x^q} - \left\{ {s \atop kq} \right\}A^j_{sl} - \left\{ {s \atop lq} \right\}A^j_{ks} + \left\{ {j \atop qs} \right\}A^s_{kl}$

(e) $A^{jkl}_{mn,q} = \dfrac{\partial A^{jkl}_{mn}}{\partial x^q} - \left\{ {s \atop mq} \right\}A^{jkl}_{sn} - \left\{ {s \atop nq} \right\}A^{jkl}_{ms} + \left\{ {j \atop qs} \right\}A^{skl}_{mn} + \left\{ {k \atop qs} \right\}A^{jsl}_{mn} + \left\{ {l \atop qs} \right\}A^{jks}_{mn}$

8.54. Prove that the covariant derivatives of the following are zero: (a) g_{jk}, (b) g^{jk}, (c) δ^j_k.

Solution

(a) $g_{jk,q} = \dfrac{\partial g_{jk}}{\partial x^q} - \left\{ {s \atop jq} \right\}g_{sk} - \left\{ {s \atop kq} \right\}g_{js} = \dfrac{\partial g_{jk}}{\partial x^q} - [jq, k] - [kq, j] = 0$ by Problem 8.45(a).

(b) $g^{jk}_{,q} = \dfrac{\partial g^{jk}}{\partial x^q} + \left\{ {j \atop qs} \right\}g^{sk} + \left\{ {k \atop qs} \right\}g^{js} = 0$ by Problem 8.45(b).

(c) $\delta^j_{k,q} = \dfrac{\partial \delta^j_k}{\partial x^q} - \left\{ {s \atop kq} \right\}\delta^j_s + \left\{ {j \atop qs} \right\}\delta^s_k = 0 - \left\{ {j \atop kq} \right\} + \left\{ {j \atop qk} \right\} = 0.$

8.55. Find the covariant derivative of $A^j_k B^{lm}_n$ with respect to x^q.

Solution

$$\left(A^j_k B^{lm}_n \right)_{,q} = \frac{\partial \left(A^j_k B^{lm}_n \right)}{\partial x^q} - \left\{ {s \atop kq} \right\}A^j_s B^{lm}_n - \left\{ {s \atop nq} \right\}A^j_k B^{lm}_s + \left\{ {j \atop qs} \right\}A^s_k B^{lm}_n + \left\{ {l \atop qs} \right\}A^j_k B^{sm}_n + \left\{ {m \atop qs} \right\}A^j_k B^{ls}_n$$

$$= \left(\frac{\partial A^j_k}{\partial x^q} - \left\{ {s \atop kq} \right\}A^j_s + \left\{ {j \atop qs} \right\}A^s_k \right)B^{lm}_n + A^j_k\left(\frac{\partial B^{lm}_n}{\partial x^q} - \left\{ {s \atop nq} \right\}B^{lm}_s + \left\{ {l \atop qs} \right\}B^{sm}_n + \left\{ {m \atop qs} \right\}B^{ls}_n \right)$$

$$= A^j_{k,q}B^{lm}_n + A^j_k B^{lm}_{n,q}$$

This illustrates the fact that the covariant derivatives of a product of tensors obey rules like those of ordinary derivatives of products in elementary calculus.

8.56. Prove $\left(g_{jk}A_n^{km}\right)_{,q} = g_{jk}A_{n,q}^{km}$.

Solution

$$\left(g_{jk}A_n^{km}\right)_{,q} = g_{jk,q}A_n^{km} + g_{jk}A_{n,q}^{km} = g_{jk}A_{n,q}^{km}$$

since $g_{jk,q} = 0$ by Problem 8.54(a). In covariant differentiation, g_{jk}, g^{jk}, and δ_k^j can be treated as constants.

Gradient, Divergence and Curl in Tensor Form

8.57. Prove that $\operatorname{div} A^p = \dfrac{1}{\sqrt{g}}\dfrac{\partial}{\partial x^k}(\sqrt{g}A^k)$.

Solution

The divergence of A^p is the contraction of the covariant derivative of A^p, that is, the contraction of $A_{,q}^p$ or $A_{,p}^p$. Then, using Problem 8.45(c),

$$\operatorname{div} A^p = A_{,p}^p = \frac{\partial A^k}{\partial x^k} + \left\{\begin{matrix} p \\ pk \end{matrix}\right\}A^k$$

$$= \frac{\partial A^k}{\partial x^k} + \left(\frac{\partial}{\partial x^k}\ln\sqrt{g}\right)A^k = \frac{\partial A^k}{\partial x^k} + \left(\frac{1}{\sqrt{g}}\frac{\partial\sqrt{g}}{\partial x^k}\right)A^k = \frac{1}{\sqrt{g}}\frac{\partial}{\partial x^k}\left(\sqrt{g}A^k\right)$$

8.58. Prove that $\nabla^2\Phi = \dfrac{1}{\sqrt{g}}\dfrac{\partial}{\partial x^k}\left(\sqrt{g}g^{kr}\dfrac{\partial\Phi}{\partial x^r}\right)$.

Solution

The gradient of Φ is $\operatorname{grad}\Phi = \nabla\Phi = \partial\Phi/\partial x^r$, a covariant tensor of rank one (see Problem 8.6(b)) defined as the covariant derivative of Φ, written $\Phi_{,r}$. The contravariant tensor of rank one associated with $\Phi_{,r}$ is $A^k = g^{kr}\partial\Phi/\partial x^r$. Then, from Problem 8.57,

$$\nabla^2\Phi = \operatorname{div}\left(g^{kr}\frac{\partial\Phi}{\partial x^r}\right) = \frac{1}{\sqrt{g}}\frac{\partial}{\partial x^k}\left(\sqrt{g}g^{kr}\frac{\partial\Phi}{\partial x^r}\right)$$

8.59. Prove that $A_{p,q} - A_{q,p} = \dfrac{\partial A_p}{\partial x^q} - \dfrac{\partial A_q}{\partial x^p}$.

Solution

$$A_{p,q} - A_{q,p} = \left(\frac{\partial A_p}{\partial x^q} - \left\{\begin{matrix} s \\ pq \end{matrix}\right\}A_s\right) - \left(\frac{\partial A_q}{\partial x^p} - \left\{\begin{matrix} s \\ qp \end{matrix}\right\}A_s\right) = \frac{\partial A_p}{\partial x^q} - \frac{\partial A_q}{\partial x^p}$$

This tensor of rank two is defined to be the curl of A_p.

8.60. Express the divergence of a vector A^p in terms of its physical components for (a) cylindrical coordinates, (b) spherical coordinates.

Solution

(a) For cylindrical coordinates $x^1 = \rho$, $x^2 = \phi$, $x^3 = z$,

$$g = \begin{vmatrix} 1 & 0 & 0 \\ 0 & \rho^2 & 0 \\ 0 & 0 & 1 \end{vmatrix} = \rho^2 \text{ and } \sqrt{g} = \rho \quad \text{(see Problem 8.30(a))}$$

The physical components, denoted by A_ρ, A_ϕ, A_z are given by

$$A_\rho = \sqrt{g_{11}}A^1 = A^1, \quad A_\phi = \sqrt{g_{22}}A^2 = \rho A^2, \quad A_z = \sqrt{g_{33}}A^3 = A^3$$

Then

$$\text{div } A^P = \frac{1}{\sqrt{g}}\frac{\partial}{\partial x^k}\left(\sqrt{g}A^k\right) = \frac{1}{\rho}\left[\frac{\partial}{\partial\rho}(\rho A_\rho) + \frac{\partial}{\partial\phi}(A_\phi) + \frac{\partial}{\partial z}(\rho A_z)\right]$$

(b)   For spherical coordinates $x^1 = r$, $x^2 = \theta$, $x^3 = \phi$,

$$g = \begin{vmatrix} 1 & 0 & 0 \\ 0 & r^2 & 0 \\ 0 & 0 & r^2\sin^2\theta \end{vmatrix} = r^4\sin^2\theta \quad \text{and} \quad \sqrt{g} = r^2\sin\theta \quad \text{(see Problem 8.30(b))}$$

The physical components, denoted by A_r, A_θ, A_ϕ are given by

$$A_r = \sqrt{g_{11}}A^1 = A^1, \quad A_\theta = \sqrt{g_{22}}A^2 = rA^2, \quad A_\phi = \sqrt{g_{33}}A^3 = r\sin\theta\, A^3$$

Then

$$\text{div } A^P = \frac{1}{\sqrt{g}}\frac{\partial}{\partial x^k}(\sqrt{g}A^k)$$

$$= \frac{1}{r^2\sin\theta}\left[\frac{\partial}{\partial r}(r^2\sin\theta A_r) + \frac{\partial}{\partial\theta}(r\sin\theta A_\theta) + \frac{\partial}{\partial\phi}(rA_\phi)\right]$$

$$= \frac{1}{r^2}\frac{\partial}{\partial r}(r^2A_r) + \frac{1}{r\sin\theta}\frac{\partial}{\partial\theta}(\sin\theta\, A_\theta) + \frac{1}{r\sin\theta}\frac{\partial A_\phi}{\partial\phi}$$

8.61. Express the Laplacian of Φ, $\nabla^2\Phi$, in (a) cylindrical coordinates, (b) spherical coordinates.

Solution

(a)   In cylindrical coordinates $g^{11} = 1$, $g^{22} = 1/\rho^2$, $g^{33} = 1$ (see Problem 8.35(a)). Then from Problem 8.58,

$$\nabla^2\Phi = \frac{1}{\sqrt{g}}\frac{\partial}{\partial x^k}\left(\sqrt{g}g^{kr}\frac{\partial\Phi}{\partial x^r}\right)$$

$$= \frac{1}{\rho}\left[\frac{\partial}{\partial\rho}\left(\rho\frac{\partial\Phi}{\partial\rho}\right) + \frac{\partial}{\partial\phi}\left(\frac{1}{\rho}\frac{\partial\Phi}{\partial\phi}\right) + \frac{\partial}{\partial z}\left(\rho\frac{\partial\Phi}{\partial z}\right)\right]$$

$$= \frac{1}{\rho}\frac{\partial}{\partial\rho}\left(\rho\frac{\partial\Phi}{\partial\rho}\right) + \frac{1}{\rho^2}\frac{\partial^2\Phi}{\partial\phi^2} + \frac{\partial^2\Phi}{\partial z^2}$$

(b)   In spherical coordinates $g^{11} = 1$, $g^{22} = 1/r^2$, $g^{33} = 1/r^2\sin^2\theta$ (see Problem 8.35(b)). Then

$$\nabla^2\Phi = \frac{1}{\sqrt{g}}\frac{\partial}{\partial x^k}\left(\sqrt{g}\,g^{kr}\frac{\partial\Phi}{\partial x^r}\right)$$

$$= \frac{1}{r^2\sin\theta}\left[\frac{\partial}{\partial r}\left(r^2\sin\theta\frac{\partial\Phi}{\partial r}\right) + \frac{\partial}{\partial\theta}\left(\sin\theta\frac{\partial\Phi}{\partial\theta}\right) + \frac{\partial}{\partial\phi}\left(\frac{1}{\sin\theta}\frac{\partial\Phi}{\partial\phi}\right)\right]$$

$$= \frac{1}{r^2}\frac{\partial}{\partial r}\left(r^2\frac{\partial\Phi}{\partial r}\right) + \frac{1}{r^2\sin\theta}\frac{\partial}{\partial\theta}\left(\sin\theta\frac{\partial\Phi}{\partial\theta}\right) + \frac{1}{r^2\sin^2\theta}\frac{\partial^2\Phi}{\partial\phi^2}$$

Intrinsic Derivatives

8.62. Calculate the intrinsic derivatives of each of the following tensors, assumed to be differentiable functions of t: (a) an invariant Φ, (b) A^j, (c) A^j_k, (d) A^{jk}_{lmn}.

Solution

(a) $\dfrac{\delta\Phi}{\delta t} = \Phi_{,q}\dfrac{dx^q}{dt} = \dfrac{\partial\Phi}{\partial x^q}\dfrac{dx^q}{dt} = \dfrac{d\Phi}{dt}$, the ordinary derivative.

(b) $\dfrac{\delta A^j}{\delta t} = A^j_{,q}\dfrac{dx^q}{dt} = \left(\dfrac{\partial A^j}{\partial x^q} + \left\{{j \atop qs}\right\}A^s\right)\dfrac{dx^q}{dt} = \dfrac{\partial A^j}{\partial x^q}\dfrac{dx^q}{dt} + \left\{{j \atop qs}\right\}A^s\dfrac{dx^q}{dt}$

$\qquad = \dfrac{dA^j}{dt} + \left\{{j \atop qs}\right\}A^s\dfrac{dx^q}{dt}$

(c) $\dfrac{\delta A^j_k}{\delta t} = A^j_{k,q}\dfrac{dx^q}{dt} = \left(\dfrac{\partial A^j_k}{\partial x^q} - \left\{{s \atop kq}\right\}A^j_s + \left\{{j \atop qs}\right\}A^s_k\right)\dfrac{dx^q}{dt}$

$\qquad = \dfrac{dA^j_k}{dt} - \left\{{s \atop kq}\right\}A^j_s\dfrac{dx^q}{dt} + \left\{{j \atop qs}\right\}A^s_k\dfrac{dx^q}{dt}$

(d) $\dfrac{\delta A^{jk}_{lmn}}{\delta t} = A^{jk}_{lmn,q}\dfrac{dx^q}{dt} = \left(\dfrac{\partial A^{jk}_{lmn}}{\partial x^q} - \left\{{s \atop lq}\right\}A^{jk}_{smn} - \left\{{s \atop mq}\right\}A^{jk}_{lsn}\right.$

$\qquad\qquad \left. - \left\{{s \atop nq}\right\}A^{jk}_{lms} + \left\{{j \atop qs}\right\}A^{sk}_{lmn} + \left\{{k \atop qs}\right\}A^{js}_{lmn}\right)\dfrac{dx^q}{dt}$

$\qquad = \dfrac{dA^{jk}_{lmn}}{dt} - \left\{{s \atop lq}\right\}A^{jk}_{smn}\dfrac{dx^q}{dt} - \left\{{s \atop mq}\right\}A^{jk}_{lsn}\dfrac{dx^q}{dt} - \left\{{s \atop nq}\right\}A^{jk}_{lms}\dfrac{dx^q}{dt}$

$\qquad\quad + \left\{{j \atop qs}\right\}A^{sk}_{lmn}\dfrac{dx^q}{dt} + \left\{{k \atop qs}\right\}A^{js}_{lmn}\dfrac{dx^q}{dt}$

8.63. Prove the intrinsic derivatives of g_{jk}, g^{jk}, and δ^j_k are zero.

Solution

$$\dfrac{\delta g_{jk}}{\delta t} = (g_{jk,q})\dfrac{dx^q}{dt} = 0, \quad \dfrac{\delta g^{jk}}{\delta t} = g^{jk}_{,q}\dfrac{dx^q}{dt} = 0, \quad \dfrac{\delta\delta^j_k}{\delta t} = \delta^j_{k,q}\dfrac{dx^q}{dt} = 0 \text{ by Problem 8.54.}$$

Relative Tensors

8.64. Let A^p_q and B^{rs}_t be relative tensors of weights w_1 and w_2, respectively. Show that their inner and outer products are relative tensors of weight $w_1 + w_2$.

Solution

By hypothesis,

$$\overline{A}^j_k = J^{w_1}\dfrac{\partial\overline{x}^j}{\partial x^p}\dfrac{\partial x^q}{\partial\overline{x}^k}A^p_q, \quad \overline{B}^{lm}_n = J^{w_2}\dfrac{\partial\overline{x}^l}{\partial x^r}\dfrac{\partial\overline{x}^m}{\partial x^s}\dfrac{\partial x^t}{\partial\overline{x}^n}B^{rs}_t$$

The outer product is

$$\overline{A}^j_k\overline{B}^{lm}_n = J^{w_1+w_2}\dfrac{\partial\overline{x}^j}{\partial x^p}\dfrac{\partial x^q}{\partial\overline{x}^k}\dfrac{\partial\overline{x}^l}{\partial x^r}\dfrac{\partial\overline{x}^m}{\partial x^s}\dfrac{\partial x^t}{\partial\overline{x}^n}A^p_qB^{rs}_t$$

a relative tensor of weight $w_1 + w_2$. Any inner product, which is a contraction of the outer product, is also a relative tensor of weight $w_1 + w_2$.

8.65. Prove that $\sqrt{g}$ is a relative tensor of weight one, i.e. a tensor density.

Solution

The elements of determinant g given by g_{pq} transform according to $\bar{g}_{jk} = \dfrac{\partial x^p}{\partial \bar{x}^j} \dfrac{\partial x^q}{\partial \bar{x}^k} g_{pq}$.

Taking determinants of both sides, $\bar{g} = \left| \dfrac{\partial x^p}{\partial \bar{x}^j} \right| \left| \dfrac{\partial x^q}{\partial \bar{x}^k} \right| g = J^2 g$ or $\sqrt{\bar{g}} = J\sqrt{g}$, which shows that $\sqrt{g}$ is a relative tensor of weight one.

8.66. Prove that $dV = \sqrt{g}\, dx^1 dx^2 \cdots dx^N$ is an invariant.

Solution

By Problem 8.65,

$$d\bar{V} = \sqrt{\bar{g}}\, d\bar{x}^1 d\bar{x}^2 \cdots d\bar{x}^N = \sqrt{g}J\, d\bar{x}^1 d\bar{x}^2 \cdots d\bar{x}^N$$

$$= \sqrt{g} \left| \dfrac{\partial x}{\partial \bar{x}} \right| d\bar{x}^1 d\bar{x}^2 \cdots d\bar{x}^N = \sqrt{g}\, dx^1 dx^2 \cdots dx^N = dV$$

From this it follows that if Φ is an invariant, then

$$\int \cdots \int_{\bar{V}} \bar{\Phi}\, d\bar{V} = \int \cdots \int_{V} \Phi\, dV$$

for any coordinate systems where the integration is performed over a volume in N-dimensional space. A similar statement can be made for surface integrals.

Miscellaneous Applications

8.67. Express in tensor form (a) the velocity and (b) the acceleration of a particle.

Solution

(a)   If the particle moves along a curve $x^k = x^k(t)$ where t is the parameter time, then $v^k = \dfrac{dx^k}{dt}$ is its velocity and is a contravariant tensor of rank one (see Problem 8.9).

(b)   The quantity $\dfrac{dv^k}{dt} = \dfrac{d^2 x^k}{dt^2}$ is not in general a tensor and so cannot represent the physical quantity acceleration in all coordinate systems. We define the acceleration a^k as the intrinsic derivative of the velocity, that is $a^k = \dfrac{\delta v^k}{\delta t}$ which is a contravariant tensor of rank one.

8.68. Write Newton's law in tensor form.

Solution

Assume the mass M of the particle to be an invariant independent of time t. Then, $Ma^k = F^k$, a contravariant tensor of rank one, is called the *force* on the particle. Thus Newton's law can be written

$$F^k = Ma^k = M\dfrac{\delta v^k}{\delta t}$$

8.69. Prove that $a^k = \dfrac{\delta v^k}{\delta t} = \dfrac{d^2 x^k}{dt^2} + \left\{ \begin{matrix} k \\ pq \end{matrix} \right\} \dfrac{dx^p}{dt}\dfrac{dx^q}{dt}$.

Solution

Since v^k is a contravariant tensor, we have by Problem 8.62(b)

$$\dfrac{\delta v^k}{\delta t} = \dfrac{dv^k}{dt} + \left\{ \begin{matrix} k \\ qs \end{matrix} \right\} v^s \dfrac{dx^q}{dt} = \dfrac{d^2 x^k}{dt^2} + \left\{ \begin{matrix} k \\ qp \end{matrix} \right\} v^p \dfrac{dx^q}{dt} = \dfrac{d^2 x^k}{dt^2} + \left\{ \begin{matrix} k \\ pq \end{matrix} \right\} \dfrac{dx^p}{dt}\dfrac{dx^q}{dt}$$

8.70. Find the physical components of (a) the velocity and (b) the acceleration of a particle in cylindrical coordinates.

Solution

(a) From Problem 8.67(a), the contravariant components of the velocity are

$$\frac{dx^1}{dt} = \frac{d\rho}{dt}, \quad \frac{dx^2}{dt} = \frac{d\phi}{dt} \quad \text{and} \quad \frac{dx^3}{dt} = \frac{dz}{dt}$$

Then the physical components of the velocity are

$$\sqrt{g_{11}}\frac{dx^1}{dt} = \frac{d\rho}{dt}, \quad \sqrt{g_{22}}\frac{dx^2}{dt} = \rho\frac{d\phi}{dt} \quad \text{and} \quad \sqrt{g_{33}}\frac{dx^3}{dt} = \frac{dz}{dt}$$

using $g_{11} = 1$, $g_{22} = \rho^2$, $g_{33} = 1$.

(b) From Problems 8.69 and 8.49(b), the contravariant components of the acceleration are

$$a^1 = \frac{d^2x^1}{dt^2} + \left\{\begin{matrix} 1 \\ 22 \end{matrix}\right\}\frac{dx^2}{dt}\frac{dx^2}{dt} = \frac{d^2\rho}{dt^2} - \rho\left(\frac{d\phi}{dt}\right)^2$$

$$a^2 = \frac{d^2x^2}{dt^2} + \left\{\begin{matrix} 2 \\ 12 \end{matrix}\right\}\frac{dx^1}{dt}\frac{dx^2}{dt} + \left\{\begin{matrix} 2 \\ 21 \end{matrix}\right\}\frac{dx^2}{dt}\frac{dx^1}{dt} = \frac{d^2\phi}{dt^2} + \frac{2}{\rho}\frac{d\rho}{dt}\frac{d\phi}{dt}$$

and

$$a^3 = \frac{d^2x^3}{dt^2} = \frac{d^2z}{dt^2}$$

Then the physical components of the acceleration are

$$\sqrt{g_{11}}a^1 = \ddot{\rho} - \rho\dot{\phi}^2, \quad \sqrt{g_{22}}a^2 = \rho\ddot{\phi} + 2\dot{\rho}\dot{\phi} \quad \text{and} \quad \sqrt{g_{33}}a^3 = \ddot{z}$$

where dots denote differentiations with respect to time.

8.71. Suppose the kinetic energy T of a particle of constant mass M moving with velocity having magnitude v is given by $T = \frac{1}{2}Mv^2 = \frac{1}{2}Mg_{pq}\dot{x}^p\dot{x}^q$. Prove that

$$\frac{d}{dt}\left(\frac{\partial T}{\partial \dot{x}^k}\right) - \frac{\partial T}{\partial x^k} = Ma_k$$

where a_k denotes the covariant components of the acceleration.

Solution

Since $T = \frac{1}{2}Mg_{pq}\dot{x}^p\dot{x}^q$, we have

$$\frac{\partial T}{\partial x^k} = \frac{1}{2}M\frac{\partial g_{pq}}{\partial x^k}\dot{x}^p\dot{x}^q, \quad \frac{\partial T}{\partial \dot{x}^k} = Mg_{kq}\dot{x}^q \quad \text{and} \quad \frac{d}{dt}\left(\frac{\partial T}{\partial \dot{x}^k}\right) = M\left(g_{kq}\ddot{x}^q + \frac{\partial g_{kq}}{\partial x^j}\dot{x}^j\dot{x}^q\right)$$

Then

$$\frac{d}{dt}\left(\frac{\partial T}{\partial \dot{x}^k}\right) - \frac{\partial T}{\partial x^k} = M\left(g_{kq}\ddot{x}^q + \frac{\partial g_{kq}}{\partial x^j}\dot{x}^j\dot{x}^q - \frac{1}{2}\frac{\partial g_{pq}}{\partial x^k}\dot{x}^p\dot{x}^q\right)$$

$$= M\left(g_{kq}\ddot{x}^q + \frac{1}{2}\left(\frac{\partial g_{kq}}{\partial x^p} + \frac{\partial g_{kp}}{\partial x^q} - \frac{\partial g_{pq}}{\partial x^k}\right)\dot{x}^p\dot{x}^q\right)$$

$$= M(g_{kq}\ddot{x}^q + [pq, k]\dot{x}^p\dot{x}^q)$$

$$= Mg_{kr}\left(\ddot{x}^r + \left\{\begin{matrix} r \\ pq \end{matrix}\right\}\dot{x}^p\dot{x}^q\right) = Mg_{kr}a^r = Ma_k$$

using Problem 8.69. The result can be used to express the acceleration in different coordinate systems.

8.72. Use Problem 8.71 to find the physical components of the acceleration of a particle in cylindrical coordinates.

Solution

Since $ds^2 = d\rho^2 + \rho^2\, d\phi^2 + dz^2$, $v^2 = (ds/dt)^2 = \dot{\rho}^2 + \rho^2\dot{\phi}^2 + \dot{z}^2$ and $T = \frac{1}{2}Mv^2 = \frac{1}{2}M(\dot{\rho}^2 + \rho^2\dot{\phi}^2 + \dot{z}^2)$.

From Problem 8.71 with $x^1 = \rho$, $x^2 = \phi$, $x^3 = z$, we find

$$a_1 = \ddot{\rho} - \rho\dot{\phi}^2, \quad a_2 = \frac{d}{dt}(\rho^2\dot{\phi}), \quad a_3 = \ddot{z}$$

Then the physical components are given by

$$\frac{a_1}{\sqrt{g_{11}}}, \quad \frac{a_2}{\sqrt{g_{22}}}, \quad \frac{a_3}{\sqrt{g_{33}}} \quad \text{or} \quad \ddot{\rho} - \rho\dot{\phi}^2, \ \frac{1}{\rho}\frac{d}{dt}(\rho^2\dot{\phi}), \ \ddot{z}$$

since $g_{11} = 1$, $g_{22} = \rho^2$, $g_{33} = 1$. Compare with Problem 8.70.

8.73. Suppose the covariant force acting on a particle is given by $F_k = -\dfrac{\partial V}{\partial x^k}$ where $V(x^1, \ldots, x^N)$ is the potential energy. Show that $\dfrac{d}{dt}\left(\dfrac{\partial L}{\partial \dot{x}^k}\right) - \dfrac{\partial L}{\partial x^k} = 0$ where $L = T - V$.

Solution

From $L = T - V$, $\dfrac{\partial L}{\partial \dot{x}^k} = \dfrac{\partial T}{\partial \dot{x}^k}$ since V is independent of $\dot{x}^k$. Then, from Problem 8.71,

$$\frac{d}{dt}\left(\frac{\partial T}{\partial \dot{x}^k}\right) - \frac{\partial T}{\partial x^k} = Ma_k = F_k = -\frac{\partial V}{\partial x^k} \quad \text{and} \quad \frac{d}{dt}\left(\frac{\partial L}{\partial \dot{x}^k}\right) - \frac{\partial L}{\partial x^k} = 0$$

The function L is called the *Lagrangean*. The equations involving L, called the *Lagrange equations*, are important in mechanics. By Problem 8.50, it follows that the results of this problem are equivalent to the statement that a particle moves in such a way that $\int_{t_1}^{t_2} L\, dt$ is an extremum. This is called *Hamilton's principle*.

8.74. Express the divergence theorem in tensor form.

Solution

Let A^k define a tensor field of rank one and let v_k denote the outward drawn unit normal to any point of a closed surface S bounding a volume V. Then the divergence theorem states that

$$\iiint_V A^k_{,k}\, dV = \iint_S A^k v_k\, dS$$

For N-dimensional space, the triple integral is replaced by an N tuple integral, and the double integral by an $N-1$ tuple integral. The invariant $A^k_{,k}$ is the divergence of A^k (see Problem 8.57). The invariant $A^k v_k$ is the scalar product of A^k and v_k, analogous to $\mathbf{A} \cdot \mathbf{n}$ in the vector notation of Chapter 2.

We have been able to express the theorem in tensor form; hence it is true for all coordinate systems since it is true for rectangular systems (see Chapter 6). Also see Problem 8.66.

8.75. Express in tensor form Maxwell's equations: (a) div $\mathbf{B} = 0$, (b) div $\mathbf{D} = 4\pi\rho$,
(c) $\nabla \times \mathbf{E} = -\dfrac{1}{c}\dfrac{\partial \mathbf{B}}{\partial t}$, (d) $\nabla \times \mathbf{H} = \dfrac{4\pi I}{c}$

Solution

Define the tensors B^k, D^k, E_k, H_k, I^k and suppose that ρ and c are invariants. Then the equations can be written

(a) $B^k_{,k} = 0$ and   (b) $D^k_{,k} = 4\pi\rho$

(c) $-\epsilon^{jkq}E_{k,q} = -\dfrac{1}{c}\dfrac{\partial B^j}{\partial t}$ or $\epsilon^{jkq}E_{k,q} = \dfrac{1}{c}\dfrac{\partial B^j}{\partial t}$

(d) $\quad -\epsilon^{jkq}H_{k,q} = \dfrac{4\pi I^j}{c}\quad$ or $\quad \epsilon^{jkq}H_{k,q} = -\dfrac{4\pi I^j}{c}$

These equations form the basis for *electromagnetic theory*.

8.76. (a) Prove that $A_{p,qr} - A_{p,rq} = R^n_{pqr}A_n$ where A_p is an arbitrary covariant tensor of rank one.

(b) Prove that R^n_{pqr} is a tensor. (c) Prove that $R_{pqrs} = g_{ns}R^n_{pqr}$ is a tensor.

Solution

(a) $\quad A_{p,qr} = (A_{p,q})_{,r} = \dfrac{\partial A_{p,q}}{\partial x^r} - \left\{\begin{matrix} j \\ pr \end{matrix}\right\}A_{j,q} - \left\{\begin{matrix} j \\ qr \end{matrix}\right\}A_{p,j}$

$\quad = \dfrac{\partial}{\partial x^r}\left(\dfrac{\partial A_p}{\partial x^q} - \left\{\begin{matrix} j \\ pq \end{matrix}\right\}A_j\right) - \left\{\begin{matrix} j \\ pr \end{matrix}\right\}\left(\dfrac{\partial A_j}{\partial x^q} - \left\{\begin{matrix} k \\ jq \end{matrix}\right\}A_k\right) - \left\{\begin{matrix} j \\ qr \end{matrix}\right\}\left(\dfrac{\partial A_p}{\partial x^j} - \left\{\begin{matrix} l \\ pj \end{matrix}\right\}A_l\right)$

$\quad = \dfrac{\partial^2 A_p}{\partial x^r \partial x^q} - \dfrac{\partial}{\partial x^r}\left\{\begin{matrix} j \\ pq \end{matrix}\right\}A_j - \left\{\begin{matrix} j \\ pq \end{matrix}\right\}\dfrac{\partial A_j}{\partial x^r} - \left\{\begin{matrix} j \\ pr \end{matrix}\right\}\dfrac{\partial A_j}{\partial x^q} + \left\{\begin{matrix} j \\ pr \end{matrix}\right\}\left\{\begin{matrix} k \\ jq \end{matrix}\right\}A_k$

$\quad - \left\{\begin{matrix} j \\ qr \end{matrix}\right\}\dfrac{\partial A_p}{\partial x^j} + \left\{\begin{matrix} j \\ qr \end{matrix}\right\}\left\{\begin{matrix} l \\ pj \end{matrix}\right\}A_l$

By interchanging q and r and subtracting, we find

$$A_{p,qr} - A_{p,rq} = \left\{\begin{matrix} j \\ pr \end{matrix}\right\}\left\{\begin{matrix} k \\ jq \end{matrix}\right\}A_k - \dfrac{\partial}{\partial x^r}\left\{\begin{matrix} j \\ pq \end{matrix}\right\}A_j - \left\{\begin{matrix} j \\ pq \end{matrix}\right\}\left\{\begin{matrix} k \\ jr \end{matrix}\right\}A_k + \dfrac{\partial}{\partial x^q}\left\{\begin{matrix} j \\ pr \end{matrix}\right\}A_j$$

$$= \left\{\begin{matrix} k \\ pr \end{matrix}\right\}\left\{\begin{matrix} j \\ kq \end{matrix}\right\}A_j - \dfrac{\partial}{\partial x^r}\left\{\begin{matrix} j \\ pq \end{matrix}\right\}A_j - \left\{\begin{matrix} k \\ pq \end{matrix}\right\}\left\{\begin{matrix} j \\ kr \end{matrix}\right\}A_j + \dfrac{\partial}{\partial x^q}\left\{\begin{matrix} j \\ pr \end{matrix}\right\}A_j$$

$$= R^j_{pqr}A_j$$

where

$$R^j_{pqr} = \left\{\begin{matrix} k \\ pr \end{matrix}\right\}\left\{\begin{matrix} j \\ kq \end{matrix}\right\} - \dfrac{\partial}{\partial x^r}\left\{\begin{matrix} j \\ pq \end{matrix}\right\} - \left\{\begin{matrix} k \\ pq \end{matrix}\right\}\left\{\begin{matrix} j \\ kr \end{matrix}\right\} + \dfrac{\partial}{\partial x^q}\left\{\begin{matrix} j \\ pr \end{matrix}\right\}$$

Replace j by n and the result follows.

(b) Since $A_{p,qr} - A_{p,rq}$ is a tensor, $R^n_{pqr}A_n$ is a tensor; and since A_n is an arbitrary tensor, R^n_{pqr} is a tensor by the quotient law. This tensor is called the *Riemann–Christoffel* tensor, and is sometimes written $R^n_{.pqr}$, $R^{..n}_{pqr.}$, or simply R^n_{pqr}.

(c) $R_{pqrs} = g_{ns}R^n_{pqr}$ is an associated tensor of R^n_{pqr} and thus is a tensor. It is called the *covariant curvature tensor* and is of fundamental importance in *Einstein's general theory of relativity*.

SUPPLEMENTARY PROBLEMS

8.77. Write each of the following using the summation convention.

(a) $a_1x^1x^3 + a_2x^2x^3 + \cdots + a_Nx^Nx^3$ (d) $g^{21}g_{11} + g^{22}g_{21} + g^{23}g_{31} + g^{24}g_{41}$

(b) $A^{21}B_1 + A^{22}B_2 + A^{23}B_3 + \cdots + A^{2N}B_N$ (e) $B^{121}_{11} + B^{122}_{12} + B^{221}_{21} + B^{222}_{22}$

(c) $A^j_1B^1 + A^j_2B^2 + A^j_3B^3 + \cdots + A^j_NB^N$

8.78. Write the terms in each of the following indicated sums.

(a) $\dfrac{\partial}{\partial x^k}(\sqrt{g}A^k),\ N = 3$ (b) $A^{jk}B^p_kC_j,\ N = 2$ (c) $\dfrac{\partial \bar{x}^j}{\partial x^k}\dfrac{\partial x^k}{\partial \bar{x}^m}$

8.79. What locus is represented by $a_kx^kx^k = 1$ where $x^k,\ k = 1, 2, \ldots, N$ are rectangular coordinates, a_k are positive constants and $N = 2, 3,$ or 4?

8.80. Let $N = 2$. Write the system of equations represented by $a_{pq}x^q = b_p$.

8.81. Write the law of transformation for the tensors   (a) A_k^{ij},   (b) B_m^{ijk},   (c) C_{mn},   (d) A_m.

8.82. Suppose the quantities $B(j, k, m)$ and $C(j, k, m, n)$ transform from a coordinate system x^i to another $\bar{x}^i$ according to the rules

     (a) $\bar{B}(p, q, r) = \dfrac{\partial x^j}{\partial \bar{x}^p} \dfrac{\partial x^k}{\partial \bar{x}^q} \dfrac{\partial \bar{x}^r}{\partial x^m} B(j, k, m)$ (b) $\bar{C}(p, q, r, s) = \dfrac{\partial \bar{x}^p}{\partial x^j} \dfrac{\partial x^q}{\partial \bar{x}^k} \dfrac{\partial x^m}{\partial \bar{x}^r} \dfrac{\partial x^s}{\partial \bar{x}^n} C(j, k, m, n)$. Determine whether

     they are tensors. If so, write the tensors in suitable notation and give the rank and the covariant and contravariant orders.

8.83. How many components does a tensor of rank 5 have in a space of 4 dimensions?

8.84. Prove that if the components of a tensor are zero in one coordinate system, they are zero in all coordinate systems.

8.85. Prove that if the components of two tensors are equal in one coordinate system, they are equal in all co-ordinate systems.

8.86. Show that the velocity $dx^k/dt = v^k$ of a fluid is a tensor, but that dv^k/dt is not a tensor.

8.87. Find the covariant and contravariant components of a tensor in (a) cylindrical coordinates ρ, ϕ, z, (b) spherical coordinates r, θ, ϕ if its covariant components in rectangular coordinates are $2x - z$, x^2y, yz.

8.88. The contravariant components of a tensor in rectangular coordinates are yz, 3, $2x + y$. Find its covariant components in parabolic cylindrical coordinates.

8.89. Evaluate   (a) $\delta_q^p B_p^{rs}$,   (b) $\delta_q^p \delta_s^r A^{qs}$,   (c) $\delta_q^p \delta_r^q \delta_s^r$,   (d) $\delta_q^p \delta_r^q \delta_s^r \delta_p^s$.

8.90. Suppose A_r^{pq} is a tensor. Show that A_r^{pr} is a contravariant tensor of rank one.

8.91. Show that $\delta_{jk} = \begin{cases} 1 & j = k \\ 0 & j \neq k \end{cases}$ is not a covariant tensor as the notation might indicate.

8.92. Let $\bar{A}_p = \dfrac{\partial x^q}{\partial \bar{x}^p} A_q$. Prove that $A_q = \dfrac{\partial \bar{x}^p}{\partial x^q} \bar{A}_p$.

8.93. Let $\bar{A}_r^p = \dfrac{\partial \bar{x}^p}{\partial x^q} \dfrac{\partial x^s}{\partial \bar{x}^r} A_s^q$. Prove that $A_s^q = \dfrac{\partial x^q}{\partial \bar{x}^p} \dfrac{\partial \bar{x}^r}{\partial x^s} \bar{A}_r^p$.

8.94. Suppose Φ is an invariant. Determine whether $\partial^2 \Phi / \partial x^p \partial x^q$ is a tensor.

8.95. Let A_q^p and B_r be tensors, prove that $A_q^p B^r$ and $A_q^p B^q$ are tensors and determine the rank of each.

8.96. Suppose A_{rs}^{pq} is a tensor. Show that $A_{rs}^{pq} + A_{sr}^{qp}$ is a symmetric tensor and $A_{rs}^{pq} - A_{sr}^{qp}$ is a skew-symmetric tensor.

8.97. Suppose A^{pq} and B_{rs} are skew-symmetric tensors. Show that $C_{rs}^{pq} = A^{pq} B_{rs}$ is symmetric.

8.98. Suppose a tensor is symmetric (skew-symmetric). Are repeated contractions of the tensor also symmetric (skew-symmetric)?

8.99. Prove that $A_{pq}x^p x^q = 0$ if A_{pq} is a skew-symmetric tensor.

8.100. What is the largest number of different components that a symmetric contravariant tensor of rank two can have when (a) $N = 4$, (b) $N = 6$? What is the number for any value of N?

8.101. How many distinct non-zero components, apart from a difference in sign, does a skew-symmetric covariant tensor of the third rank have?

8.102. Suppose A_{rs}^{pq} is a tensor. Prove that a double contraction yields an invariant.

8.103. Prove that a necessary and sufficient condition for a tensor of rank R to become an invariant by repeated contraction is that R be even and that the number of covariant and contravariant indices be equal to $R/2$.

8.104. Given A_{pq} and B^{rs} are tensors. Show that the outer product is a tensor of rank four and that two inner products can be formed of rank two and zero, respectively.

8.105. Let $A(p, q)B_q = C^p$ where B_q is an arbitrary covariant tensor of rank one and C^p is a contravariant tensor of rank one. Show that $A(p, q)$ must be a contravariant tensor of rank two.

8.106. Let A^p and B_q be arbitrary tensors. Show that if $A^p B_q C(p, q)$ is an invariant, then $C(p, q)$ is a tensor that can be written C_p^q.

8.107. Find the sum $S = A + B$, difference $D = A - B$, and products $P = AB$ and $Q = BA$, where A and B are the matrices

(a) $A = \begin{bmatrix} 3 & -1 \\ 2 & 4 \end{bmatrix}$, $B = \begin{bmatrix} 4 & 3 \\ -2 & -1 \end{bmatrix}$, (b) $A = \begin{bmatrix} 2 & 0 & 1 \\ -1 & -2 & 2 \\ -1 & 3 & -1 \end{bmatrix}$, $B = \begin{bmatrix} 1 & -1 & 2 \\ 3 & 2 & -4 \\ -1 & -2 & 2 \end{bmatrix}$

8.108. Find $(3A - 2B)(2A - B)$, where A and B are the matrices in the preceding problem.

8.109. (a) Verify that $\det(AB) = \{\det A\}\{\det B\}$ for the matrices in Problem 8.107.

(b) Is $\det(AB) = \det(BA)$?

8.110. Let $A = \begin{bmatrix} 3 & -1 & 2 \\ 4 & 2 & 3 \end{bmatrix}$, $B = \begin{bmatrix} -3 & 2 & -1 \\ 1 & 3 & -2 \\ 2 & 1 & 2 \end{bmatrix}$.

Show that (a) AB is defined and find it, (b) BA and $A + B$ are not defined.

8.111. Find x, y, and z such that $\begin{bmatrix} 2 & -1 & 3 \\ 1 & 2 & -4 \\ -1 & 3 & -2 \end{bmatrix} \begin{bmatrix} x \\ y \\ z \end{bmatrix} = \begin{bmatrix} 1 \\ -3 \\ 6 \end{bmatrix}$

8.112. The inverse of a square matrix A, written A^{-1} is defined by the equation $AA^{-1} = I$, where I is the unit matrix having ones down the main diagonal and zeros elsewhere.

Find A^{-1} if (a) $A = \begin{bmatrix} 3 & -2 \\ -5 & 4 \end{bmatrix}$, (b) $A = \begin{bmatrix} 1 & -1 & 1 \\ 2 & 1 & -1 \\ 1 & -1 & 2 \end{bmatrix}$. Is $A^{-1}A = I$ in these cases?

8.113. Prove that $A = \begin{bmatrix} 2 & 1 & -2 \\ 1 & -2 & 3 \\ 4 & -3 & 4 \end{bmatrix}$ has no inverse.

8.114. Prove that $(AB)^{-1} = B^{-1}A^{-1}$, where A and B are non-singular square matrices.

8.115. Express in matrix notation the transformation equations for (a) a contravariant vector (b) a covariant tensor of rank two (c) a mixed tensor of rank two.

8.116. Given $A = \begin{bmatrix} 2 & -2 \\ -3 & 1 \end{bmatrix}$, determine the values of the constant λ such that $AX = \lambda X$, for some nonzero matrix X (depending on λ). These values of λ are called *characteristic values* or *eigenvalues* of the matrix A.

8.117. The equation $F(\lambda) = 0$ of the previous problem for determining the characteristic values of a matrix A is called the *characteristic equation* for A. Show that $F(A) = O$, where $F(A)$ is the matrix obtained by replacing λ by A in the characteristic equation and where the constant term c is replaced by the matrix cI, and O is a matrix whose elements are zero (called the null matrix). The result is a special case of the *Hamilton–Cayley theorem*, which states that a matrix satisfies its own characteristic equation.

8.118. Prove that $(AB)^T = B^T A^T$.

8.119. Determine the metric tensor and conjugate metric tensor in (a) parabolic cylindrical and (b) elliptic cylindrical coordinates.

8.120. Consider the affine transformation $\bar{x}^r = a_p^r x^p + b^r$, where a_p^r and b^r are constants such that $a_r^p a_q^r = \delta_q^p$. Prove that there is no distinction between the covariant and contravariant components of a tensor. In the special case where the transformations are from one rectangular coordinate system to another, the tensors are called *Cartesian tensors*.

8.121. Find g and g^{jk} corresponding to $ds^2 = 3(dx^1)^2 + 2(dx^2)^2 + 4(dx^3)^2 - 6(dx^1 dx^3)$.

8.122. Let $A^k = g^{jk} A_j$. Show that $A_j = g_{jk} A^k$ and conversely.

8.123. Express the relationship between the associated tensors

(a) A^{pq} and $A_j^{\cdot q}$,   (b) $A_{\cdot q}^{p \cdot r}$ and A_{jql},   (c) $A_{pq}^{\cdot \cdot r}$ and $A_{\cdot \cdot l}^{jk}$.

8.124. Show that (a) $A_p^{\cdot q} B_{\cdot rs}^p = A^{pq} B_{prs}$, (b) $A_{\cdot \cdot r}^{pq} B_p^{\cdot r} = A_{p \cdot r}^{\cdot q} B^{pr} = A_p^{\cdot qr} B_{\cdot r}^p$. Hence demonstrate the general result that a dummy symbol in a term may be lowered from its upper position and raised from its lower position without changing the value of the term.

8.125. Show that if $A_{\cdot qr}^p = B_{\cdot q}^p C_r$, then $A_{pqr} = B_{pq} C_r$ and $A_p^{\cdot qr} = B_p^{\cdot q} C^r$. Hence demonstrate the result that a free index in a tensor equation may be raised or lowered without affecting the validity of the equation.

8.126. Show that the tensors g_{pq}, g^{pq} and δ_q^p are associated tensors.

8.127. Prove   (a) $\bar{g}_{jk} \dfrac{\partial \bar{x}^j}{\partial x^p} = g_{pq} \dfrac{\partial x^q}{\partial \bar{x}^k}$,   (b) $\bar{g}^{jk} \dfrac{\partial x^p}{\partial \bar{x}^j} = g^{pq} \dfrac{\partial \bar{x}^k}{\partial x^q}$.

8.128. Let A^p be a vector field. Find the corresponding unit vector.

8.129. Show that the cosines of the angles which the three-dimensional unit vector U^i make with the coordinate curves are given by $\dfrac{U_1}{\sqrt{g_{11}}}, \quad \dfrac{U_2}{\sqrt{g_{22}}}, \quad \dfrac{U_3}{\sqrt{g_{33}}}$.

8.130. Determine the Christoffel symbols of the first kind in (a) rectangular, (b) cylindrical, and (c) spherical coordinates.

8.131. Determine the Christoffel symbols of the first and second kinds in (a) parabolic cylindrical, (b) elliptic cylindrical coordinates.

8.132. Find differential equations for the geodesics in (a) cylindrical, (b) spherical coordinates.

8.133. Show that the geodesics on a plane are straight lines.

8.134. Show that the geodesics on a sphere are arcs of great circles.

8.135. Write the Christoffel symbols of the second kind for the metric

$$ds^2 = (dx^1)^2 + \left[(x^2)^2 - (x^1)^2\right](dx^2)^2$$

and the corresponding geodesic equations.

8.136. Write the covariant derivative with respect to x^q of each of the following tensors:

(a) A_l^{jk},   (b) A_{lm}^{jk},   (c) A_{klm}^j,   (d) A_m^{jkl},   (e) A_{lmn}^{jk}.

8.137. Find the covariant derivative of (a) $g_{jk} A^k$, (b) $A^j B_k$, and (c) $\delta_k^j A_j$ with respect to x^q.

8.138. Use the relation $A^j = g^{jk} A_k$ to obtain the covariant derivative of A^j from the covariant derivative of A_k.

8.139. Suppose Φ is an invariant. Prove that $\Phi_{,pq} = \Phi_{,qp}$; that is, the order of covariant differentiation of an invariant is immaterial.

8.140. Show that ϵ_{pqr} and ϵ^{pqr} are covariant and contravariant tensors, respectively.

8.141. Express the divergence of a vector A^p in terms of its physical components for (a) parabolic cylindrical, (b) paraboloidal coordinates.

8.142. Find the physical components of grad Φ in (a) parabolic cylindrical, (b) elliptic cylindrical coordinates.

8.143. Find $\nabla^2 \Phi$ in parabolic cylindrical coordinates.

8.144. Using the tensor notation, show that (a) div curl $A^r = 0$, (b) curl grad $\Phi = 0$.

8.145. Calculate the intrinsic derivatives of the following tensor fields, assumed to be differentiable functions of t:

(a) A_k, (b) A^{jk}, (c) $A_j B^k$, (d) ϕA_k^j where ϕ is an invariant.

8.146. Find the intrinsic derivative of (a) $g_{jk} A^k$, (b) $\delta_k^j A_j$, (c) $g_{jk} \delta_r^j A_p^r$.

8.147. Prove $\dfrac{d}{dt}(g^{pq} A_p A_q) = 2g^{pq} A_p \dfrac{\delta A_q}{\delta t}$.

8.148. Show that if no external force acts, a moving particle of constant mass travels along a geodesic given by

$$\frac{\delta}{\delta s}\left(\frac{dx^p}{ds}\right) = 0.$$

8.149. Prove that the sum and difference of two relative tensors of the same weight and type is also a relative tensor of the same weight and type.

8.150. Suppose A_r^{pq} is a relative tensor of weight w. Prove that $g^{-w/2} A_r^{pq}$ is an absolute tensor.

8.151. Let $A(p, q) B_r^{qs} = C_{pr}^s$, where B_r^{qs} is an arbitrary relative tensor of weight w_1 and C_{pr}^s is a known relative tensor of weight w_2. Prove that $A(p, q)$ is a relative tensor of weight $w_2 - w_1$. This is an example of the quotient law for relative tensors.

8.152. Show that the quantity $G(j, k)$ of Solved Problem 8.31 is a relative tensor of weight two.

8.153. Find the physical components of (a) the velocity and (b) the acceleration of a particle in spherical coordinates.

8.154. Let A^r and B^r be 2 vectors in 3-dimensional space. Show that if λ and μ are constants, then $C^r = \lambda A^r + \mu B^r$ is a vector lying in the plane of A^r and B^r. What is the interpretation in higher dimensional space?

8.155. Show that a vector normal to the surface $\phi(x^1, x^2, x^3) = $ constant is given by $A^p = g^{pq} \dfrac{\partial \phi}{\partial x^q}$. Find the corresponding unit normal.

8.156. The equation of continuity is given by $\nabla \cdot (\sigma v) + \dfrac{\partial \sigma}{\partial t} = 0$ where σ is the density and v is the velocity of a fluid. Express the equation in tensor form.

8.157. Express the continuity equation in (a) cylindrical and (b) spherical coordinates.

8.158. Express Stokes' theorem in tensor form.

8.159. Prove that the covariant curvature tensor R_{pqrs} is skew-symmetric in (a) p and q, (b) r and s, (c) q and s.

8.160. Prove $R_{pqrs} = R_{rspq}$.

8.161. Prove (a) $R_{pqrs} + R_{psqr} + R_{prsq} = 0$, (b) $R_{pqrs} + R_{rqps} + R_{rspq} + R_{psrq} = 0$.

8.162. Prove that covariant differentiation in a Euclidean space is commutative. Thus show that the Riemann–Christoffel tensor and curvature tensor are zero in a Euclidean space.

8.163. Let $T^p = \dfrac{dx^p}{ds}$ be the tangent vector to curve C whose equation is $x^p = x^p(s)$ where s is the arc length. (a) Show that $g_{pq} T^p T^q = 1$. (b) Prove that $g_{pq} T^p \dfrac{\delta T^q}{\delta s} = 0$ and thus show that $N^q = \dfrac{1}{\kappa} \dfrac{\delta T^q}{\delta s}$ is a unit normal to C for suitable κ. (c) Prove that $\dfrac{\delta N^q}{\delta s}$ is orthogonal to N^q.

8.164. With the notation of the previous problem, prove:

(a) $g_{pq} T^p N^q = 0$, (b) $g_{pq} T^p \dfrac{\delta N^q}{\delta s} = -\kappa$ or $g_{pq} T^p \left(\dfrac{\delta N^q}{\delta s} + \kappa T^q\right) = 0$.

Hence show that $B^r = \dfrac{1}{\tau}\left(\dfrac{\delta N^r}{\delta s} + \kappa T^r\right)$ is a unit vector for suitable τ orthogonal to both T^p and N^q.

8.165. Prove the Frenet–Serret formulas

$$\frac{\delta T^p}{\delta s} = \kappa N^p, \qquad \frac{\delta N^p}{\delta s} = \tau B^p - \kappa T^p, \qquad \frac{\delta B^p}{\delta s} = -\tau N^p$$

where T^p, N^p, and B^p are the unit tangent, unit normal, and unit binormal vectors to C, and κ and τ are the curvature and torsion of C.

8.166. Show that $ds^2 = c^2(dx^4)^2 - dx^k dx^k$ $(N = 3)$ is invariant under the linear (affine) transformation

$$\bar{x}^1 = \gamma(x^1 - vx^4), \quad \bar{x}^2 = x^2, \quad \bar{x}^3 = x^3, \quad \bar{x}^4 = \gamma\left(x^4 - \frac{\beta}{c}x^1\right)$$

where γ, β, c, and v are constants, $\beta = v/c$ and $\gamma = (1 - \beta^2)^{-1/2}$. This is the *Lorentz transformation* of special relativity. Physically, an observer at the origin of the x^i system sees an event occurring at position x^1, x^2, x^3 at time x^4 while an observer at the origin of the $\bar{x}^i$ system sees the same event occurring at position $\bar{x}^1$, $\bar{x}^2$, $\bar{x}^3$ at time $\bar{x}^4$. It is assumed that (1) the two systems have the x^1 and $\bar{x}^1$ axes coincident, (2) the positive x^2 and x^3 axes are parallel respectively to the positive $\bar{x}^2$ and $\bar{x}^3$ axes, (3) the $\bar{x}^i$ system moves with velocity v relative to the x^i system, and (4) the velocity of light c is a constant.

8.167. Show that to an observer fixed in the $x^i(\bar{x}^i)$ system, a rod fixed in the $\bar{x}^i(x^i)$ system lying parallel to the $\bar{x}^1(x^1)$ axis and of length L in this system appears to have the reduced length $L\sqrt{1 - \beta^2}$. This phenomena is called the *Lorentz–Fitzgerald contraction*.

ANSWERS TO SUPPLEMENTARY PROBLEMS

8.77. (a) $a_k x^k x^3$ (b) $A^{2j} B_j$ (c) $A^j_k B^k$ (d) $g^{2q} g_{q1}, N = 4$ (e) $B^{p2r}_{pr}, N = 2$

8.78. (a) $\dfrac{\partial}{\partial x^1}(\sqrt{g}A^1) + \dfrac{\partial}{\partial x^2}(\sqrt{g}A^2) + \dfrac{\partial}{\partial x^3}(\sqrt{g}A^3)$ (b) $A^{11}B^p_1 C_1 + A^{21}B^p_1 C_2 + A^{12}B^p_2 C_1 + A^{22}B^p_2 C_2$

(c) $\dfrac{\partial \bar{x}^j}{\partial x^1}\dfrac{\partial x^1}{\partial \bar{x}^m} + \dfrac{\partial \bar{x}^j}{\partial x^2}\dfrac{\partial x^2}{\partial \bar{x}^m} + \cdots + \dfrac{\partial \bar{x}^j}{\partial x^N}\dfrac{\partial x^N}{\partial \bar{x}^m}$

8.79. Ellipse for $N = 2$, ellipsoid for $N = 3$, hyperellipsoid for $N = 4$.

8.80. $\begin{cases} a_{11}x^1 + a_{12}x^2 = b_1 \\ a_{21}x^1 + a_{22}x^2 = b_2 \end{cases}$

8.81. (a) $\bar{A}^{pq}_r = \dfrac{\partial \bar{x}^p}{\partial x^i}\dfrac{\partial \bar{x}^q}{\partial x^j}\dfrac{\partial x^k}{\partial \bar{x}^r} A^{ij}_k$ (b) $\bar{B}^{pqr}_s = \dfrac{\partial \bar{x}^p}{\partial x^i}\dfrac{\partial \bar{x}^q}{\partial x^j}\dfrac{\partial \bar{x}^r}{\partial x^k}\dfrac{\partial x^m}{\partial \bar{x}^s} B^{ijk}_m$ (c) $\bar{C}_{pq} = \dfrac{\partial x^m}{\partial \bar{x}^p}\dfrac{\partial x^n}{\partial \bar{x}^q} C_{mn}$ (d) $\bar{A}_p = \dfrac{\partial x^m}{\partial \bar{x}^p} A_m$

8.82. (a) $B(j, k, m)$ is a tensor of rank three and is covariant of order two and contravariant of order one. If can be written B^m_{jk}. (b) $C(j, k, m, n)$ is not a tensor.

8.83. $4^5 = 1024$

8.87. (a) $2\rho \cos^2 \phi - z \cos \phi + \rho^3 \sin^2 \phi \cos^2 \phi, -2\rho^2 \sin \phi \cos \phi + \rho z \sin \phi + \rho^4 \sin \phi \cos^3 \phi, \rho z \sin \phi.$

(b) $2r \sin^2 \theta \cos^2 \phi - r \sin \theta \cos \theta \cos \phi + r^3 \sin^4 \theta \sin^2 \phi \cos^2 \phi + r^2 \sin \theta \cos^2 \theta \sin \phi,$

$2r^2 \sin \theta \cos \theta \cos^2 \phi - r^2 \cos^2 \theta \cos \phi + r^4 \sin^3 \theta \cos \theta \sin^2 \phi \cos^2 \phi - r^3 \sin^2 \theta \cos \theta \sin \phi,$

$-2r^2 \sin^2 \theta \sin \phi \cos \phi + r^2 \sin \theta \cos \theta \sin \phi + r^4 \sin^4 \theta \sin \phi \cos^3 \phi$

8.88. $u^2 vz + 3v, 3u - uv^2 z, u^2 + uv - v^2$

8.89. (a) B^{rs}_q, (b) A^{pr}, (c) δ^p_s, (d) N **8.98.** Yes.

8.94. It is not a tensor. **8.100.** (a) 10, (b) 21, (c) $N(N + 1)/2$

8.95. Rank 3 and rank 1, respectively. **8.101.** $N(N - 1)(N - 2)/6$

8.107. (a) $S = \begin{bmatrix} 7 & 2 \\ 0 & 3 \end{bmatrix}$, $D = \begin{bmatrix} -1 & -4 \\ 4 & 5 \end{bmatrix}$, $P = \begin{bmatrix} 14 & 10 \\ 0 & 2 \end{bmatrix}$, $Q = \begin{bmatrix} 18 & 8 \\ -8 & -2 \end{bmatrix}$

(b) $S = \begin{bmatrix} 3 & -1 & 3 \\ 2 & 0 & -2 \\ -2 & 1 & 1 \end{bmatrix}$, $D = \begin{bmatrix} 1 & 1 & -1 \\ -4 & -4 & 6 \\ 0 & 5 & -3 \end{bmatrix}$, $P = \begin{bmatrix} 1 & -4 & 6 \\ -9 & -7 & 10 \\ 9 & 9 & -16 \end{bmatrix}$,

$Q = \begin{bmatrix} 1 & 8 & -3 \\ 8 & -16 & 11 \\ -2 & 10 & -7 \end{bmatrix}$

8.108. (a) $\begin{bmatrix} -52 & -86 \\ 104 & 76 \end{bmatrix}$ (b) $\begin{bmatrix} 3 & -16 & 20 \\ 9 & 163 & -136 \\ -61 & -135 & 132 \end{bmatrix}$ **8.110.** $\begin{bmatrix} -6 & 5 & 3 \\ -4 & 17 & -2 \end{bmatrix}$

8.111. $x = -1, y = 3, z = 2$ **8.112.** (a) $\begin{bmatrix} 2 & 1 \\ 5/2 & 3/2 \end{bmatrix}$ (b) $\begin{bmatrix} 1/3 & 1/3 & 0 \\ -5/3 & 1/3 & 1 \\ -1 & 0 & 1 \end{bmatrix}$. Yes

8.115. (a) $\begin{bmatrix} \overline{A}^1 \\ \overline{A}^2 \\ \overline{A}^3 \end{bmatrix} = \begin{bmatrix} \dfrac{\partial \overline{x}^1}{\partial x^1} & \dfrac{\partial \overline{x}^1}{\partial x^2} & \dfrac{\partial \overline{x}^1}{\partial x^3} \\ \dfrac{\partial \overline{x}^2}{\partial x^1} & \dfrac{\partial \overline{x}^2}{\partial x^2} & \dfrac{\partial \overline{x}^2}{\partial x^3} \\ \dfrac{\partial \overline{x}^3}{\partial x^1} & \dfrac{\partial \overline{x}^3}{\partial x^2} & \dfrac{\partial \overline{x}^3}{\partial x^3} \end{bmatrix} \begin{bmatrix} A^1 \\ A^2 \\ A^3 \end{bmatrix}$

(b) $\begin{bmatrix} \overline{A}_{11} & \overline{A}_{12} & \overline{A}_{13} \\ \overline{A}_{21} & \overline{A}_{22} & \overline{A}_{23} \\ \overline{A}_{31} & \overline{A}_{32} & \overline{A}_{33} \end{bmatrix} = \begin{bmatrix} \dfrac{\partial x^1}{\partial \overline{x}^1} & \dfrac{\partial x^2}{\partial \overline{x}^1} & \dfrac{\partial x^3}{\partial \overline{x}^1} \\ \dfrac{\partial x^1}{\partial \overline{x}^2} & \dfrac{\partial x^2}{\partial \overline{x}^2} & \dfrac{\partial x^3}{\partial \overline{x}^2} \\ \dfrac{\partial x^1}{\partial \overline{x}^3} & \dfrac{\partial x^2}{\partial \overline{x}^3} & \dfrac{\partial x^3}{\partial \overline{x}^3} \end{bmatrix} \begin{bmatrix} A_{11} & A_{12} & A_{13} \\ A_{21} & A_{22} & A_{23} \\ A_{31} & A_{32} & A_{33} \end{bmatrix} \begin{bmatrix} \dfrac{\partial x^1}{\partial \overline{x}^1} & \dfrac{\partial x^1}{\partial \overline{x}^2} & \dfrac{\partial x^1}{\partial \overline{x}^3} \\ \dfrac{\partial x^2}{\partial \overline{x}^1} & \dfrac{\partial x^2}{\partial \overline{x}^2} & \dfrac{\partial x^2}{\partial \overline{x}^3} \\ \dfrac{\partial x^3}{\partial \overline{x}^1} & \dfrac{\partial x^3}{\partial \overline{x}^2} & \dfrac{\partial x^3}{\partial \overline{x}^3} \end{bmatrix}$

(c) $\begin{bmatrix} \overline{A}_1^1 & \overline{A}_2^1 & \overline{A}_3^1 \\ \overline{A}_1^2 & \overline{A}_2^2 & \overline{A}_3^2 \\ \overline{A}_1^3 & \overline{A}_2^3 & \overline{A}_3^3 \end{bmatrix} = \begin{bmatrix} \dfrac{\partial \overline{x}^1}{\partial x^1} & \dfrac{\partial \overline{x}^1}{\partial x^2} & \dfrac{\partial \overline{x}^1}{\partial x^3} \\ \dfrac{\partial \overline{x}^2}{\partial x^1} & \dfrac{\partial \overline{x}^2}{\partial x^2} & \dfrac{\partial \overline{x}^2}{\partial x^3} \\ \dfrac{\partial \overline{x}^3}{\partial x^1} & \dfrac{\partial \overline{x}^3}{\partial x^2} & \dfrac{\partial \overline{x}^3}{\partial x^3} \end{bmatrix} \begin{bmatrix} A_1^1 & A_2^1 & A_3^1 \\ A_1^2 & A_2^2 & A_3^2 \\ A_1^3 & A_2^3 & A_3^3 \end{bmatrix} \begin{bmatrix} \dfrac{\partial x^1}{\partial \overline{x}^1} & \dfrac{\partial x^1}{\partial \overline{x}^2} & \dfrac{\partial x^1}{\partial \overline{x}^3} \\ \dfrac{\partial x^2}{\partial \overline{x}^1} & \dfrac{\partial x^2}{\partial \overline{x}^2} & \dfrac{\partial x^2}{\partial \overline{x}^3} \\ \dfrac{\partial x^3}{\partial \overline{x}^1} & \dfrac{\partial x^3}{\partial \overline{x}^2} & \dfrac{\partial x^3}{\partial \overline{x}^3} \end{bmatrix}$

8.116. $\lambda = 4, -1$

8.119. (a) $\begin{bmatrix} u^2 + v^2 & 0 & 0 \\ 0 & u^2 + v^2 & 0 \\ 0 & 0 & 1 \end{bmatrix}$, $\begin{bmatrix} \dfrac{1}{u^2 + v^2} & 0 & 0 \\ 0 & \dfrac{1}{u^2 + v^2} & 0 \\ 0 & 0 & 1 \end{bmatrix}$

(b) $\begin{bmatrix} a^2(\sinh^2 u + \sin^2 v) & 0 & 0 \\ 0 & a^2(\sinh^2 u + \sin^2 v) & 0 \\ 0 & 0 & 1 \end{bmatrix}$, $\begin{bmatrix} \dfrac{1}{a^2(\sinh^2 u + \sin^2 v)} & 0 & 0 \\ 0 & \dfrac{1}{a^2(\sinh^2 u + \sin^2 v)} & 0 \\ 0 & 0 & 1 \end{bmatrix}$

8.121. $g = 6$, $(g^{jk}) = \begin{bmatrix} 4/3 & 0 & 1 \\ 0 & 1/2 & 0 \\ 1 & 0 & 1 \end{bmatrix}$

8.123. (a) $A^{pq} = g^{pj} A_j^{\cdot q}$, (b) $A_{\cdot q}^{p \cdot r} = g^{pj} g^{rl} A_{jql}$, (c) $A_{pq}^{\cdot \cdot r} = g_{pj} g_{qk} g^{rl} A_{\cdot \cdot l}^{jk}$ **8.128.** $\dfrac{A^p}{\sqrt{A^p A_p}}$ or $\dfrac{A^p}{\sqrt{g_{pq} A^p A^q}}$

8.130. (a) They are all zero.   (b) $[22, 1] = -\rho$, $[12, 2] = [21, 2] = \rho$. All others are zero.

    (c) $[22, 1] = -r$, $[33, 1] = -r \sin^2 \theta$, $[33, 2] = -r^2 \sin \theta \cos \theta$

      $[21, 2] = [12, 2] = r$, $[31, 3] = [13, 3] = r \sin^2 \theta$

      $[32, 3] = [23, 3] = r^2 \sin \theta \cos \theta$.   All others are zero.

8.131. (a) $[11, 1] = u$, $[22, 2] = v$, $[11, 2] = -v$, $[22, 1] = -u$,

      $[12, 1] = [21, 1] = v$, $[21, 2] = [12, 2] = u$.

$$\begin{Bmatrix} 1 \\ 11 \end{Bmatrix} = \frac{u}{u^2 + v^2}, \quad \begin{Bmatrix} 2 \\ 22 \end{Bmatrix} = \frac{v}{u^2 + v^2}, \quad \begin{Bmatrix} 1 \\ 22 \end{Bmatrix} = \frac{-u}{u^2 + v^2}, \quad \begin{Bmatrix} 2 \\ 11 \end{Bmatrix} = \frac{-v}{u^2 + v^2},$$

$$\begin{Bmatrix} 1 \\ 21 \end{Bmatrix} = \begin{Bmatrix} 1 \\ 12 \end{Bmatrix} = \frac{v}{u^2 + v^2}, \quad \begin{Bmatrix} 2 \\ 21 \end{Bmatrix} = \begin{Bmatrix} 2 \\ 12 \end{Bmatrix} = \frac{u}{u^2 + v^2}. \quad \text{All others are zero.}$$

    (b) $[11, 1] = 2a^2 \sinh u \cosh u$, $[22, 2] = 2a^2 \sin v \cos v$, $[11, 2] = -2a^2 \sin v \cos v$

      $[22, 1] = -2a^2 \sinh u \cosh u$, $[12, 1] = [21, 1] = 2a^2 \sin v \cos v$, $[21, 2] = [12, 2] = 2a^2 \sinh u \cosh u$

$$\begin{Bmatrix} 1 \\ 11 \end{Bmatrix} = \frac{\sinh u \cosh u}{\sinh^2 u + \sin^2 v}, \quad \begin{Bmatrix} 2 \\ 22 \end{Bmatrix} = \frac{\sin v \cos v}{\sinh^2 u + \sin^2 v}, \quad \begin{Bmatrix} 1 \\ 22 \end{Bmatrix} = \frac{-\sinh u \cosh u}{\sinh^2 u + \sin^2 v},$$

$$\begin{Bmatrix} 2 \\ 11 \end{Bmatrix} = \frac{-\sin v \cos v}{\sinh^2 u + \sin^2 v}, \quad \begin{Bmatrix} 1 \\ 21 \end{Bmatrix} = \begin{Bmatrix} 1 \\ 12 \end{Bmatrix} = \frac{\sin v \cos v}{\sinh^2 u + \sin^2 v},$$

$$\begin{Bmatrix} 2 \\ 21 \end{Bmatrix} = \begin{Bmatrix} 2 \\ 12 \end{Bmatrix} = \frac{\sinh u \cosh u}{\sinh^2 u + \sin^2 v}. \quad \text{All others are zero.}$$

8.132. (a) $\dfrac{d^2 \rho}{ds^2} - \rho \left(\dfrac{d\phi}{ds} \right)^2 = 0$, $\dfrac{d^2 \phi}{ds^2} + \dfrac{2}{\rho} \dfrac{d\rho}{ds} \dfrac{d\phi}{ds} = 0$, $\dfrac{d^2 z}{ds^2} = 0$

    (b) $\dfrac{d^2 r}{ds^2} - r \left(\dfrac{d\theta}{ds} \right)^2 - r \sin^2 \theta \left(\dfrac{d\phi}{ds} \right)^2 = 0$, $\quad \dfrac{d^2 \theta}{ds^2} + \dfrac{2}{r} \dfrac{dr}{ds} \dfrac{d\theta}{ds} - \sin \theta \cos \theta \left(\dfrac{d\phi}{ds} \right)^2 = 0$

      $\dfrac{d^2 \phi}{ds^2} + \dfrac{2}{r} \dfrac{dr}{ds} \dfrac{d\phi}{ds} + 2 \cot \theta \dfrac{d\theta}{ds} \dfrac{d\phi}{ds} = 0$

8.135. $\begin{Bmatrix} 1 \\ 22 \end{Bmatrix} = x^1$, $\begin{Bmatrix} 2 \\ 12 \end{Bmatrix} = \begin{Bmatrix} 2 \\ 21 \end{Bmatrix} = \dfrac{x^1}{(x^1)^2 - (x^2)^2}$, $\begin{Bmatrix} 2 \\ 22 \end{Bmatrix} = \dfrac{x^2}{(x^2)^2 - (x^1)^2}$. All others are zero.

$$\frac{d^2 x^1}{ds^2} + x^1 \left(\frac{dx^2}{ds} \right)^2 = 0, \quad \frac{d^2 x^2}{ds^2} + \frac{2x^1}{(x^1)^2 - (x^2)^2} \frac{dx^1}{ds} \frac{dx^2}{ds} + \frac{x^2}{(x^2)^2 - (x^1)^2} \left(\frac{dx^2}{ds} \right)^2 = 0$$

8.136. (a) $A_{l,q}^{jk} = \dfrac{\partial A_l^{jk}}{\partial x^q} - \begin{Bmatrix} s \\ lq \end{Bmatrix} A_s^{jk} + \begin{Bmatrix} j \\ qs \end{Bmatrix} A_l^{sk} + \begin{Bmatrix} k \\ qs \end{Bmatrix} A_l^{js}$

    (b) $A_{lm,q}^{jk} = \dfrac{\partial A_{lm}^{jk}}{\partial x^q} - \begin{Bmatrix} s \\ lq \end{Bmatrix} A_{sm}^{jk} - \begin{Bmatrix} s \\ mq \end{Bmatrix} A_{ls}^{jk} + \begin{Bmatrix} j \\ qs \end{Bmatrix} A_{lm}^{sk} + \begin{Bmatrix} k \\ qs \end{Bmatrix} A_{lm}^{js}$

    (c) $A_{klm,q}^{j} = \dfrac{\partial A_{klm}^{j}}{\partial x^q} - \begin{Bmatrix} s \\ kq \end{Bmatrix} A_{slm}^{j} - \begin{Bmatrix} s \\ lq \end{Bmatrix} A_{ksm}^{j} - \begin{Bmatrix} s \\ mq \end{Bmatrix} A_{kls}^{j} + \begin{Bmatrix} j \\ qs \end{Bmatrix} A_{klm}^{s}$

    (d) $A_{m,q}^{jkl} = \dfrac{\partial A_m^{jkl}}{\partial x^q} - \begin{Bmatrix} s \\ mq \end{Bmatrix} A_s^{jkl} + \begin{Bmatrix} j \\ qs \end{Bmatrix} A_m^{skl} + \begin{Bmatrix} k \\ qs \end{Bmatrix} A_m^{jsl} + \begin{Bmatrix} l \\ qs \end{Bmatrix} A_m^{jks}$

(e) $A^{jk}_{lmn,q} = \dfrac{\partial A^{jk}_{lmn}}{\partial x^q} - \left\{{s \atop lq}\right\} A^{jk}_{smn} - \left\{{s \atop mq}\right\} A^{jk}_{lsn} - \left\{{s \atop nq}\right\} A^{jk}_{lms} + \left\{{j \atop qs}\right\} A^{sk}_{lmn} + \left\{{k \atop qs}\right\} A^{js}_{lmn}$

8.137. (a) $g_{jk} A^k_{,q}$, (b) $A^j_{,q} B_k + A^j B_{k,q}$, (c) $\delta^j_k A_{j,q}$

8.141. (a) $\dfrac{1}{u^2 + v^2}\left[\dfrac{\partial}{\partial u}(\sqrt{u^2 + v^2} A_u) + \dfrac{\partial}{\partial v}(\sqrt{u^2 + v^2} A_v)\right] + \dfrac{\partial A_z}{\partial z}$

(b) $\dfrac{1}{uv(u^2 + v^2)}\left[\dfrac{\partial}{\partial u}(uv\sqrt{u^2 + v^2} A_u) + \dfrac{\partial}{\partial v}(uv\sqrt{u^2 + v^2} A_v)\right] + \dfrac{1}{uv}\dfrac{\partial^2 A_z}{\partial z^2}$

8.142. (a) $\dfrac{1}{\sqrt{u^2 + v^2}}\dfrac{\partial \Phi}{\partial u}\mathbf{e}_u + \dfrac{1}{\sqrt{u^2 + v^2}}\dfrac{\partial \Phi}{\partial v}\mathbf{e}_v + \dfrac{\partial \Phi}{\partial z}\mathbf{e}_z$

(b) $\dfrac{1}{a\sqrt{\sinh^2 u + \sin^2 v}}\left(\dfrac{\partial \Phi}{\partial u}\mathbf{e}_u + \dfrac{\partial \Phi}{\partial v}\mathbf{e}_v\right) + \dfrac{\partial \Phi}{\partial z}\mathbf{e}_z$

where $\mathbf{e}_u$, $\mathbf{e}_v$ and $\mathbf{e}_z$ are unit vectors in the directions of increasing u, v, and z, respectively.

8.143. $\dfrac{1}{u^2 + v^2}\left[\dfrac{\partial^2 \Phi}{\partial u^2} + \dfrac{\partial^2 \Phi}{\partial v^2} + (u^2 + v^2)\Phi\right]$

8.145. (a) $\dfrac{\delta A_k}{\delta t} = A_{k,\,q}\dfrac{dx^q}{dt} = \left(\dfrac{\partial A_k}{\partial x^q} - \left\{{s \atop kq}\right\} A_s\right)\dfrac{dx^q}{dt} = \dfrac{dA_k}{dt} - \left\{{s \atop kq}\right\} A_s\dfrac{dx^q}{dt}$

(b) $\dfrac{\delta A^{jk}}{\delta t} = \dfrac{dA^{jk}}{dt} + \left\{{j \atop qs}\right\} A^{sk}\dfrac{dx^q}{dt} + \left\{{k \atop qs}\right\} A^{js}\dfrac{dx^q}{dt}$

(c) $\dfrac{\delta}{\delta t}(A_j B^k) = \dfrac{\delta A_j}{\delta t}B^k + A_j\dfrac{\delta B^k}{\delta t} = \left(\dfrac{dA_j}{dt} - \left\{{s \atop jq}\right\} A_s\dfrac{dx^q}{dt}\right)B^k + A_j\left(\dfrac{dB^k}{dt} + \left\{{k \atop qs}\right\} B^s\dfrac{dx^q}{dt}\right)$

(d) $\dfrac{\delta}{\delta t}(\Phi A^j_k) = \Phi\dfrac{\delta A^j_k}{\delta t} + \dfrac{\delta \Phi}{\delta t}A^j_k = \Phi\left(\dfrac{dA^j_k}{dt} + \left\{{j \atop qs}\right\} A^s_k\dfrac{dx^q}{dt} - \left\{{s \atop kq}\right\} A^j_s\dfrac{dx^q}{dt}\right) + \dfrac{d\Phi}{dt}A^j_k$

8.146. (a) $g_{jk}\dfrac{\delta A^k}{\delta t} = g_{jk}\left(\dfrac{dA^k}{dt} + \left\{{k \atop qs}\right\} A^s\dfrac{dx^q}{dt}\right)$

(b) $\delta^j_k\dfrac{\delta A_j}{\delta t} = \delta^j_k\left(\dfrac{dA_j}{dt} - \left\{{s \atop jq}\right\} A_s\dfrac{dx^q}{dt}\right) = \dfrac{dA_k}{dt} - \left\{{s \atop kq}\right\} A_s\dfrac{dx^q}{dt}$

(c) $g_{jk}\delta^j_r\dfrac{\delta A^r_p}{\delta t} = g_{rk}\left(\dfrac{dA^r_p}{dt} - \left\{{s \atop pq}\right\} A^r_s\dfrac{dx^q}{dt} + \left\{{r \atop qs}\right\} A^s_p\dfrac{dx^q}{dt}\right)$

8.153. (a) $\dot{r}$, $r\dot\theta$, $r\sin\theta\,\dot\phi$ (b) $\ddot{r} - r\dot\theta^2 - r\sin^2\theta\,\dot\phi^2$, $\dfrac{1}{r}\dfrac{d}{dt}(r^2\dot\theta) - r\sin\theta\cos\theta\,\dot\phi^2$, $\dfrac{1}{r\sin\theta}\dfrac{d}{dt}(r^2\sin^2\theta\,\dot\phi)$

8.156. $\dfrac{\partial(\sigma v^q)}{\partial x^q} + \dfrac{\sigma v^q}{2g}\dfrac{\partial g}{\partial x^q} + \dfrac{\partial \sigma}{\partial t} = 0$ where v^q are the contravariant components of the velocity.

8.157. (a) $\dfrac{\partial}{\partial \rho}(\sigma v^1) + \dfrac{\partial}{\partial \phi}(\sigma v^2) + \dfrac{\partial}{\partial z}(\sigma v^3) + \dfrac{\sigma v^1}{\rho} + \dfrac{\partial \sigma}{\partial t} = 0$

(b) $\dfrac{\partial}{\partial r}(\sigma v^1) + \dfrac{\partial}{\partial \theta}(\sigma v^2) + \dfrac{\partial}{\partial \phi}(\sigma v^3) + \sigma\left(\dfrac{2v^1}{r} + v^2\cot\theta\right) + \dfrac{\partial \sigma}{\partial t} = 0$ where v^1, v^2, and v^3 are the contravariant components of the velocity.

8.158. $\displaystyle\int_C A_p\dfrac{dx^p}{ds}ds = -\iint_S \epsilon^{pqr} A_{q,r} v_p\,dS$ where $\dfrac{dx^p}{ds}$ is the unit tangent vector to the closed curve C and v^p is the positive unit normal to the surface S, which has C as boundary.

Index